T0289879

Handbook of Heat Transfer and Fluid Flow

Handbook of Heat Transfer and Fluid Flow

Edited by
Emilia Lloyd

www.willfordpress.com

Published by Willford Press,
118-35 Queens Blvd., Suite 400,
Forest Hills, NY 11375, USA

ISBN: 978-1-64728-535-7

Cataloging-in-Publication Data

Handbook of heat transfer and fluid flow / edited by Emilia Lloyd.
    p. cm.
Includes bibliographical references and index.
ISBN 978-1-64728-535-7
1. Heat--Transmission. 2. Fluid dynamics. 3. Energy transfer. 4. Fluid mechanics. I. Lloyd, Emilia.
QC320 .H36 2023
621.402 2--dc23

For information on all Willford Press publications
visit our website at www.willfordpress.com

Contents

Permissions

List of Contributors

Index

Preface

Fluid flow refers to the motion of a fluid under the influence of various unbalanced forces. It primarily belongs to a branch of fluid mechanics that is typically concerned with the dynamics of fluids. Heat transfer is a field of thermal engineering that focuses on the consumption, exchange, generation and conversion of thermal energy among physical systems. It is divided into several mechanisms, including thermal radiation, thermal conduction, thermal convection and transmission of energy through phase changes. Fluid flow and heat transfer are critical components of engineering process design. Heat transfer into a fluid leads to the development of an equilibrium temperature under constant flow, which requires the solution of the Navier-Stokes equations. This book includes some of the vital pieces of works being conducted across the world on fluid flow and heat transfer. It will also provide interesting topics for research, which interested readers can take up. This book is a resource guide for experts as well as students.

The researches compiled throughout the book are authentic and of high quality, combining several disciplines and from very diverse regions from around the world. Drawing on the contributions of many researchers from diverse countries, the book's objective is to provide the readers with the latest achievements in the area of research. This book will surely be a source of knowledge to all interested and researching the field.

In the end, I would like to express my deep sense of gratitude to all the authors for meeting the set deadlines in completing and submitting their research chapters. I would also like to thank the publisher for the support offered to us throughout the course of the book. Finally, I extend my sincere thanks to my family for being a constant source of inspiration and encouragement.

Editor

1

Numerical Investigation on Forced Hybrid Nanofluid Flow and Heat Transfer Inside a Three-Dimensional Annulus Equipped with Hot and Cold Rods: Using Symmetry Simulation

Aysan Shahsavar Goldanlou [1,2], Mohammad Badri [3], Behzad Heidarshenas [4][iD],
Ahmed Kadhim Hussein [5][iD], Sara Rostami [6,7][iD] and Mostafa Safdari Shadloo [8,*][iD]

[1] Institute of Research and Development, Duy Tan University, Da Nang 550000, Vietnam;
aysanshahsavargoldanlou@duytan.edu.vn
[2] Faculty of Electrical—Electronic Engineering, Duy Tan University, Da Nang 550000, Vietnam
[3] Department of Mechanical Engineering, University of Kashan, Kashan 8731753153, Iran;
badri_m@grad.kashanu.ac.ir
[4] College of Mechanical of Electrical Engineering, Nanjing University of Aeronautics and Astronautics,
Nanjing 210016, China; behzadheidarshenas@nuaa.edu.cn
[5] College of Engineering—Mechanical Engineering Department, University of Babylon, Babylon 51001, Iraq;
ahmed.hussein.eng@uobabylon.edu.iq
[6] Laboratory of Magnetism and Magnetic Materials, Advanced Institute of Materials Science,
Ton Duc Thang University, Ho Chi Minh City 758307, Vietnam; sara.rostami@tdtu.edu.vn
[7] Faculty of Applied Sciences, Ton Duc Thang University, Ho Chi Minh City 758307, Vietnam
[8] CORIA-UMR 6614, CNRS & INSA of Rouen, Normandie University, 76000 Rouen, France
* Correspondence: msshadloo@coria.fr

Abstract: A 3D computational fluid dynamics method is used in the current study to investigate the hybrid nanofluid (HNF) flow and heat transfer in an annulus with hot and cold rods. The chief goal of the current study is to examine the influences of dissimilar Reynolds numbers, emissivity coefficients, and dissimilar volume fractions of nanoparticles on hydraulic and thermal characteristics of the studied annulus. In this way, the geometry is modeled using a symmetry scheme. The heat transfer fluid is a water, ethylene–glycol, or water/ethylene–glycol mixture-based Cu-Al$_2$O$_3$ HNF, which is a Newtonian NF. According to the findings for the model at Re = 3000 and ϕ_1 = 0.05, all studied cases with different base fluids have similar behavior. ϕ_1 and ϕ_2 are the volume concentration of Al$_2$O$_3$ and Cu nanoparticles, respectively. For all studied cases, the total average Nusselt number (Nu_{ave}) reduces firstly by an increment of the volume concentrations of Cu nanoparticles until ϕ_2 = 0.01 or 0.02 and then, the total Nu_{ave} rises by an increment of the volume concentrations of Cu nanoparticles. Additionally, for the case with water as the base fluid, the total Nu_{ave} at ϕ_2 = 0.05 is higher than the values at ϕ_2 = 0.00. On the other hand, for the other cases, the total Nu_{ave} at ϕ_2 = 0.05 is lower than the values at ϕ_2 = 0.00. For all studied cases, the case with water as the base fluid has the maximum Nu_{ave}. Plus, for the model at Re = 4000 and ϕ_1 = 0.05, all studied cases with different base fluids have similar behavior. For all studied cases, the total Nu_{ave} reduces firstly by an increment of the volume concentrations of Cu nanoparticles until ϕ_2 = 0.01 and then, the total Nu_{ave} rises by an increment of the volume concentrations of Cu nanoparticles. The Nu_{ave} augments are found by an increment of Reynolds numbers. Higher emissivity values should lead to higher radiation heat transfer, but the portion of radiative heat transfer in the studied annulus is low and therefore, has no observable increment in HNF flow and heat transfer.

Keywords: steady-state solution; forced convection; hybrid nanofluid; Nusselt number; streamlines

1. Introduction

Heat exchangers are types of equipment that have many industrial applications and this factor makes their performance improvement an attractive subject for researchers. Extensive research in this field has led to the development of new methods to improve the performance of heat exchangers, some of the most important of which are the use of corrugated surfaces [1–3], turbulators [4–6], fins [7–9], magnetic fields [10–12], and the replacement of common coolants with nanofluids (NFs) [13–15]. The suspension of nanoparticles in conventional coolants improves their cooling performance, and this fact has been confirmed in many laboratory-based and numerical studies [16–18].

All of the above methods, in turn, improve the performance of heat exchangers, but researchers are always looking for ideas to improve the effectiveness of these methods. A group of researchers posed the question: if adding one nanoparticle to a coolant improves its cooling performance, does adding two types of nanoparticles to the coolants not lead to a further improvement in cooling performance? Numerous experimental [19–21] and numerical [22–24] studies were conducted to answer this question, and it was finally found that the cooling performance of hybrid nanofluids (HNFs) was generally better than that of conventional NFs. Li et al. [25] investigated numerically different features of convection heat transfer inside a cavity equipped with a twin-web turbine disk and pin fins. Their obtained results indicate that there is a clockwise circular fluid flow inside the studied cavity which influences the front and back web for the improvement of the local convection heat transfer coefficients. The back web, Nu, is strangely higher than the Nu of the front web, and the back web temperature is perceptibly lower than the front web temperature. Chorin et al. [26] studied experimentally free convective heat transfer in a differentially heated cavity equipped with localized turbulators. The turbulator was located in the warm boundary coating of the cavity fluid flow. Experiments were carried out in heat transfers, temperature profiles, and velocity fields terms. The effect of the vertical location and the length for a conducting turbulator and an insulator was investigated. For the insulator turbulator, a fluid flow part diverged inside the colder turbulator region, leading to a downstream heat transfer upsurge. Giwa et al. [27] investigated experimentally and numerically different uniform magnetic induction effects on heat transfer presentation of a rectangular cavity equipped with aqueous HNF. The aqueous HNF's thermal properties had been studied for different nanoparticle volume fractions and various temperature ranges. Their obtained results show that by employing the vertically magnetic field on the cavity-side walls, the maximum Nu improvement was attained in compression with the model which did not have a magnetic field. Furthermore, it was realized that an increment in the magnetic field can improve the heat transfer characteristics significantly. Mansouri et al. [28] considered numerically conjugate conduction–convection heat transfer in a cavity filled with air and equipped with a rhombus conducting block exposed to subdivision, such as opposing and cooperating roles. The temperature and flow results were obtained by employing the Lattice Boltzmann technique with the finite volume method and the multi-relaxation time collision scheme. The influences of the initial solid block thermal conductivity on heat transfer and fluid flow inside the cavity were studied. Their obtained results present that the initial block subdivision leads to a reduction in heat transfer inside the cavity. Thiers et al. [29] studied numerical heat transfer improvement inside a rectangular differentially heated cavity with different thermal perturbations on cold and hot walls. The different effects of wave characteristics such as phase shift, frequency, and amplitude, and also the disturbance area's vertical location, were studied in the range of different Rayleigh numbers. The key aim of their study was to find the optimum location of differentially heated sources for heat transfer improvement. Ataei-Dadavi et al. [30] considered experimental fluid flow and heat transfer characteristics in a differential side-heated coarse porous media cavity. Their findings illustrated that the presence of a porous medium in the cavity leads to heat transfer reduction compared to the cavity without porous media. They attained a novel correlation method to predict the Nu for coarse porous media-based cavities. Farsani et al. [31] investigated heat transfer improvement in cavities using baffles melting, which were filled with gallium. They modeled the process of the phase using the fixed grid-based enthalpy–porosity technique coupled with the Semi-Implicit Method for

Pressure Linked Equations (SIMPLE) algorithm. They reported the streamlines and isotherm lines, and also, Nu values, on hot walls to investigate and analyze the problem. Vishnu et al. [32] examined numerically different conditions of heat transfer inside an oblique cavity located in supersonic flow. They simulated a 2D compressible unsteady turbulent flow field using the finite volume method and the Harten-Lax-van Leer-Contact (HLLC) scheme. They also validated their numerical procedure with available empirical and mathematical data. Sadaghiani et al. [33] examined parametrically and experimentally different influences of bubble coalescence on critical heat flux and pool boiling heat transfer in a cavity. They examined different surface wettability effects on the structured surfaces' performance by employing a Teflon film with a thickness of 50 nm. Additionally, they examined bubble dynamics by employment of a high-speed camera. Pan et al. [34] experimentally considered the heat transfer features of a heat exchanger equipped with microchannel and fan-shaped cavities. Their obtained data illustrated that the microchannel heat exchanger's performance, which is equipped with fan-shaped cavities, is sharply better than the heat exchangers without fan-shaped cavities and also, their pressure drop is lower than the heat exchangers without fan-shaped cavities. Balotaki et al. [35] modeled numerically, and by employing the lattice Boltzmann method, the heat transfer of free natural convections in a triangular cavity filling with different operating fluids, including water, TiO_2-H_2O, Al-H_2O, and Cu-H_2O, and equipped with double distribution functions. They reported various interesting results such as average predicted Nu, heat lines, an isothermal line, streamlines, and entropy generation values. Liu and Huang [36] studied numerically different effects using different linear thermal forcing models on fluid flow and convective heat transfer in a rectangular finned enclosure. The results showed fluctuations in the flow above the fin. Seo et al. [37] numerically analyzed the performance of a naturally cooled rectangular enclosure with a sinusoidal cylinder. The outcomes revealed the significant impact of cylinder diameter and Rayleigh number on the performance features of the enclosure. Razzaghpanah and Sarunac [38] simulated the free convection from a bundle of heated cylinders submerged in molten salt. They proposed a correlation set for Nusselt number in terms of Rayleigh number. Krakov and Nikiforov [39] employed numerical data to reveal the impact of interior cylinder shape on thermomagnetic convection heat transfer through a horizontal annulus. It was demonstrated that the shape of the interior cylinder can increase the heat transfer by 40–50%. Pawar et al. [40] numerically investigated convection heat transfer and forced laminar steady state flow inside a cylindrical finned cavity at various incidence angles. Due to analyzing aerodynamic characteristics, factors such as local time-averaged flux and vorticity, length of recirculation, Strouhal number and moment, and lift and drag coefficients were reported by them. Alam et al. [41] considered the flow around a cylinder with variable cross-sections and analyzed the dependence of velocity and temperature field on corner radius and attack angle. The dramatic role of boundary layers in improving heat transfer was reported. The main reason for this behavior is that the primary wake bubbles are mostly linked to the forces. Vyas et al. [42] conducted an experimental numerical assessment on fluid flow in the wake region of a cylinder submerged in a conduit, considering the impacts of blockage ratio. It was realized that the highest downstream distance traveled by the vortices was at a blockage ratio of 0.38. Hadžiabdić et al. [43] simulated the flow and heat transfer around a rotating cylinder. They reported the high local rates of heat transfer in the cylinder. Furthermore, they found that the overall mean predicted Nu values of rotational cylinder did not have a significant variation inside the cylinder but its time-average presented some changes in comparison with the basic cylinder.

The literature review clarifies that even though the influence of using NF for electrical heat exchangers was evaluated, to the best knowledge of the authors, there is not any research which examines the symmetry simulation of forced HNF flow and heat transfer in a three-dimensional annulus equipped with hot and cold rods. The chief goal of the current study is to examine the influences of dissimilar Reynolds numbers, emissivity coefficients, and dissimilar volume fractions of nanoparticles on hydraulic and thermal characteristics of the studied annulus. Moreover, in this study, the geometry is modeled using symmetry scheme.

2. Methodology

2.1. Problem Statement and Governing Equations

Figure 1 demonstrates a basic and simulated three-dimensional cylindrical shape cavity (annulus) equipped with hot and cold rods in the present study. In the present paper, this geometry is analyzed using a symmetry boundary condition. Figure 1 shows, on the right side, a schematic diagram of the studied configuration. The height of the cylinder is changed from 300 mm. Additionally, the height of the square is changed from 46.03 to 70.71 mm, the length of the square is 1000 mm, and the gravity influences are determined in z direction. This is a hypothetical geometry and was designed by the authors of the article. However, it has many applications in electrical components and electronics industries. The system is made of an insulator in the center and outer sides of the cylinder, with a boundary condition of zero heat flux. However, this material is presumed with two dissimilar emissivity values of 0.2 and 0.5 for analyzing its influences on radiation heat transfer. Moreover, two dissimilar Reynolds numbers, Re = 3000 and 4000, are in transient regime. The initial NF temperature is $T_{initial}$ = 350 K and the cold side temperature is T_c = 300 K. The hot side temperature can be T_h = 430 K. The heat transfer fluid is a water-, ethylene–glycol-, or water/ethylene–glycol mixture-based Cu-Al$_2$O$_3$ HNF, which creates a Newtonian NF. To attain the most proficient Newtonian NF in the current research, Cu and Al$_2$O$_3$ nanoparticles were added to the base fluid in different volume concentrations of 0.01 to 0.05 with diameters of 24 nm. Table 1 presents the thermophysical features of the base fluid and nanoparticles.

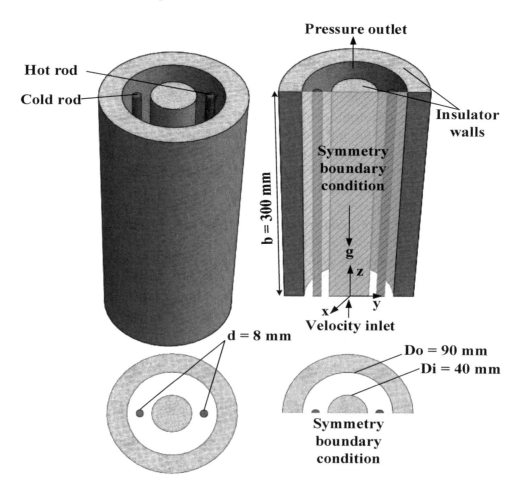

Figure 1. Schematic diagrams of the basic (left) and computation domain (right) geometries in the present paper.

Table 1. Base fluid and nanoparticle numerical values for thermophysical properties (T = 298 K) [44].

Thermophysical Properties	k (W/m·K)	c_p (J/kg·K)	ρ (kg/m^3)
Cu	400	385	8933
Al$_2$O$_3$	40	765	3970
Ethylene-glycol	0.252	2415	1114.4
H$_2$O	0.613	4179	997.1
H$_2$O/ethylene-glycol mixture	0.3799	3300	1067.5

The equations presented in Table 2 have been used to calculate the thermophysical properties of NF. The density (ρ_{hnf}) and specific heat ($c_{P,hnf}$) of the HNF at each section temperature (T_m) were calculated using the equations presented in Table 2 [44], where ϕ_1 is the volume concentration of Al$_2$O$_3$ and ϕ_2 is the volume concentration of Cu nanoparticles. Moreover, the effective dynamic viscosity and thermal conductivity of HNF were obtained by the equations in Table 2.

Table 2. Useful equations to model the studied nanofluid (NF) in the present work [44].

Density	$\rho_{hnf} = (1 - \phi_2)\left[(1 - \phi_1)\rho_{bf} + \phi_1\rho_{np,1}\right] + \phi_2\rho_{np,2}$
Specific heat	$(\rho c_p)_{hnf} = (1 - \phi_2)\left[(1 - \phi_1)(\rho c_p)_{bf} + \phi_1(\rho c_p)_{np,1}\right] + \phi_2(\rho c_p)_{np,2}$
Thermal conductivity	$k_{hnf} = k_{nf}\left[k_{np,2} + 2k_{nf} - 2\phi_2\frac{k_{nf}-k_{np,2}}{k_{np,2}} + 2k_{nf} + \phi_2(k_{nf} - k_{np,2})\right]$
	$k_{nf} = k_{bf}\left[k_{np,1} + 2k_{bf} - \frac{2\phi_1(k_{bf}-k_{np,1})}{k_{np,1}} + 2k_{bf} + \phi_1(k_{bf} - k_{np,1})\right]$
Dynamic viscosity	$\mu_{hnf} = \frac{\mu_{bf}}{(1-\phi_1)^{2.5}(1-\phi_2)^{2.5}}$

A 3D numerical analysis has been used in the current research to examine the transient HNF flow and heat transfer in an annulus with hot and cold rods. Given that the Reynolds number is more than 2300, the flow regime is considered transient. The Reynolds-averaged Navier–Stokes (RANS) equations with the shear-stress (SST) k–ω turbulence model were used for simulating the turbulence regime. The used equations can be defined as reported in Table 3. As mentioned earlier, in the current modeling, the SST k–ω turbulence model [45] has been presumed as the turbulence model [46–53]. This is used to have a better forecast of the flow conditions, such as large normal strain, strong acceleration, reverse pressure gradients, and separating flow, which can occur in the HNF flow appearances adjacent to the rods and walls. Based on the SST k–ω model, the turbulence kinetic energy (k) and specific dissipation rate (ω) can be attained by the equations reported in Table 3, where Ω_{ij} is the mean rate-of-rotation tensor and F_1 and F_2 are the blending functions. Additionally, D_ω^+ is the positive portion of the cross-diffusion term. The constant values reported in Table 3 are defined in Table 4. All equations were resolved until the maximum residual of grid control volume became smaller than 10^{-7}.

The Monte Carlo method has been used for modeling the radiation in the annulus gap. In this model, the radiation has been determined to affect the domain by heating its surface, without radiant energy transfer directly to the domain (Surface-to-Surface transfer mode (S2S)). The spectral dependency of the radiative heat transfer equation was estimated using the Gray Model (GM), which determines whether all radiation amounts are closely uniform through the spectrum. The inner rod and outer wall are determined as insulators and no heat transfer occurs in these sections.

It was also predictable that the radiation mechanism could not play a decisive role in heat transfer. However, the goal was to model the problem with the least assumptions and the most accurate details. Therefore, despite the low share of radiation in heat transfer, its effect was considered.

In order to simulate the nanofluid flow inside the cylinder, the single-phase technique was employed. Just the thermophysical characteristics were calculated at different temperatures and nanoparticle volume fractions and have been entered into software.

Table 3. Governing equations and parameters of interest [45].

Continuity equation	$\frac{\partial}{\partial x_i}(\rho u_i) = 0$
Momentum equation	$\frac{\partial}{\partial x_i}(\rho u_i u_j) = -\frac{\partial p}{\partial x_i} + \frac{\partial \tau_{ij}}{\partial x_j} + \rho g_i$
	$\tau_{ij} = \frac{\mu^* \partial u_i}{\partial x_j}$
	$\mu^* = \mu + \mu_t$
Energy equation	$\frac{\partial}{\partial x_i}(\rho c_p u_i T) = \frac{\partial}{\partial x_i}\left(\frac{\lambda^* \partial T}{\partial x_i}\right) + Qi$
	$\lambda^* = \lambda + \lambda_t$
	$\lambda_t = \frac{c_p \mu_t}{\sigma_t}$
Turbulence equations	$\frac{\partial}{\partial x_i}(\rho u_i k) = \frac{\partial}{\partial x_i}\left[\left(\mu + \frac{\mu_t}{\sigma_k}\right)\frac{\partial k}{\partial x_i}\right] + G_k - Y_k$
	$\frac{\partial}{\partial x_i}(\rho u_i \omega) = \frac{\partial}{\partial x_i}\left[\left(\mu + \frac{\mu_t}{\sigma_\omega}\right)\frac{\partial k}{\partial x_i}\right] + G_\omega - Y_\omega + D_\omega$
	$\mu_t = \frac{\rho k}{\omega}\frac{1}{max\left\{\frac{1}{\alpha^*}, \frac{\Omega F_2}{a_1 \omega}\right\}}$
	$\Omega = \sqrt{2\Omega_{ij}\Omega_{ij}}$
	$\Omega_{ij} = \frac{1}{2}\left(\frac{\partial u_i}{\partial x_j} - \frac{\partial u_j}{\partial x_i}\right)$
	$\sigma_k = \frac{1}{\frac{F_1}{\sigma_{k,1}} + \frac{1-F_1}{\sigma_{k,2}}}$
	$\sigma_\omega = \frac{1}{\frac{F_1}{\sigma_{\omega,1}} + \frac{1-F_1}{\sigma_{\omega,2}}}$
	$F_1 = tanh\left(\Phi_1^4\right)$
	$F_2 = tanh\left(\Phi_2^2\right)$
	$\Phi_1 = min\left\{max\left\{\frac{\sqrt{k}}{0.09\omega y}, \frac{500\mu}{\rho y^2 \omega}\right\}, \frac{4\rho k}{\sigma_{\omega,2}D_\omega^+ y^2}\right\}$
	$\Phi_2 = max\left\{\frac{2\sqrt{k}}{0.09\omega y}, \frac{500\mu}{\rho y^2 \omega}\right\}$
	$D_\omega^+ = max\left\{\frac{2\rho}{\omega \rho_{,2}}\frac{\partial k}{\partial x_i}\frac{\partial \omega}{\partial x_i}, 10^{-20}\right\}$
	$\alpha^* = \alpha_\infty^*\left(\frac{\alpha_0^* + \frac{Re_t}{R_k}}{1 + \frac{Re_t}{R_k}}\right), \quad \alpha_0^* = \frac{\beta_i}{3}, \quad Re_t = \frac{\rho k}{\mu \omega}, \quad \beta_i = F_1 \beta_{i,1} + (1-F_1)\beta_{i,2}$
	$G_k = \tau_{t,ij}\frac{\partial u_i}{\partial x_j}, \tau_{t,ij} = \mu_t\left(\frac{\partial u_i}{\partial x_j} + \frac{\partial u_j}{\partial x_i}\right) - \frac{2}{3}\rho k\delta_{ij}$
	$Y_k = \rho\beta^* k\omega, G_\omega = \frac{\rho\alpha}{\mu_t}G_k, Y_\infty = \rho\beta_i\omega^2$
	$\alpha = \frac{\alpha_\infty}{\alpha^*} = \left(\frac{\alpha_0^* + \frac{Re_t}{R_k}}{1 + \frac{Re_t}{R_k}}\right), \alpha_\infty = F_1\alpha_{\infty,1} + (1-F_1)\alpha_{\infty,2}$
	$\alpha_{\infty,1} = \frac{\beta_{i,1}}{\beta_\infty^*} - \frac{\kappa^2}{\sigma_{\omega,1}\sqrt{\beta_\infty^*}}, \alpha_{\infty,2} = \frac{\beta_{i,2}}{\beta_\infty^*} - \frac{\kappa^2}{\sigma_{\omega,2}\sqrt{\beta_\infty^*}}$
Parameters of interest	$\alpha_{nf} = \frac{k_{nf}}{(\rho c_p)_{nf}}, Re = \frac{U_{ref}D_h}{v_f}, Nu = \frac{k_{nf}}{k_f}\int\frac{\partial T}{\partial x}$

Table 4. Constant values of the shear-stress (SST) $k-\omega$ turbulence model [45].

σ_t	$\alpha_{k,1}$	$\alpha_{k,2}$	$\alpha_{\omega,1}$	$\alpha_{\omega,2}$	a_1	$\beta_{i,1}$	$\beta_{i,2}$	β_∞^*	κ	R_k
0.85	1.176	1.000	2.000	1.168	0.31	0.0750	0.0828	0.0900	0.41	6

2.2. Validation

As displayed in Figure 2, a grid independency study was implemented for the studied cavity containing Al_2O_3-Cu/water HNF to examine the influences of mesh sizes on the findings. For the validation, the emissivity of the cylinder is constant and equal to 0.2. Additionally, the volume fraction of the Al_2O_3 nanoparticles is 0.05. A non-uniform structured mesh was used for the channel. Moreover, the near-wall mesh was fine enough to be able to resolve the viscous sublayer ($y+ \leq 1$). As observed, several meshes have been tested versus different Reynolds numbers and volume concentrations of Cu nanoparticles. Finally, it was determined that the mesh configurations, including 113,285 nodes, had acceptable precision with a maximum error of 3%.

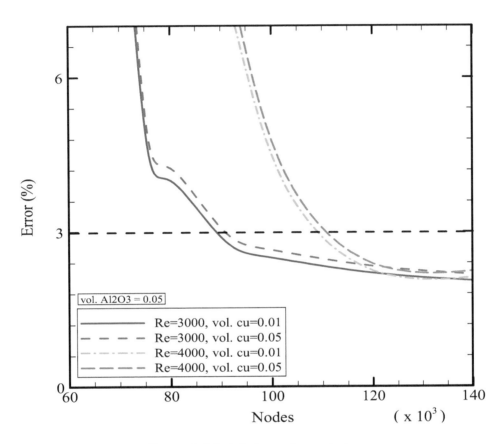

Figure 2. Mesh independency test.

Figure 3 shows code validation in the present study. Figure 3a illustrates dimensionless temperature dissimilarity versus dimensionless distance to validate the obtained results in the present work and the empirical data of Kuehn and Goldstein [54]. Due the lack of empirical results for natural heat transfer in a cylinder, the numerical process has been validated with the presented empirical data of Kuehn and Goldstein [54] for $\Delta T = 0.371$ K, Pr = 5.45, $Ra_L = 2.09 \times 10^5$, $L/2ri = 0.8$, and $T_{avg} = 303.18$ K, providing $T_i > T_o$. The judgments for different temperature profiles at various angles are indicated in Figure 3. It is seen clearly that there is a remarkable coincidence between the obtained results from reference [54] and the numerical data in the present investigation.

In addition, Figure 3b illustrates Nusselt number variation versus Reynolds number (2500 < Re < 3900) to validate the obtained results in the present work and the numerical data of Liu and Yu [55] in the transient flow regime. It is seen clearly that there is a good agreement between the obtained results from reference [55] and the numerical data in the present investigation.

Due to the specific geometric conditions of the problem, the boundary conditions, and the explanations in the manual of ANSYS-Fluent software, using the symmetry boundary condition in the present work is a very effective and appropriate method for modeling. In order to validate the boundary condition, Figure 3c is plotted. In this diagram, the results of modeling a full cylinder and a half cylinder modeled with the symmetry condition are presented. The good agreement of the answers is well observed, which indicates the correctness of the boundary conditions used.

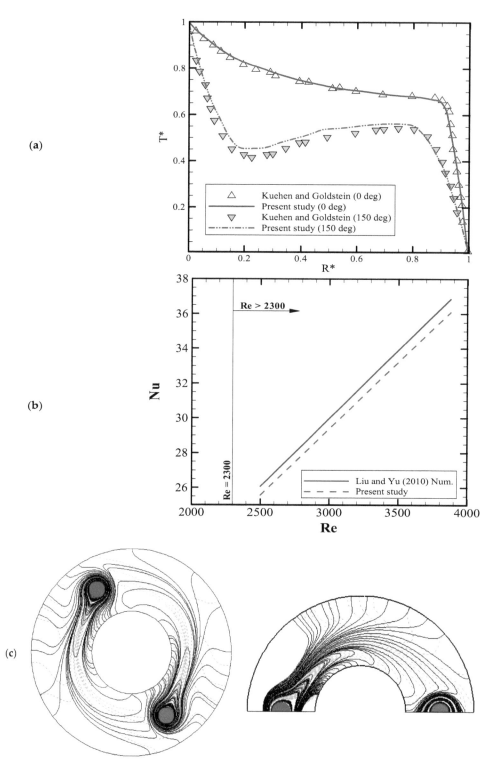

Figure 3. (**a**) Dimensionless temperature dissimilarity versus dimensionless distance to validate the obtained results in the present work and the empirical data of Kuehn and Goldstein [46]. (**b**) Code validation with numerical data of Liu and Yu [47] in transient regime flow. (**c**) Code validation for symmetry boundary condition.

3. Results and Discussion

Figure 4 illustrates streamlines and isotherm lines at different cross-sections of the studied cylinder for the case filled with Al_2O_3-Cu/water HNF at $\phi_1 = 0.05$, $\phi_2 = 0.01$, $\varepsilon = 0.2$, and Re = 3000. As seen in this figure, HNF flow and heat transfer through the z-direction of the cylinder are developed.

Additionally, the calculations show that different vortexes are through the annulus. The main heat transfer mechanism in the annulus is convection and the next one is radiation. However, the portion of radiation heat transfer is sharply lower than the convective mechanism. In addition, the presence of rods leads to more heat transfer because of vortexes being made. Rods are like obstacles that destroy the laminar sublayers and make local turbulence which enhance the convective heat transfer coefficient in the annulus and intensify the Nu. Figure 5 illustrates streamlines and isotherm lines at different cross-sections of the studied cylinder for the case filled with Al_2O_3-Cu/water HNF at $\phi_1 = 0.05$, $\phi_2 = 0.01$, $\varepsilon = 0.2$, and Re = 4000. As seen in this figure, HNF flow and heat transfer through the z-direction of the cylinder are developed. Additionally, the calculations show that different vortexes are through the annulus. The main heat transfer mechanism in the annulus is convection and the next one is radiation. However, the portion of radiation heat transfer is sharply lower than the convective mechanism. Additionally, the presence of rods leads to more heat transfer because of vortexes being made. Rods are like obstacles that destroy the laminar sublayers and make local turbulence which enhance the convective heat transfer coefficient in the annulus and augment the Nu. Additionally, it is realized that higher Reynolds numbers lead to more heat transfer coefficients because of more vortexes and turbulence through the cylindrical annulus. By comparing the streamlines and isotherm lines in Figures 4 and 5, some similarities between them are clearly seen.

Figure 6 illustrates the streamlines and isotherm lines at different cross-sections of the studied cylinder for the case filled with Al_2O_3-Cu/water HNF at $\phi_1 = 0.05$, $\phi_2 = 0.05$, $\varepsilon = 0.2$, and Re = 4000. As seen in this figure, HNF flow and heat transfer through the z-direction of cylinder are developed. Additionally, the calculations show that different vortexes are through the annulus. The main heat transfer mechanism in the annulus is convection and the next one is radiation. However, the portion of radiation heat transfer is sharply lower than the convective mechanism. In addition, the presence of rods leads to more heat transfer because of vortexes being made. Rods are like obstacles that destroy the laminar sublayers and make local turbulence which enhance the convective heat transfer coefficient in the annulus and augment the Nu. Furthermore, it is realized that higher volume concentrations of nanoparticles lead to more heat transfer coefficients because of the higher thermal conductivity of HNF flow and also more vortexes and turbulence through the cylindrical annulus because of more dynamic viscosity values. By comparing the streamlines and isotherm lines in Figures 5 and 6, some similarities between them are clearly seen.

Figure 7 illustrates the streamlines and isotherm lines at different cross-sections of the studied cylinder for the case filled with Al_2O_3-Cu/water HNF at $\phi_1 = 0.05$, $\phi_2 = 0.01$, $\varepsilon = 0.7$, and Re = 3000. As seen in this figure, HNF flow and heat transfer through the z-direction of the cylinder are developed. Additionally, the calculations show that different vortexes are through the annulus. The main heat transfer mechanism in the annulus is convection and the next one is radiation. However, the portion of radiation heat transfer is sharply lower than the convective mechanism. In addition, the presence of rods leads to more heat transfer because of vortexes being made. Rods are like obstacles that destroy the laminar sublayers and make local turbulence which enhance the convective heat transfer coefficient in the annulus and augment the Nu. Figure 8 illustrates streamlines and isotherm lines at different cross-sections of the studied cylinder for the case filled with Al_2O_3-Cu/water HNF at $\phi_1 = 0.05$, $\phi_2 = 0.01$, $\varepsilon = 0.7$, and Re = 4000. As seen in this figure, HNF flow and heat transfer through the z-direction of cylinder are developed. Additionally, the calculations show that different vortexes are through the annulus. The main heat transfer mechanism in the annulus is convection and the next one is radiation. However, the portion of radiation heat transfer is sharply lower than the convective mechanism. Additionally, the presence of rods leads to more heat transfer because of vortexes being made. Rods are like obstacles that destroy the laminar sublayers and make local turbulence which enhance the convective heat transfer coefficient in the annulus and augment the Nu. Additionally, it is realized that higher Reynolds numbers lead to more heat transfer coefficients because of more vortexes and turbulence through the cylindrical annulus. By comparing the streamlines and isotherm lines in Figures 7 and 8, some similarities between them are clearly seen. As noted previously, the portion of radiation heat transfer inside the studied cylinder is significantly lower than the convection heat transfer. Therefore, variations in emissivity values do not have an important influence on HNF flow and heat transfer distribution inside the cylindrical annulus, which is shown clearly by comparing Figures 4–8 with each other. Higher emissivity values should lead to higher radiation heat transfer, but the portion of radiative heat transfer in the studied annulus is low and therefore, does not have an observable increment in HNF flow and heat transfer.

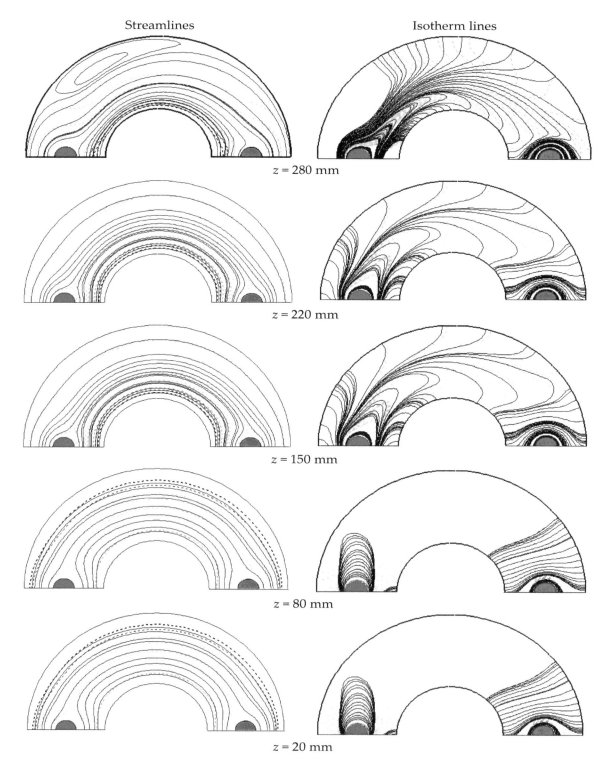

Figure 4. Streamlines and isotherm lines at different cross-sections of the studied cylinder for the case filled with Al$_2$O$_3$-Cu/water HNF at $\phi_1 = 0.05$, $\phi_2 = 0.01$, $\varepsilon = 0.2$, and Re = 3000.

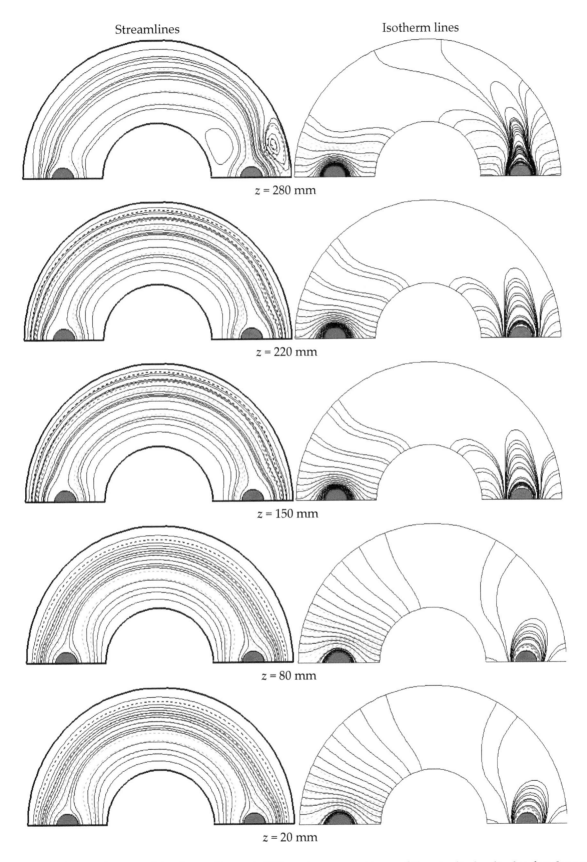

Figure 5. Streamlines and isotherm lines at different cross-sections of the studied cylinder for the case filled with Al_2O_3-Cu/water HNF at $\phi_1 = 0.05$, $\phi_2 = 0.01$, $\varepsilon = 0.2$, and Re = 4000.

Streamlines   Isotherm lines

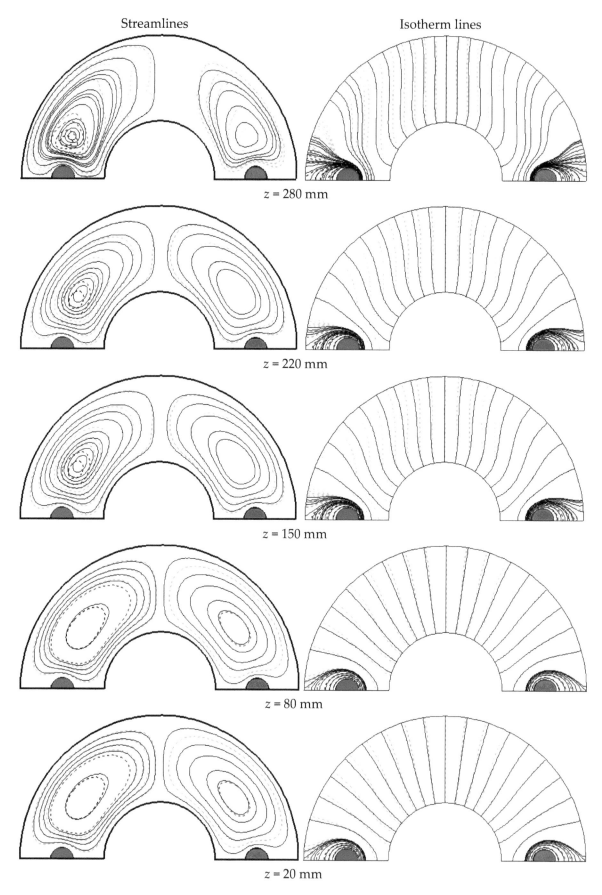

z = 280 mm

z = 220 mm

z = 150 mm

z = 80 mm

z = 20 mm

Figure 6. Streamlines and isotherm lines at different cross-sections of the studied cylinder for the case filled with Al_2O_3-Cu/water HNF at $\phi_1 = 0.05$, $\phi_2 = 0.05$, $\varepsilon = 0.2$, and Re = 4000.

Streamlines   Isotherm lines

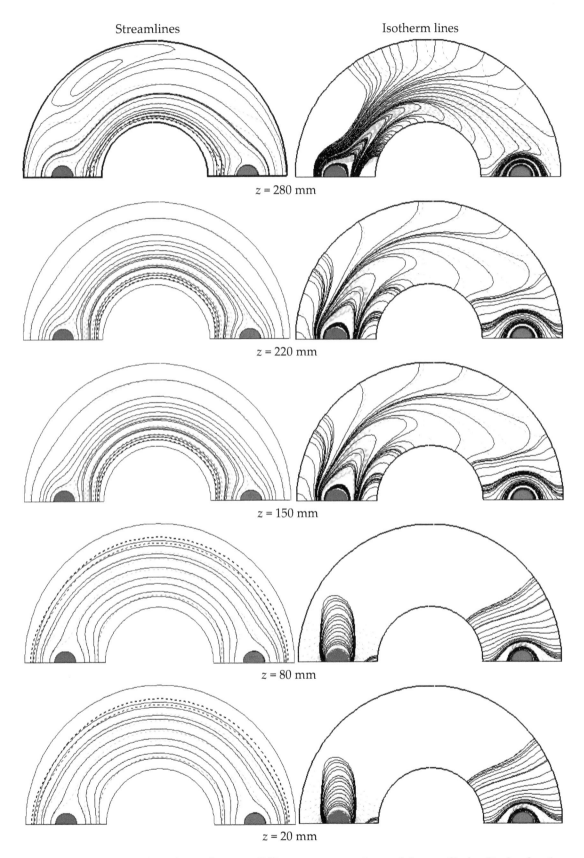

$z = 280$ mm

$z = 220$ mm

$z = 150$ mm

$z = 80$ mm

$z = 20$ mm

Figure 7. Streamlines and isotherm lines at different cross-sections of the studied cylinder for the case filled with Al_2O_3-Cu/water HNF at $\phi_1 = 0.05$, $\phi_2 = 0.01$, $\varepsilon = 0.7$, and Re = 3000.

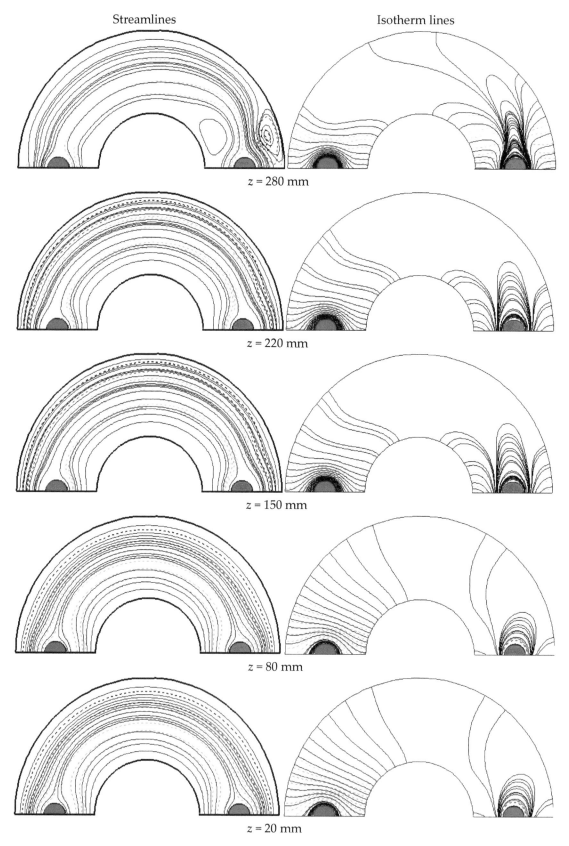

Figure 8. Streamlines and isotherm lines at different cross-sections of the studied cylinder for the case filled with Al_2O_3-Cu/water HNF at $\phi_1 = 0.05$, $\phi_2 = 0.01$, $\varepsilon = 0.7$, and Re = 4000.

Figure 9 illustrates streamlines and isotherm lines at different cross-sections of the studied cylinder for the case filled with Al_2O_3-Cu/water HNF at $\phi_1 = 0.05$, $\phi_2 = 0.05$, $\varepsilon = 0.7$ and Re = 4000. As seen in this figure, HNF flow and heat transfer through the z-direction of the cylinder are developed. Additionally, the calculations show different vortexes are through the annulus. The main heat transfer mechanism in the annulus is convection and the next one is radiation. However, the portion of radiation heat transfer is sharply lower than the convective mechanism. Additionally, the presence of rods leads to more heat transfer because of vortexes being made. Rods are like obstacles that destroy the laminar sublayers and make local turbulence which enhance the convective heat transfer coefficient in the annulus and augment the Nu. In addition, it is realized that higher volume concentrations of nanoparticles lead to more heat transfer coefficients because of more thermal conductivity of HNF flow and also, more vortexes and turbulence through the cylindrical annulus because of more dynamic viscosity values. By comparing the streamlines and isotherm lines in Figures 6 and 9, some similarities between them are clearly seen. As noted previously, the portion of radiation heat transfer inside the studied cylinder is significantly lower than the convection heat transfer. Therefore, variations in emissivity values do not have an important influence on HNF flow and heat transfer distribution inside the cylindrical annulus, which is shown clearly by comparing Figures 6 and 9 with each other. Higher emissivity values should lead to higher radiation heat transfer, but the portion of radiative heat transfer in the studied annulus is low and therefore, does not have an observable increment in HNF flow and heat transfer.

Figure 10 demonstrates the effects of using different base fluids on variation of predicted Nu_{ave} versus different volume fractions of Cu nanoparticles for the studied cylinder for the case filled with Al_2O_3-Cu HNF at $\phi_1 = 0.05$ or 0.10, $\varepsilon = 0.2$, and Re = 3000 or 4000. As seen in Figure 10a, for the model at Re = 3000 and $\phi_1 = 0.05$, all studied cases with different base fluids have similar behavior. For all studied cases, the total Nu_{ave} reduces firstly by an increment of the volume concentrations of Cu nanoparticles until $\phi_2 = 0.01$ or 0.02 and then, the total Nu_{ave} rises by an increment of the volume concentrations of Cu nanoparticles. Additionally, it is seen that for the case with water as the base fluid, the total Nu_{ave} at $\phi_2 = 0.05$ is more than the values at $\phi_2 = 0.00$, while for the other cases, the total Nu_{ave} at $\phi_2 = 0.05$ is less than the values at $\phi_2 = 0.00$. Furthermore, for all studied cases, the case with water as the base fluid has the maximum Nu_{ave}. As seen in Figure 10b, for the model at Re = 4000 and $\phi_1 = 0.05$, all studied cases with different base fluids have similar behavior. For all studied cases, the total Nu_{ave} reduces firstly by an increment of the volume concentrations of Cu nanoparticles until $\phi_2 = 0.01$ and then, the total Nu_{ave} rises by an increment of the volume concentrations of Cu nanoparticles. Additionally, it is seen that for the case with water as the base fluid, the total Nu_{ave} at $\phi_2 = 0.05$ is more than the values at $\phi_2 = 0.00$, while for the other cases, the total Nu_{ave} at $\phi_2 = 0.05$ is less than the values at $\phi_2 = 0.00$. Additionally, it is seen that by increment of Reynolds numbers, the Nu_{ave} augments, as found by the comparison of Figure 10a,b.

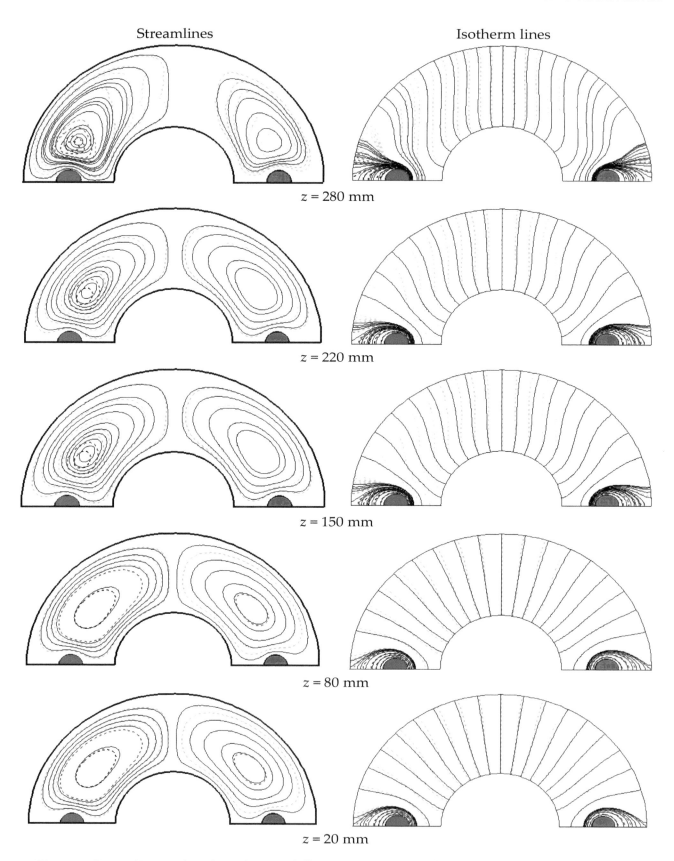

Figure 9. Streamlines and isotherm lines at different cross-sections of the studied cylinder for the case filled with Al_2O_3-Cu/water HNF at $\phi_1 = 0.05$, $\phi_2 = 0.05$, $\varepsilon = 0.7$, and Re = 4000.

As seen in Figure 10c, for the model at Re = 3000 and ϕ_1 = 0.10, all studied cases with different base fluids have similar behavior. For all studied cases, the total Nu_{ave} increases firstly by an increment of the volume concentrations of Cu nanoparticles until ϕ_2 = 0.01, 0.02, or 0.03 and then, the total Nu_{ave} diminishes by the intensification of volume concentrations of Cu nanoparticles. Additionally, it is seen that for the case with water as the base fluid, the total Nu_{ave} at ϕ_2 = 0.05 is more than the values at ϕ_2 = 0.00, while for the other cases, the total Nu_{ave} at ϕ_2 = 0.05 is less than the values at ϕ_2 = 0.00. Furthermore, for all studied cases, the highest Nu_{ave} belongs to the case with water as the base fluid. As seen in Figure 10d, for the model at Re = 4000 and ϕ_1 = 0.10, all studied cases with different base fluids have similar behavior. For all studied cases, the total Nu_{ave} reduces firstly by an increment of the volume concentrations of Cu nanoparticles until ϕ_2 = 0.01, 0.02, and 0.03, and then, the total Nu_{ave} rises by an increment of the volume concentrations of Cu nanoparticles. Additionally, it is seen that for the case with water as the base fluid, the total Nu_{ave} at ϕ_2 = 0.05 is more than the values at ϕ_2 = 0.00, while for the other cases, the total Nu_{ave} at ϕ_2 = 0.05 is less than the values at ϕ_2 = 0.00. Furthermore, for all studied cases, the case with water as the base fluid has the maximum Nu_{ave}. Additionally, it is seen that by increment of Reynolds numbers, the Nu_{ave} augments, as found by the comparison of Figure 10c,d.

Figure 11 shows different portions of the predicted Nu versus different volume fractions of Cu nanoparticles of the studied cylinder for the case filled with ethylene glycol-based HNF at ϕ_1 = 0.05, ε = 0.2, and Re = 3000. It is seen that just about 5 to 7% of the total average Nusselt number is related to radiation heat transfer and a huge portion is related to convection heat transfer. This is why variations of emissivity coefficient do not have a significant effect on the final results.

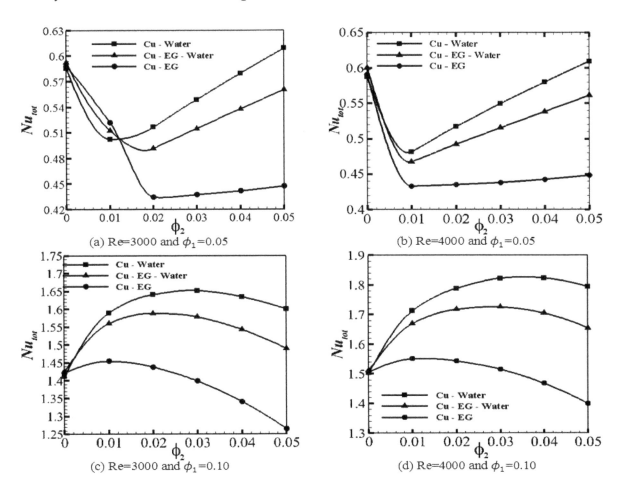

Figure 10. Effects of using different base fluids on variation of predicted Nu_{ave} versus different volume fractions of Cu nanoparticles of the studied cylinder for the case filled with Al$_2$O$_3$-Cu HNF at ϕ_1 = 0.05 or 0.10, ε = 0.2, and Re = 3000 or 4000.

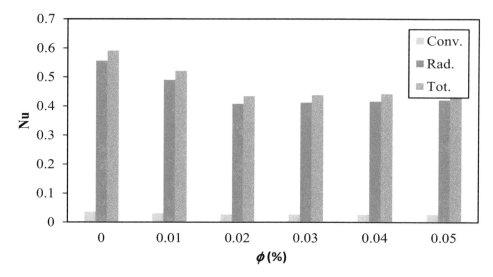

Figure 11. Different portions of the predicted Nu versus different volume fractions of Cu nanoparticles of the studied cylinder for the case filled with EG-based HNF at $\phi_1 = 0.05$, $\varepsilon = 0.2$, and Re = 3000.

4. Conclusions

The present paper investigates symmetry simulation of transient forced HNF flow and heat transfer in a three-dimensional annulus equipped with hot and cold rods. The chief goal of the current study is to model the geometry using a symmetry scheme and also examine the influences of dissimilar Reynolds numbers, emissivity coefficients, and dissimilar volume fractions of nanoparticles on the thermal and hydraulic characteristics of the studied annulus. The height of cylinder is changed from 300 mm. Additionally, the height of square is changed from 46.03 to 70.71 mm, the length of square is 1000 mm, and the gravity influences are determined in z direction. The system is made of an insulator in the center and outer sides of the cylinder, with a boundary condition of zero heat flux. However, this material is assumed with two different emissivity values, $\varepsilon = 0.2$ and $\varepsilon = 0.5$, in order to analyze the effects of emissivity values on radiation heat transfer, in addition to two different Reynolds numbers, Re = 3000 and 4000. For all studied models, the initial NF temperature is $T_{initial} = 350$ K and the cold side temperature is $T_c = 300$ K. The hot side temperature can be $T_h = 430$ K. The heat transfer fluid is a water-, ethylene–glycol- or water/ethylene–glycol mixture-based Cu-Al$_2$O$_3$ HNF, which makes a Newtonian NF. The RANS equations with the shear-stress (SST) k–ω turbulence model have been used for simulating the turbulence regime. The most important obtained results are as follows:

- For the model at Re = 3000 and $\phi_1 = 0.05$, all studied cases with different base fluids have similar behavior. For all studied cases, the total Nu_{ave} reduces firstly by an increment of the volume concentrations of Cu nanoparticles until $\phi_2 = 0.01$ or 0.02 and then, the total Nu_{ave} rises by an increment of the volume concentrations of Cu nanoparticles.

- For the case with water as the base fluid, the total Nu_{ave} at $\phi_2 = 0.05$ is more than the values at $\phi_2 = 0.00$, while for the other cases, the total Nu_{ave} at $\phi_2 = 0.05$ is less than the values at $\phi_2 = 0.00$.

- For all studied cases, the case with water as the base fluid has the maximum Nu_{ave}.

- For the model at Re = 4000 and $\phi_1 = 0.05$, all studied cases with different base fluids have similar behavior. For all studied cases, the total Nu_{ave} reduces firstly by an increment of the volume concentrations of Cu nanoparticles until $\phi_2 = 0.01$ and then, the total Nu_{ave} rises by an increment of the volume concentrations of Cu nanoparticles.

- By increment of Reynolds numbers, the Nu_{ave} augments.

- Higher emissivity values should lead to higher radiation heat transfer, but the portion of radiative heat transfer in the studied annulus is low and therefore, does not have an observable increment in HNF flow and heat transfer.

Author Contributions: Conceptualization, A.S.G.; methodology, M.B.; software, A.S.G. and B.H.; validation, B.H.; formal analysis, A.K.H.; investigation, A.K.H.; resources, A.K.H.; data curation, S.R.; writing—original draft preparation, S.R.; writing—review and editing, S.R.; visualization, M.B.; supervision, M.S.S.; project administration, M.S.S. All authors have read and agreed to the published version of the manuscript.

Acknowledgments: M. S. Shadloo acknowledges the access to French HPC resources provided by the French regional computing center of Normandy CRIANN (2017002).

Nomenclature

c_p	Specific heat (J/kg·K)
Cu	Copper
d_{np}	Nanoparticles size
$\vec{g}$	Gravity
k	Thermal conductivity (W/m·K)
Ra	Rayleigh number
M	Molecular weight of the base fluid
N	Avogadro number
Nu	Nusselt number
$\vec{P}$	Pressure
T_c	Cold side temperature
T_{fr}	liquid freezing point of base fluid
T_h	Hot side temperature
$T_{initial}$	Initial nanofluid temperature
T_m	Each section-temperature
$\vec{U}_{dr,bf}$	Nanoparticles drift velocity
$\vec{U}_{dr,s}$	Base fluid drift velocity
u_B	Average Brownian velocity
Greek Symbols	
$\vec{\alpha}$	Acceleration
ρ	Density (kg/m^3)
ε	emissivity
ϕ	volume concentration of nanoparticles
μ	Dynamic viscosity

References

1. Shahsavar, A.; Shaham, A.; Talebizadehsardari, P. Wavy channels triple-tube LHS unit with sinusoidal variable wavelength in charging/discharging mechanism. *Int. Commun. Heat Mass Transf.* **2019**, *107*, 93–105. [CrossRef]

2. Al-Rashed, A.A.; Shahsavar, A.; Rasooli, O.; Moghimi, M.; Karimipour, A.; Tran, M.D. Numerical assessment into the hydrothermal and entropy generation characteristics of biological water-silver nano-fluid in a wavy walled microchannel heat sink. *Int. Commun. Heat Mass Transf.* **2019**, *104*, 118–126. [CrossRef]

3. Shahsavar, A.; Al-Rashed, A.A.; Entezari, S.; Talebizadehsardari, P. Melting and solidification characteristics of a double-pipe latent heat storage system with sinusoidal wavy channels embedded in a porous medium. *Energy* **2019**, *171*, 751–769. [CrossRef]

4. Rostami, S.; Shahsavar, A.; Kefayati, G.; Goldanlou, A.S. Energy and Exergy Analysis of Using Turbulator in a Parabolic Trough Solar Collector Filled with Mesoporous Silica Modified with Copper Nanoparticles Hybrid Nanofluid. *Energies* **2020**, *13*, 2946. [CrossRef]

5. Yan, S.-R.; Moria, H.; Pourhedayat, S.; Hashemian, M.; Asaadi, S.; Dizaji, H.S.; Jermsittiparsert, K. A critique of effectiveness concept for heat exchangers; theoretical-experimental study. *Int. J. Heat Mass Transf.* **2020**, *159*, 120160. [CrossRef]

6.   Yi, Y.; Xie, X.; Jiang, Y. Optimization of solution flow rate and heat transfer area allocation in the two-stage absorption heat exchanger system based on a complete heat and mass transfer simulation model. *Appl. Therm. Eng.* **2020**, *178*, 115616. [CrossRef]

7.   Yildiz, C.; Arici, M.; Nizetic, S.; Shahsavar, A. Numerical investigation of natural convection behavior of molten PCM in an enclosure having rectangular and tree-like branching fins. *Energy* **2020**, *207*, 118223. [CrossRef]

8.   Shahsavar, A.; Rashidi, M.; Mosghani, M.M.; Toghraie, D.; Talebizadehsardari, P. A numerical investigation on the influence of nanoadditive shape on the natural convection and entropy generation inside a rectangle-shaped finned concentric annulus filled with boehmite alumina nanofluid using two-phase mixture model. *J. Therm. Anal. Calorim.* **2019**, *141*, 915–930. [CrossRef]

9.   Ma, Y.; Shahsavar, A.; Talebizadehsardari, P. Two-phase mixture simulation of the effect of fin arrangement on first and second law performance of a bifurcation microchannels heatsink operated with biologically prepared water-Ag nanofluid. *Int. Commun. Heat Mass Transf.* **2020**, *114*, 104554. [CrossRef]

10.   Zhang, R.; Aghakhani, S.; Pordanjani, A.H.; Vahedi, S.M.; Shahsavar, A.; Afrand, M. Investigation of the entropy generation during natural convection of Newtonian and non-Newtonian fluids inside the L-shaped cavity subjected to magnetic field: Application of lattice Boltzmann method. *Eur. Phys. J. Plus* **2020**, *135*, 184. [CrossRef]

11.   Liu, W.; Shahsavar, A.; Barzinjy, A.A.; Al-Rashed, A.A.; Afrand, M. Natural convection and entropy generation of a nanofluid in two connected inclined triangular enclosures under magnetic field effects. *Int. Commun. Heat Mass Transf.* **2019**, *108*, 104309. [CrossRef]

12.   Alsarraf, J.; Rahmani, R.; Shahsavar, A.; Afrand, M.; Wongwises, S. Effect of magnetic field on laminar forced convective heat transfer of MWCNT–Fe3O4/water hybrid nanofluid in a heated tube. *J. Therm. Anal. Calorim.* **2019**, *137*, 1809–1825. [CrossRef]

13.   Zheng, Y.; Shahsavar, A.; Afrand, M. Sonication time efficacy on Fe3O4-liquid paraffin magnetic nanofluid thermal conductivity: An experimental evaluation. Ultrason. *Sonochemistry* **2020**, *64*, 105004. [CrossRef] [PubMed]

14.   Alsarraf, J.; Shahsavar, A.; Khaki, M.; Ranjbarzadeh, R.; Karimipour, A.; Afrand, M. Numerical investigation on the effect of four constant temperature pipes on natural cooling of electronic heat sink by nanofluids: A multifunctional optimization. *Adv. Powder Technol.* **2020**, *31*, 416–432. [CrossRef]

15.   Al-Rashed, A.A.A.; Rahimi-Nasrabadi, M.; Aghaei, A.; Monfared, F.; Shahsavar, A.; Afrand, M. Effect of a porous medium on flow and mixed convection heat transfer of nanofluids with variable properties in a trapezoidal enclosure. *J. Therm. Anal. Calorim.* **2019**, *139*, 741–754. [CrossRef]

16.   Chen, Z.; Shahsavar, A.; Al-Rashed, A.A.; Afrand, M. The impact of sonication and stirring durations on the thermal conductivity of alumina-liquid paraffin nanofluid: An experimental assessment. *Powder Technol.* **2020**, *360*, 1134–1142. [CrossRef]

17.   Shahsavar, A.; Baseri, M.M.; Al-Rashed, A.A.; Afrand, M. Numerical investigation of forced convection heat transfer and flow irreversibility in a novel heatsink with helical microchannels working with biologically synthesized water-silver nano-fluid. Int. Commun. *Heat Mass Transf.* **2019**, *108*, 104324. [CrossRef]

18.   Liu, W.; Al-Rashed, A.A.; AlSagri, A.S.; Mahmoudi, B.; Afrand, M.; Afrand, M. Laminar forced convection performance of non-Newtonian water-CNT/Fe$_3$O$_4$ nano-fluid inside a minichannel hairpin heat exchanger: Effect of inlet temperature. *Powder Technol.* **2019**, *354*, 247–258. [CrossRef]

19.   Yang, L.; Ji, W.; Mao, M.; Huang, J.-N. An updated review on the properties, fabrication and application of hybrid-nanofluids along with their environmental effects. *J. Clean. Prod.* **2020**, *257*, 120408. [CrossRef]

20.   Geng, Y.; Al-Rashed, A.A.A.; Mahmoudi, B.; AlSagri, A.S.; Shahsavar, A.; Sardari, P.T. Characterization of the nanoparticles, the stability analysis and the evaluation of a new hybrid nano-oil thermal conductivity. *J. Therm. Anal. Calorim.* **2019**, *139*, 1553–1564. [CrossRef]

21.   Wu, H.; Al-Rashed, A.A.; Barzinjy, A.A.; Shahsavar, A.; Karimi, A.; Talebizadehsardari, P. Curve-fitting on experimental thermal conductivity of motor oil under influence of hybrid nano additives containing multi-walled carbon nanotubes and zinc oxide. *Phys. A Stat. Mech. Appl.* **2019**, *535*, 122128. [CrossRef]

22.   Shahsavar, A.; Talebizadeh, P.; Toghraie, D. Free convection heat transfer and entropy generation analysis of water-Fe$_3$O$_4$/CNT hybrid nanofluid in a concentric annulus. *Int. J. Numer. Methods Heat Fluid Flow* **2019**, *29*, 915–934. [CrossRef]

23. Shahsavar, A.; Godini, A.; Sardari, P.T.; Toghraie, D.; Salehipour, H. Impact of variable fluid properties on forced convection of Fe$_3$O$_4$/CNT/water hybrid nanofluid in a double-pipe mini-channel heat exchanger. *J. Therm. Anal. Calorim.* **2019**, *137*, 1031–1043. [CrossRef]

24. Salman, S.; Talib, A.R.A.; Saadon, S.; Sultan, M.T.H. Hybrid nanofluid flow and heat transfer over backward and forward steps: A review. *Powder Technol.* **2020**, *363*, 448–472. [CrossRef]

25. Li, L.; Tang, Z.; Li, H.; Gao, W.; Yue, Z.; Xie, G. Convective heat transfer characteristics of twin-web turbine disk with pin fins in the inner cavity. *Int. J. Therm. Sci.* **2020**, *152*, 106303. [CrossRef]

26. Chorin, P.; Moreau, F.; Saury, D. Heat transfer modification of a natural convection flow in a differentially heated cavity by means of a localized obstacle. *Int. J. Therm. Sci.* **2020**, *151*, 106279. [CrossRef]

27. Giwa, S.; Sharifpur, M.; Meyer, J. Effects of uniform magnetic induction on heat transfer performance of aqueous hybrid ferrofluid in a rectangular cavity. *Appl. Therm. Eng.* **2020**, *170*, 115004. [CrossRef]

28. El Mansouri, A.; Hasnaoui, M.; Amahmid, A.; Alouah, M. Numerical analysis of conjugate convection-conduction heat transfer in an air-filled cavity with a rhombus conducting block subjected to subdivision: Cooperating and opposing roles. *Int. J. Heat Mass Transf.* **2020**, *150*, 119375. [CrossRef]

29. Thiers, N.; Gers, R.; Skurtys, O. Heat transfer enhancement by localised time varying thermal perturbations at hot and cold walls in a rectangular differentially heated cavity. *Int. J. Therm. Sci.* **2020**, *151*, 106245. [CrossRef]

30. Ataei-Dadavi, I.; Rounaghi, N.; Chakkingal, M.; Kenjeres, S.; Kleijn, C.R.; Tummers, M.J. An experimental study of flow and heat transfer in a differentially side heated cavity filled with coarse porous media. *Int. J. Heat Mass Transf.* **2019**, *143*, 118591. [CrossRef]

31. Farsani, R.Y.; Mahmoudi, A.; Jahangiri, M. How a conductive baffle improves melting characteristic and heat transfer in a rectangular cavity filled with gallium. *Therm. Sci. Eng. Prog.* **2020**, *16*, 100453. [CrossRef]

32. Vishnu, A.; Aravind, G.; Deepu, M.; Sadanandan, R. Effect of heat transfer on an angled cavity placed in supersonic flow. *Int. J. Heat Mass Transf.* **2019**, *141*, 1140–1151. [CrossRef]

33. Sadaghiani, A.K.; Altay, R.; Noh, H.; Kwak, H.J.; Şendur, K.; Mısırlıoğlu, B.; Park, H.S.; Koşar, A. Effects of bubble coalescence on pool boiling heat transfer and critical heat flux—A parametric study based on artificial cavity geometry and surface wettability. *Int. J. Heat Mass Transf.* **2020**, *147*, 118952. [CrossRef]

34. Pan, M.; Wang, H.; Zhong, Y.; Hu, M.; Zhou, X.; Dong, G.; Huang, P. Experimental investigation of the heat transfer performance of microchannel heat exchangers with fan-shaped cavities. *Int. J. Heat Mass Transf.* **2019**, *134*, 1199–1208. [CrossRef]

35. Balotaki, H.K.; Havaasi, H.; KhakRah, H.; Hooshmand, P.; Ross, D. WITHDRAWN: Modelling of free convection heat transfer in a triangular cavity equipped using double distribution functions (DDF) lattice Boltzmann method (LBM). *Therm. Sci. Eng. Prog.* **2020**, 100495. [CrossRef]

36. Liu, Y.; Huang, H. Effect of three modes of linear thermal forcing on convective flow and heat transfer in rectangular cavities. *Int. J. Heat Mass Transf.* **2020**, *147*, 118951. [CrossRef]

37. Seo, Y.M.; Luo, K.; Ha, M.Y.; Park, Y.G. Direct numerical simulation and artificial neural network modeling of heat transfer characteristics on natural convection with a sinusoidal cylinder in a long rectangular enclosure. *Int. J. Heat Mass Transf.* **2020**, *152*, 119564. [CrossRef]

38. Razzaghpanah, Z.; Sarunac, N. Natural convection heat transfer from a bundle of in-line heated circular cylinders immersed in molten solar salt. *Int. J. Heat Mass Transf.* **2020**, *148*, 119032. [CrossRef]

39. Krakov, M.; Nikiforov, I. Influence of the shape of the inner boundary on thermomagnetic convection in the annulus between horizontal cylinders: Heat transfer enhancement. *Int. J. Therm. Sci.* **2020**, *153*, 106374. [CrossRef]

40. Pawar, A.P.; Sarkar, S.; Saha, S.K. Forced convective flow and heat transfer past an unconfined blunt headed cylinder at different angles of incidence. *Appl. Math. Model.* **2020**, *82*, 888–915. [CrossRef]

41. Alam, M.; Abdelhamid, T.; Sohankar, A. Effect of cylinder corner radius and attack angle on heat transfer and flow topology. *Int. J. Mech. Sci.* **2020**, *175*, 105566. [CrossRef]

42. Vyas, A.; Mishra, B.; Srivastava, A. Investigation of the effect of blockage ratio on flow and heat transfer in the wake region of a cylinder embedded in a channel using whole field dynamic measurements. *Int. J. Therm. Sci.* **2020**, *153*, 106322. [CrossRef]

43. Hadžiabdić, M.; Palkin, E.; Mullyadzhanov, R.; Hanjalić, K. Heat transfer in flow around a rotary oscillating cylinder at a high subcritical Reynolds number: A computational study. *Int. J. Heat Fluid Flow* **2019**, *79*, 108441. [CrossRef]

44. Aladdin, N.A.L.; Bachok, N.; Pop, I. Cu-Al$_2$O$_3$/water hybrid nanofluid flow over a permeable moving surface in presence of hydromagnetic and suction effects. *Alex. Eng. J.* **2020**, *59*, 657–666. [CrossRef]

45. Tseng, Y.; Ferng, Y.; Lin, C. Investigating flow and heat transfer characteristics in a fuel bundle with split-vane pair grids by CFD methodology. *Ann. Nucl. Energy* **2014**, *64*, 93–99. [CrossRef]

46. Chowdhury, D.; Neogi, S. Thermal performance evaluation of traditional walls and roof used in tropical climate using guarded hot box. *Constr. Build. Mater.* **2019**, *218*, 73–89. [CrossRef]

47. Jamil, B.; Akhtar, N. Effect of specific height on the performance of a single slope solar still: An experimental study. *Desalination* **2017**, *414*, 73–88. [CrossRef]

48. Qin, H.; Wang, C.; Zhang, D.; Tian, W.; Su, G.; Qiu, S. Parametric investigation of radiation heat transfer and evaporation characteristics of a liquid droplet radiator. *Aerosp. Sci. Technol.* **2020**, *106*, 106214. [CrossRef]

49. Kan, K.; Zheng, Y.; Chen, H.; Zhou, D.; Dai, J.; Binama, M.; Yu, A. Numerical simulation of transient flow in a shaft extension tubular pump unit during runaway process caused by power failure. *Renew. Energy* **2020**, *154*, 1153–1164. [CrossRef]

50. Naung, S.W.; Rahmati, M.; Farokhi, H. Direct Numerical Simulation of Interaction between Transient Flow and Blade Structure in a Modern Low-Pressure Turbine. *Int. J. Mech. Sci.* **2020**, 106104. [CrossRef]

51. Kan, K.; Chen, H.; Zheng, Y.; Zhou, D.; Binama, M.; Dai, J. Transient characteristics during power-off process in a shaft extension tubular pump by using a suitable numerical model. *Renew. Energy* **2021**, *164*, 109–121. [CrossRef]

52. Abdelhafiz, M.M.; Hegele, L.A.; Oppelt, J.F. Numerical transient and steady state analytical modeling of the wellbore temperature during drilling fluid circulation. *J. Pet. Sci. Eng.* **2020**, *186*, 106775. [CrossRef]

53. Ma, Y.; Mohebbi, R.; Rashidi, M.; Yang, Z. MHD convective heat transfer of Ag-MgO/water hybrid nanofluid in a channel with active heaters and coolers. *Int. J. Heat Mass Transf.* **2019**, *137*, 714–726. [CrossRef]

54. Kuehn, T.; Goldstein, R. Numerical solution to the Navier-Stokes equations for laminar natural convection about a horizontal isothermal circular cylinder. *Int. J. Heat Mass Transf.* **1980**, *23*, 971–979. [CrossRef]

55. Liu, D.; Yu, L. Experimental investigation of single-phase convective heat transfer of nanofluids in a minichannel. In Proceedings of the 14th International Heat Transfer Conference, IHTC14, Washington, DC, USA, 8–13 August 2010.

Numerical Simulations of the Flow of a Dense Suspension Exhibiting Yield-Stress and Shear-Thinning Effects

Meng-Ge Li [1], Feng Feng [1], Wei-Tao Wu [1] and Mehrdad Massoudi [2],* ⓘD

[1] School of Mechanical Engineering, Nanjing University of Science & Technology, Nanjing 210094, China;
menggeli@njust.edu.cn (M.-G.L.); nust203@aliyun.com (F.F.); weitaowwtw@njust.edu.cn (W.-T.W.)

[2] U.S. Department of Energy, National Energy Technology Laboratory (NETL), Pittsburgh, PA 15236, USA

* Correspondence: Mehrdad.Massoudi@NETL.DOE.GOV

Abstract: Many types of dense suspensions are complex materials exhibiting both solid-like and fluid-like behavior. These suspensions are usually considered to behave as non-Newtonian fluids and the rheological characteristics such as yield stress, thixotropy and shear-thinning/thickening can have significant impact on the flow and the engineering applications of these materials. Therefore, it is important to understand the rheological features of these fluids. In this paper, we study the flow of a nonlinear fluid which exhibits yield stress and shear-thinning effects. The geometries of interests are a straight channel, a channel with a crevice and a pipe with a contraction; we assume the fluid behaves as a Herschel-Bulkley fluid. The numerical simulations indicate that for flows with low Reynolds number and high Bingham number an unyielded plug may form in the center of the channel. In the case of a channel with a crevice, the fluid in the deep portion of the crevice is at an extremely high level of viscosity, forming a plug which is hard to yield. For the pipe with a contraction, near the pipe neck the unyielded region is smaller due to the enhanced flow disturbance.

Keywords: suspension; shear-thinning fluids; dense suspension; yield stress; non-Newtonian

1. Introduction

Fluid–solid dense suspension with yield stress are complex materials used in many industrial applications. These fluids usually show both solid-like and fluid-like behaviour [1–3]. In many situations, these complex fluids are modelled as non-linear fluids where their material properties can depend on temperature, shear rate, shear history and so forth [4,5]. The complex rheological behaviour can influence the design of the system [6]. The high viscosity and the yield stress increase the frictional losses, while the shear dependent viscosity and fluid–solid transition caused by the yield stress make it more difficult to control the operation of these complex fluids. Therefore, understanding the rheological behaviour of these non-linear fluids is of great importance in these industrial applications.

Complex fluids with yield stress show solid-like behavior, when the local stress is lower than a critical value, called the yield stress [7]. Many experiments have been designed and performed to investigate the behavior of these types of fluids. Coussot et al. (2002) [8] show the "abrupt" solid–fluid transition in gels and clay suspensions. They indicate that such an abrupt solid–fluid transition is associated with a bifurcation of the rheological property of the materials: for small stresses lower than the critical value, the apparent viscosity of the material increases in time, with the material stopping to flow eventually. Once the stresses become larger than the critical value, the apparent viscosity begins to decrease continuously (due to the change of the material microstructure) and as a result the flow accelerates. According to Stickel (2005) [6], at zero shear rate, dense suspensions may exhibit yield stress behavior, while general suspensions are Newtonian. As the shear rate increases, the viscosity of the suspensions is shear thinning until it reaches a Newtonian plateau, followed by a short region of shear thickening; however, the viscosity variation beyond this region is still inconclusive. Zhu et al. (2002) [9] measured the yield stress of suspensions using a slotted-plate technique, where multiple slots are opened and dragged through the suspensions. As a result, a shearing motion is created to eliminate the effects of the wall slip and the value of the yield stress could be determined by the point on the experimentally measured force versus time curve that deviates from linearity. Qian et al. (2018) [10] indicated that the dynamic yield stress can be obtained by fitting the Bingham model, where the equilibrium flow curve is measured by a shear rate controlled steady-state protocol. On the other hand, they suggested that the static yield stress is related to the creep stress where a viscosity bifurcation occurs in a stress-controlled creep recovery measurement. Although the concept of yield stress has been widely used in many engineering applications and investigated in numerous experiments, there still exits some uncertainty about accurate measurement of the yield stress [see Barnes and Walters (1985) [7], Barnes (1999) [5] and Putz and Burghelea (2009) [4]]. In 2017, Malkin et al. [11] discussed the recent developments in yield stress. They pointed out that yielding is now widely regarded as a transition that extends to a certain range of stress and this occurs over time, rather than the collapse of structure at the transition to fluid material. Therefore, the yield stress should be characterized by the durability depending on the stress.

The flow behavior of a yield stress fluid is difficult to predict, as it usually involves solid to fluid transition and shear thinning effects; these effects cannot be simply predicted a priori without the help of numerical simulations [12]. Based on the experimental tests, Gomes et al. [13] developed a finite volume model to study the flow of a Bingham plastic and a Herschel-Bulkley fluid in annular and jetting regions. Their numerical simulations agree with the results presented in the literature, indicating that numerical simulations supported by reliable experimental data can effectively promote the study of flow under extreme conditions. Saeid et al. [14] numerically investigated the effects of operating conditions and geometries on the flow of a suspension in a vertical well; the suspension is modeled as a water-based mixture exhibiting shear-thinning behavior. They found that increasing the number of jets and rotational speed as well as decreasing of Reynolds number both help reduce the pressure drop. In 2000, Subramanian [15] experimentally studied the pressure drop of five different suspension in pipe and annular flows and moreover, they compared the measured data with the results predicted by the numerical models using the Bingham plastic, power-law and yield power-law models. Their results indicated that for most of the suspension tested, the yield power-law model, namely the Herschel-Bulkley fluid, predicts better than the others but for different specific situations, the accuracy of the models needs further study. Ovarlez et al. (2015) [16] carried out experimental and theoretical investigations on the rheology of noncolloidal spheres suspended in yield stress fluids. Their results indicate that the Hershel-Bulkley model with the index of the interstitial fluid can be used to study suspensions. On the specific usage of Herschel-Bulkley model, Saasen et al. (2020) [17] suggested that accurate and proper parameters should be selected carefully and the dimensionless shear rate is better to use in the Herschel-Bulkley model for wider applications.

In this paper, we study the flow of a yield-stress and shear-thinning fluid in three representative geometries used in many engineering delivery applications. In Section 2, we discuss the governing equations and the constitutive relation. We present the geometries and a brief discussion of the boundary conditions in Section 3. And the numerical results are presented and discussed in Section 4.

2. Mathematical Framework

Flows of fluids infused with small macroscopic solid particles can be modeled in a variety of ways. These are multi-component materials, especially when the suspension is considered to be dense, that is, a large number of particles are present in the fluid. In general, at least three different ways of modeling, based on the methods in continuum mechanics, are available. The *first* and perhaps the most advanced approach is to use Mixture Theory where governing equations are presented for each component. In this approach, for a simple two-component case, three constitutive relations are needed; namely the stress tensor for each component and one vector equation for the interaction forces. In this approach, the coupled equations have to be solved numerically [see Soo (1990) [18]; Raj and Tao (1995) [19], Johnson et al. (1991) [20], Massoudi (2010) [21]]. In the *second* approach, the mixture is considered to be a non-homogeneous single component fluid; in this approach the governing equations for the single component fluid also includes a convection-diffusion equation for the particle motion. Two constitutive relations are needed: one for the stress tensor and one for the concentration flux [see Phillips et al. (1992) [22]; Wu et al. (2016) [23]; Tao et al. (2019) [24]]. This approach is simpler than the first one, both from a modeling point of view and also from a computational perspective. Finally, in the *third* approach, the suspension is treated as a single component homogenous fluid where the basic equations of motion are needed and only one constitutive relation is necessary, namely for the stress tensor. This is by far the simplest and most often used methods in engineering applications. In this paper, we use this approach. Other methods include numerical simulations, statistical mechanics and experimental or phenomenological approaches [see Peker and Helvaci, (2011) [25]].

2.1. Governing Equations

If thermo-chemical and electro-magnetic effects are ignored, the governing equations for a single component non-linear fluid, are the conservation equations for mass and linear momentum [26].

2.1.1. Conservation of Mass

The conservation of mass is:

$$\frac{\partial \rho}{\partial t} + div(\rho \boldsymbol{v}) = 0, \tag{1}$$

where ρ is the density of the fluid; $\partial / \partial t$ is the partial derivative with respect to time; div is the divergence operator and $\boldsymbol{v}$ is the velocity vector. For an incompressible fluid, the equation is simplified to:

$$div \, \boldsymbol{v} = 0. \tag{2}$$

2.1.2. Conservation of Linear Momentum

The conservation of linear momentum is:

$$\rho \frac{d\boldsymbol{v}}{dt} = div\boldsymbol{T} + \boldsymbol{b}, \tag{3}$$

where $\boldsymbol{T}$ is the Cauchy stress tensor, $\boldsymbol{b}$ is the body force vector, which is ignored in this paper and d / dt is the total time derivative, given by $d(.) / dt = \partial(.) / \partial t + [grad(.)]\boldsymbol{v}$, where $grad$ designates the gradient operator. The conservation of angular momentum indicates that in the absence of couple stresses the stress tensor is symmetric, that is, $\boldsymbol{T} = \boldsymbol{T}^{\mathrm{T}}$. Notice that in flows of fluid–solid suspensions, in general, the fluid is non-homogenous and particle volume fraction is variable, thus the convection-diffusion equation is necessary in many applications [27]. For alternative ways of modeling non-homogeneous fluids, we refer the reader to Massoudi and Vaidya (2008) [28]. In this paper, we ignore the presence of the particles and model the suspension as a homogenous nonlinear fluid.

2.2. Constitutive Equation for the Stress Tensor

Dense suspensions are complex multi-component fluids; they are composed of a base (host) fluid, for example, water, with other dispersed components such as sand, oil and so forth, as well as some chemicals. Both viscous stress with shear-thinning characteristic and yield-stress behavior should be considered in the constitutive relation for stress tensor of these complex fluids. We therefore assume that the total stress tensor is composed of two parts, where:

$$\boldsymbol{T} = \boldsymbol{T}_y + \boldsymbol{T}_v,$$

where T_y is the yield stress and T_v is the viscous stress. Many nonlinear fluids show distinct non-Newtonian features, such as thixotropy, yield stress, shear dependent viscosity [29–31] and so forth. In this paper, we consider the constitutive equation for a fluid which has a yield stress in addition to its viscous characteristics:

$$T = -pI + T_y + T_v \tag{4}$$

$$T_v = \eta_p(\dot{\gamma})A_1 \tag{5}$$

$$A_1 = \mathrm{grad}v + \mathrm{grad}v^T, \dot{\gamma} = \sqrt{1/2 tr(A_1)^2}, \tag{6}$$

where p is the pressure, I the identity tesnor, η_p is the shear (plastic) viscosity depending on the shear rate, tr is the trace of a tensor and $\dot{\gamma}$ is the shear rate, which is defined by the second invariant of A_1.

As observed by many experiments [2,32,33], many dense suspensions show shear-thinning behavior which can be described by a power law model:

$$\eta_p = \eta_r \dot{\gamma}^{n-1}, \tag{7}$$

where η_r is the reference viscosity, and n is the power-law exponent, indicating the intensity of the shear rate dependency; when $n < 1$, the fluid is shear-thinning, when $n > 1$ the fluid is shear-thickening, and when $n = 1$ the fluid behaves as a Newtonian fluid. The shear viscosity of non-linear fluids is, in general, a function of several parameters such as time, shear rate, temperature, pressure, concentration, material composition, molecular weight, electric field, magnetic field and so forth. In this paper, we assume that the viscosity depends only on the shear rate. That is, even though we recognize that particle concentration or in general, the microstructure of the suspension, can influence the rheological behavior of the suspension, for example, as sedimentation, in this paper, we are ignoring those factors ([see our previous work [24,27] for details).

In general, a yield-stress fluid is broadly defined as a material that behaves as a solid material below a critical applied stress and would flow like a fluid at higher stresses. One of the most popular model for a yield stress fluid is the generalized Bingham viscoplastic fluid model, known as the as Herschel-Bulkley model where the constitutive equation is given as [34,35]:

$$\begin{cases} T_y = \left(\frac{\tau_y}{\sqrt{1/2 II_{DA_1}}} + \eta_p \right)A_1, \ \sqrt{1/2 II_T} \geq \tau_y \\ A_1 = 0, \ \sqrt{1/2 II_T} < \tau_y \end{cases}, \tag{8}$$

where τ_y is the yield stress, II_{A_1} is the second invariant of the A_1 and II_T is the second invariant of the stress tensor. If η_p is constant, then the above equation reduces to the basic Bingham model [see Prager (2004), p. 136 [36]] The effective viscosity, defined as the ratio of the shear stress to the shear rate, is:

$$\eta = \frac{T}{A_1} = \eta_p + \frac{\tau_y}{\dot{\gamma}}. \tag{9}$$

If $T > \tau_y$ everywhere, then η_p is well defined and in the computational scheme we can use methods suitable for Newtonian fluids. However, when the stress approaches the yield stress, the viscosity will tend to infinity and this is physically unreasonable; therefore, we need some other methods to approximate the viscosity. The most common approach is to remove the discontinuity by regularization [34,37], which transforms the computational problem into a conventional one for a purely viscous fluid and then to vary the regularization parameter to try to obtain convergence to the solution of the discontinuous problem. At least three regularization approaches are used [34,38]:

Simple model [39]:

$$\eta = \eta_p + \tau_y \left(\frac{1}{\varepsilon + \dot{\gamma}} \right). \tag{10}$$

Bercovier and Engelman model [40]:

$$\eta = \eta_p + \tau_y \left(\frac{1}{\left[\varepsilon^2 + \dot{\gamma}^2 \right]^{1/2}} \right). \tag{11}$$

Papanastasiou model [41]:

$$\eta = \eta_p + \tau_y \left(\frac{1 - e^{-\dot{\gamma}/\varepsilon}}{\dot{\gamma}} \right). \tag{12}$$

The Bingham model should be obtained in all three formulations as $\varepsilon \to 0$. Physically, the regularized viscosity approaches a large but finite viscosity at zero shear rate. In this paper, we first test all three regularization methods and then we use the simple model combined with the power law model for the simulations. That is, we assume:

$$\eta = \eta_r \dot{\gamma}^{n-1} + \frac{\tau_y}{\varepsilon + \dot{\gamma}}. \tag{13}$$

2.3. Expanded Form of the Governing Equations

The following dimensionless form of the governing equations can be obtained:

$$div^* \boldsymbol{V} = 0 \tag{14}$$

$$\rho^* \left(\frac{\partial \boldsymbol{V}}{\partial \tau} + \boldsymbol{V} \cdot grad^* \boldsymbol{V} \right) = -grad^* P + \frac{1}{Re^*} div^* \left[\left(\dot{\gamma}^{*m} + \frac{B}{\left[\varepsilon^2 + \dot{\gamma}^{*2} \right]^{1/2}} \right) \boldsymbol{A_1}^* \right] \tag{15}$$

using the non-dimensional parameters:

$$Y = \frac{y}{H}, X = \frac{x}{H}, \boldsymbol{V} = \frac{v}{v_0}, \tau = \frac{t v_0}{H}, P = \frac{p}{\rho_r v_0^2}, \rho^* = \frac{\rho}{\rho_r}, Re^* = \frac{\rho_r v_0^{1-m} H^{1+m}}{\eta_r}$$

$$grad^*(\cdot) = H grad(\cdot), div^*(\cdot) = H div(\cdot), \boldsymbol{A_1}^* = \frac{H}{v_0} \boldsymbol{A_1}, B = \frac{\tau_y H^{1+m}}{\eta_r v_0^{1+m}},$$

where v_0 is a reference velocity and m equals $n - 1$ in Equation (13) for simplicity. As a result of this non-dimensionalization, two dimensionless numbers Re^* and B—appear, where Re^* is the ratio of the inertia force to the viscous force and is similar to the Reynolds number and B is the ratio of the yield stress to the viscous stress which is equivalent to the Bingham number. Flow of yield-stress fluids are frequently characterized by the dimensionless Bingham number, reflecting the relative importance of the yield stress and the viscous stress.

3. Problem Description

In this paper, we assume that the flow is laminar. We investigate the flow of a yield-stress shear-thinning fluid in three different geometries, see Figure 1 for the details. First, we will study the flow in a horizontal and straight channel to reveal the basic features of the fluid, see Figure 1a. Half of the height is H and the length of the channel is kept at $10H$, which is long enough to guarantee that for all the cases the flow near the crevice and the contraction will not be affected by the outlet boundary. The criterion is that the gradient of the dimensionless velocity along the axial direction is less than 10^{-3}. In order to save computational cost, we have assumed that the flow is two dimensional (axisymmetric).

The second problem is a two-dimensional channel with a crevice representing geometries commonly found in industrial products, such as the parts joints and weld crack, as shown in Figure 1b. The width of the channel is W_i and L_i is the length of the straight channel, assumed to be $3.3W_i$ (for a fully developed flow). In order to simulate the effect of crevice, the width W_c and the depth D are both set as $0.7W_i$. The length of the channel after the crevice is $Lo = 2W_i$.

Figure 1c shows the geometry of a two-dimensional pipe with a contraction, which may represent a portion of a valve in a real industrial application. We also make use of the symmetry in this problem. The radius of the pipe is R_i and the radius in the contraction part is $R_c = 0.3R_i$, the length of the contraction is R_i, the length of the part before and after the contraction are set as $5R_i$ and $4R_i$, respectively.

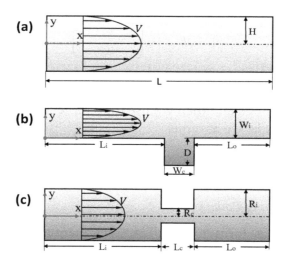

Figure 1. Schematic of (a) straight channel (b) channel with a crevice (c) pipe with a contraction.

Table 1 presented the boundary conditions. The average apparent viscosity is also defined in this paper, $\bar{\eta}_1$ for the pipe geometry and $\bar{\eta}_2$ for the channel:

$$\begin{aligned}\bar{\eta}_1 &= \tfrac{1}{R^2}\int 2\eta r\,dr \\ \bar{\eta}_2 &= \tfrac{1}{W}\int \eta\,dy.\end{aligned} \tag{16}$$

The grid convergence tests are carried out in order to determine the appropriate meshes to use. Here we only show the results of the straight channel with an inlet generalized Reynolds number of 1×10^{-2} and a Bingham number of $612\left(Re^* = 3.3\times10^{-3}, B = 15\right)$, as shown in Table 2 As a result, Grid C is chosen for further studies. The convergence criterion is that the residuals of pressure and velocity are less than 10^{-6} and 10^{-7}, respectively.

Table 1. Boundary conditions used in the numerical simulations.

Boundary Type	Boundary Conditions	
	Pressure	Velocity
Wall	Zero gradient	Fixed value (0)
Inlet	Zero gradient	Fixed value
Outlet	Fixed value (0)	Zero gradient

Table 2. Mesh dependency study.

Label	Grid Number	Mean Plastic Viscosity at Exit (10^2)
Grid A	25,000	2.12074
Grid B	43,956	2.04562
Grid C	69,104	2.00631
Grid D	100,000	1.99802

4. Results and Discussion

We define a generalized Reynolds number often used in flow of non-Newtonian fluids using the power-law type model [42] $\left(\eta = \eta_r \dot{\gamma}^{n-1}\right)$:

$$Re = \frac{\rho v^{2-n} d^n}{\eta_r \left(0.75 + \frac{0.25}{n}\right)^n 8^{n-1}},\tag{17}$$

where d is the inlet diameter. We also use the dimensionless Bingham number $Bn = \tau_y H / \eta_r v_0$ $\left(B = (H/v_0)^m Bn\right)$ to characterize the influence of the yield stress.

4.1. Convergence Properties of the Regularization Methods

For the flow of yield stress fluids, as the yield surface is approached, the presence of the $\dot{\gamma}$ in the denominator of Equation (9) makes the apparent viscosity unbounded. Moreover, while calculating the velocity field, the shape and the location of the yield surface are unknown a priori. To overcome this difficulty, several regularization methods have been proposed; these are continuous and apply to both the yielded and unyielded regions. In this section, we first test the convergence properties of three typical regularization methods, introduced in Section 2.2, on the determination of the yield surface; we do this by simulating the flow of a yield stress fluid in a straight channel for different regularization parameter, ε. We set the reference length and the reference velocity as half of the inlet height and the mean velocity, respectively; we also assume $Re = 1.18 \times 10^{-4}$, $Bn = 163$ $\left(Re^* = 3.89 \times 10^{-5}, B = 31\right)$. The yield surface for different regularization parameter using the three regularization methods are shown in Figure 2 It can be observed that the unyielded region is reduced as ε (the regularization parameter) decreases and eventually converges. In addition, when ε is small enough for convergence, the yield surface of the three regularization methods is nearly at the same position. However, the maximum value of ε required to represent the yield surface is much lower for the Simple model than that of the Papanastasiou and the Bercovier & Engelman models. Therefore, the appropriate choice of the regularization parameter should be made depending on both the flow condition and the regularization method. In the following simulations, the simple regularization method with the proper regularization parameter, ε, is used.

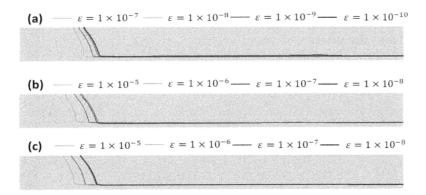

Figure 2. Yield surfaces in the straight channel with different regularization parameter using: (a) Simple; (b) Papanastasiou; (c) Bercovier & Engelman models.

4.2. Flow in a Straight Channel

We first investigate the flow characteristics of a yield-stress, shear-thinning fluid in a straight channel. The reference length and the reference velocity are half of the inlet height and the mean velocity. Typical flow fields with inlet Reynolds number of 1.18×10^{-4} and Bingham number of 612 $\left(Re^* = 3.89 \times 10^{-5}, B = 117\right)$ are shown in Figure 3 As it can be seen, the maximum value of the velocity is obtained at the center line where the shear rate is almost zero, resulting in a high plastic viscosity. The apparent viscosity depends on both the plastic viscosity and the yield stress, see Equation (9), therefore, the apparent viscosity reaches an extremely high value near the center line. In this region, the fluid has not yielded and it is flowing with the same velocity, as the plug flow. This observation is similar to Abdali's work (1992) [43] where the flow is fully developed, also see Mitsoulis (2017) [44] for a review of several benchmark problems dealing with yield stress shear-thinning fluids.

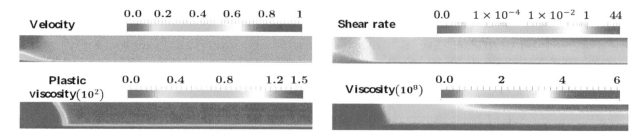

Figure 3. Velocity, shear rate, plastic viscosity and apparent viscosity fields in a straight channel with inlet Reynolds number of 1.18×10^{-4} and $B_n = 612$.

Figure 4 shows the velocity and viscosity distributions along the Y-axis for different Reynolds numbers. Here, the reference velocity, V_0, is the mean velocity at the inlet, the reference viscosity η_r is the viscosity coefficient appearing in the power-law model; the viscosity is plotted in logarithmic coordinates. It can be seen that as the Reynolds number increases, the velocity profile becomes more parabolic and near the axis, the constant-velocity profile becomes smaller, indicating that the size of the unyielded plug is reduced. That is, the shear banding becomes larger. Figure 4b indicates that the plastic viscosity gradually increases near the pipe axis, due to the decreasing of the shear rate; the gradient of the apparent viscosity becomes large near the edge of the plug flow region. The variation of the apparent viscosity is similar to the plastic viscosity; however, with the influence of the yield stress, the magnitude of the apparent viscosity is several times larger than the plastic viscosity. Furthermore, the apparent viscosity in the yield region decreases when the Reynolds number increases, while in the plug flow region, the difference is less significant. The position of high viscosity gradient gradually approaches the center line. The yield region for different Reynolds numbers are shown in Figure 5 It can be seen that the plug flow decreases for larger Reynolds number and the variation is similar to that of the apparent viscosity. The location where the apparent viscosity abruptly changes also roughly coincides with the location where the yield occurs; in addition, due to the flow disturbance the yield surface moves away from the inlet.

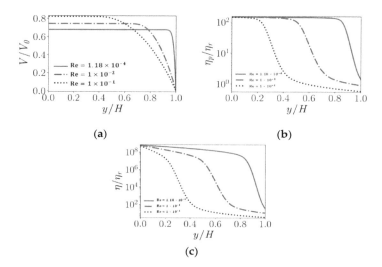

Figure 4. Profiles for: **(a)** velocity; **(b)** plastic viscosity; **(c)** apparent viscosity along the Y-axis in the straight channel for different inlet Reynolds numbers.

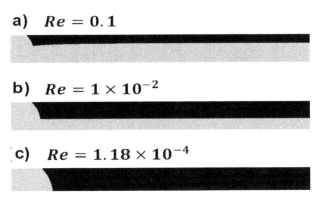

a) $Re = 0.1$

b) $Re = 1 \times 10^{-2}$

c) $Re = 1.18 \times 10^{-4}$

Figure 5. Distribution of the yield surfaces for different Reynolds numbers. The shaded regions are unyielded.

In addition, we also consider the effect of the Bingham number. The viscosity and velocity profiles along the Y-axis for different Bingham numbers are plotted in Figure 6. The Bingham number characterizes the ratio of the yield stress to the viscous stress; as the Bingham number increases, the plug size becomes larger. The viscosity increases for higher Bingham numbers, which is in contrast to the behavior for the Reynolds number; furthermore, the maximum value of the apparent viscosity changes with Bingham number. Figure 7 shows the yield region for different Bingham numbers. Similar to the previous studies, the yield region is closely related to the change of the apparent viscosity and the yield surface moves towards the wall and the inlet when the Bingham number increases.

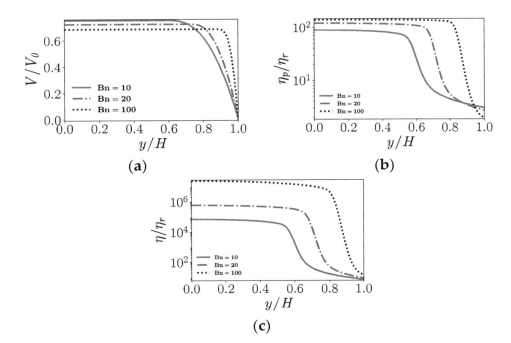

Figure 6. Profiles for: (**a**) velocity; (**b**) plastic viscosity; (**c**) the apparent viscosity along the Y-axis in the straight channel for different Bingham numbers and $Re = 1.18 \times 10^{-4}$.

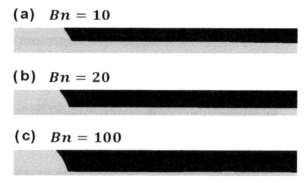

Figure 7. Distribution of the yield surfaces for different Bingham numbers and $Re = 1.18 \times 10^{-4}$. The shaded regions are unyielded.

Figure 8 shows the velocity and the viscosity profiles along the Y-axis for different values of m, which indicates the intensity of the shear dependent viscosity. Since in this paper we are assuming that the fluid behaves as a shear-thinning fluid, then the value of m is kept below zero. As observed in Figure 8, the plastic viscosity rises with the increase of the absolute value of m; on the other hand, the apparent viscosity is negatively proportional to the absolute value of m. Meanwhile, the gradient of the apparent viscosity becomes less as m decreases but the maximum value of the apparent viscosity is hardly affected by m. As the absolute value of m decrease a lower velocity near the pipe center is obtained, resulting in a larger plug flow.

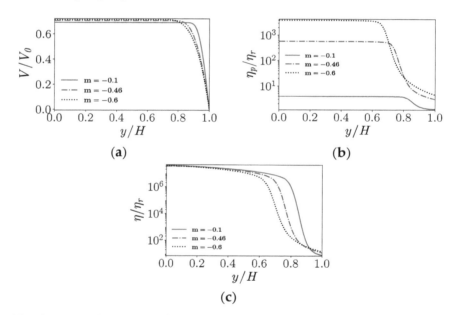

Figure 8. Profiles for: **(a)** velocity; **(b)** plastic viscosity; **(c)** apparent viscosity along the Y-axis in the straight channel for different values of the exponent m in the power law model.

4.3. Flow in a Channel with a Crevice

In this section, we study the flow in a channel with a crevice. The height of the inlet section is chosen as the reference length. Figure 9 shows the velocity fields, streamlines, the shear rate, plastic viscosity and the apparent viscosity distribution when $Re = 7.56 \times 10^{-5}$ and $B_n = 40.8$ $\left(Re^* = 3.89 \times 10^{-5}, B = 7.8\right)$. For the flow in the channel with a crevice, the velocity distribution in the main channel is similar to that in the straight channel, where a plug flow is observed near the center of the channel. In the crevice, the velocity and the shear rate remain nearly zero, resulting in a high viscosity and with little or no yielded region. It can also be seen that some of the fluid flows into the crevice; however, it is difficult for the fluid in the deep portion of the crevice to move due to the existence of the yield stress. By looking at the streamlines inside the crevice, we can see that the incoming flow forms a free shear layer near the interface of the main channel and the crevice and the small disturbances grow into vortical structures that propagate inside the crevice. Furthermore, near the crevice, the flow in the main channel is disturbed a little, resulting in a reduction of the plastic and the apparent viscosity in that region.

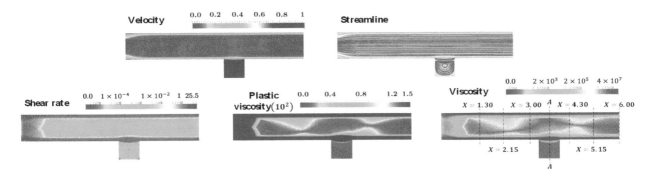

Figure 9. The velocity, streamlines, shear rate, plastic viscosity and plastic viscosity fields in the channel with a crevice with an inlet Reynolds number of 7.56×10^{-5} and $B_n = 40.8$.

The velocity and the viscosity profiles along the line A-A are shown in Figure 10. It can be seen that the velocity is nearly zero in the crevice and near the center of the main channel the velocity has a constant value. In the main channel, the higher Reynolds number leads to higher maximum velocity and smaller plug region, which indicates that the unyielded region can be suppressed by increasing the inlet Reynolds number (usually the inlet velocity); in the crevice, the fluid begins to move as the Reynolds number increases, albeit at low velocity. Along line A-A, the viscosity fluctuates greatly; this could be due to the flow disturbance caused by the presence of the crevice. Interestingly, when the Reynolds number reaches 0.01, the plastic viscosity inside the crevice drops a little and then begins to rise to reach the maximum value, which is in accordance with shape of the vortical structures observed in the streamlines in the crevice. Figures 11 and 12 show the viscosity profiles in the Y-axis at different X positions, from which we can see that the distribution is no longer consistent with the profiles of the straight channel because of the crevice; the maximum value of the viscosity is no longer in the right of the center of the channel and differs with X positions. The profiles at symmetrical positions along the crevice, such as X = 3.0 and X = 4.3, present similar patterns, see the distributions in Figures 9 and 13. The profiles for the yield region with different Reynolds numbers are shown in Figure 13. It can be seen that the crevice does not play a favorable role for yield stress fluid; as the Reynolds number increases to 0.01, even though more fluid has yielded at the top of the crevice, most of the fluid in the crevice remain stagnant.

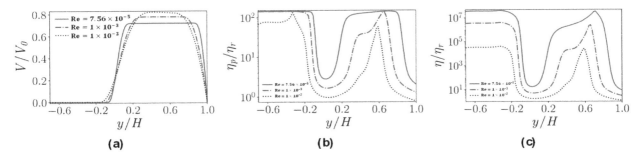

Figure 10. (**a**) Velocity; (**b**) Plastic viscosity; (**c**) Apparent viscosity profiles along line A-A for different inlet Reynolds numbers.

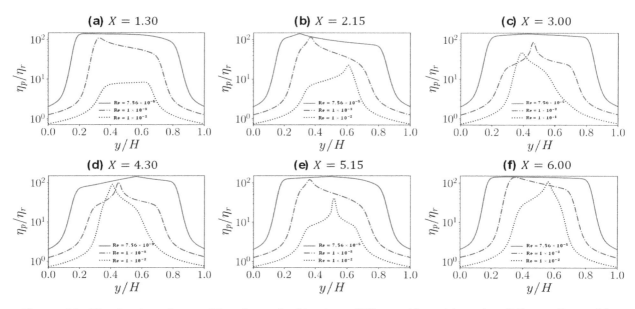

Figure 11. Plastic viscosity profiles along the Y-axis at different X positions for different Reynolds numbers ($X = x/H$).

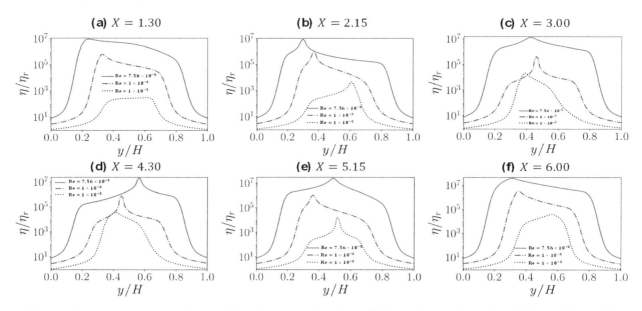

Figure 12. Apparent viscosity profiles along the Y-axis at different X positions for different Reynolds numbers ($X = x/H$).

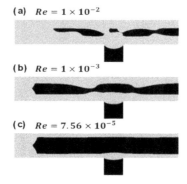

Figure 13. Distribution of the yield surfaces for different Reynolds numbers. The shaded regions are unyielded.

Figures 14–17 show the effect of the Bingham number. As the Bingham number increases, the velocity decreases, while the viscosity rises, which results in larger unyielded region. In addition, the viscosity becomes more non-linear as the Bingham number increases. The profiles for the yield region are plotted in Figure 17, which has a similar shape to that in Figure 13. Therefore, with higher Reynolds number and lower Bingham number, the size of unyielded plug is reduced.

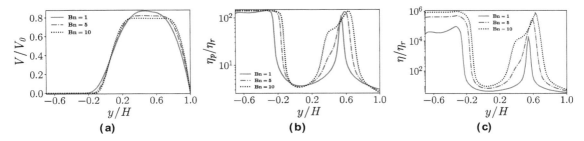

Figure 14. (a) Velocity (b) Plastic viscosity (c) Apparent viscosity profiles along A-A for different Bingham numbers and $Re = 7.56 \times 10^{-5}$.

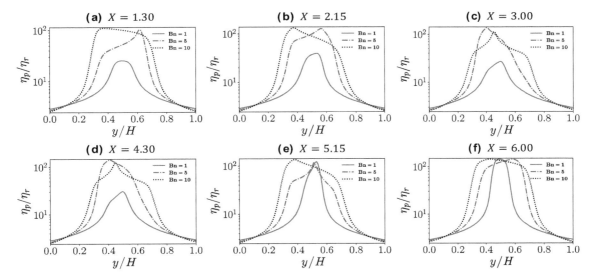

Figure 15. Plastic viscosity profiles along the Y-axis at different X positions for different Bingham numbers ($X = x/H$).

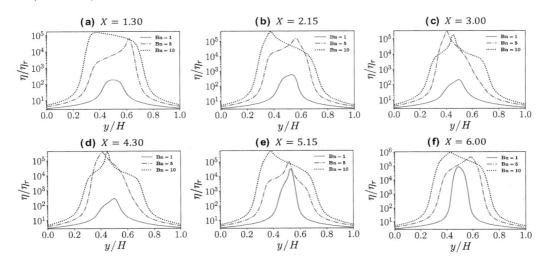

Figure 16. Apparent viscosity profiles along the Y-axis at different X positions for different Bingham numbers ($X = x/H$).

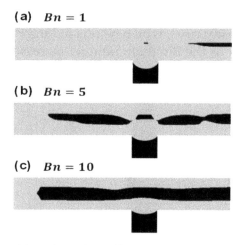

Figure 17. Distribution of the yield surfaces for different Bingham numbers and $Re = 7.56 \times 10^{-5}$. The shaded regions are unyielded.

4.4. Flow in a Pipe with a Contraction

Finally, we consider the flow of a yield stress fluid in a pipe with a contraction. The inlet radius is selected as the reference length. Figure 18 shows the profiles for the velocity, shear rate, plastic viscosity and apparent viscosity when the inlet Reynolds number is 1.18×10^{-4} and $B_n = 163$ $\left(Re^* = 3.89 \times 10^{-5}, B = 31\right)$. It is observed that the velocity rises rapidly to several times its value in the contraction, which leads to a higher shear rate and lower viscosity. The shear rate is affected by both the magnitude of the velocity and the shape of the pipe: at the corner M and N where the pipe is suddenly contracted, a minimum shear rate is observed and a higher level of the shear rate is observed at the walls of the contraction and near the two ends of the contracted segment; therefore, the viscosity is higher near the corners and the centerline away from the contraction, forming three unyielded regions in this geometry.

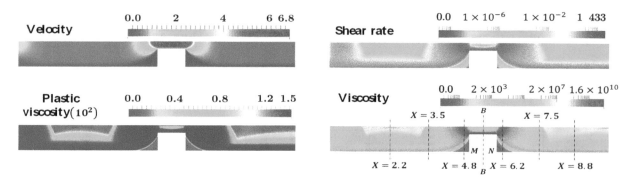

Figure 18. Velocity, shear rate, plastic viscosity and apparent viscosity fields in the pipe with a contraction when the inlet Reynolds number is 1.18×10^{-4} and $B_n = 163$.

Figure 19 shows the velocity and the viscosity profiles along line B-B for different Reynolds numbers. It can be seen that in the narrow section of the pipe there is a significant increase in the velocity, resulting in smaller plug flow; when the Reynolds number increases, the variation of the velocity profile in the contraction is not as sensitive as that in the regular section of the pipe. The viscosity increases along the radial direction but unlike the straight channel, the maximum value does not occur at the centerline; the viscosity also decreases after reaching a maximum, which indicates that the effect of the inlet and outlet of the contraction. To investigate this effect, the viscosity profiles along the radial direction at different X positions are plotted in Figures 20 and 21. It can be seen that the maximum apparent viscosity appears at the position near the center line away from the contraction and the stagnant zones at the corners. A change in the geometry would change the streamlines and lead to a higher shear rate, as a result, the viscosity decreases upstream and downstream of the contraction. Furthermore, for a yield-stress fluid, the high Reynolds number can help the fluid approach a lower viscosity.

From the yield regions shown in Figure 22, which is strongly related to the viscosity, we can see that the unyielded regions are roughly symmetric with respect to the contraction. The size of the plug flow continues to reduce while the Reynolds number keeps increasing.

The effect of the Bingham number is shown in Figures 23–26 With smaller Bingham numbers, it is easier for the fluid to flow; this results in a lower apparent viscosity and a more parabolic velocity profile. And as the Bingham number approaches smaller values, the unyielded region is observed only near the corners. Alexandrou et al. (2001) [45] showed similar results in the entry-expansion flow, where the geometry studied in their work is similar to the third case in our study. When designing a transport system, for example, a proper Bingham number (flow condition) should be considered in order to achieve an acceptable maximum unyielded region which is considered unfavorable in engineering applications. Furthermore, as the Reynolds number increases, the flow may become unstable after the channel contraction; this has been observed for the Casson, the Power-Law and the Quemada fluid models. The instability behavior depends on the specific parameters involved in each model's constitutive equation [46].

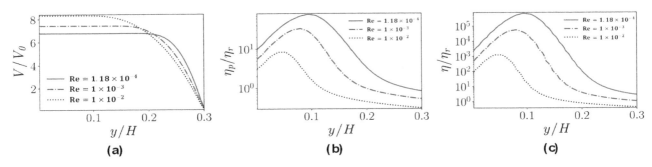

Figure 19. (a) Velocity; (b) Plastic viscosity; (c) Apparent viscosity profiles along line B-B for different Reynolds numbers.

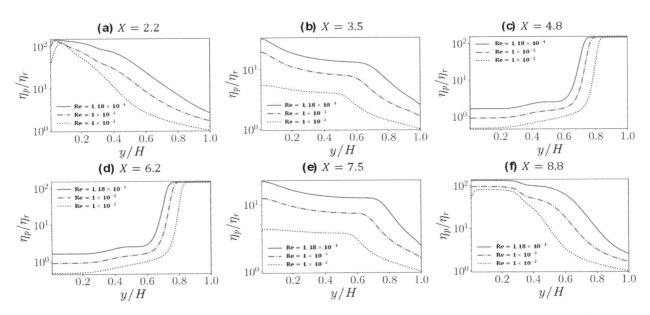

Figure 20. Plastic viscosity profiles along the radial direction at different X positions for different Reynolds numbers ($X = x/H$).

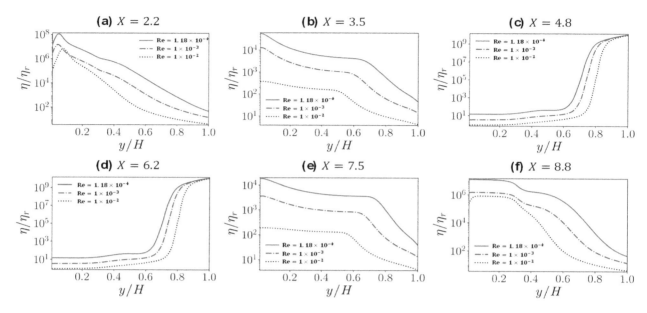

Figure 21. Apparent viscosity profiles along the radial direction at different X positions for different Reynolds numbers ($X = x/H$).

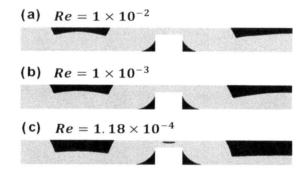

Figure 22. Distribution of the yield surfaces for different Reynolds numbers. The shaded regions are unyielded.

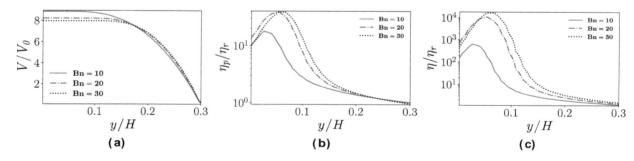

Figure 23. (a) Velocity (b) Plastic viscosity (c) Apparent viscosity profiles along B-B for different Bingham numbers and $Re = 1.18 \times 10^{-4}$.

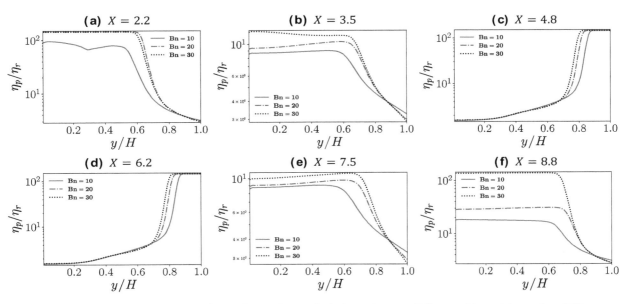

Figure 24. Plastic viscosity profiles along the radial direction at different X positions for different Bingham numbers and $Re = 1.18 \times 10^{-4}$ $(X = x/H)$.

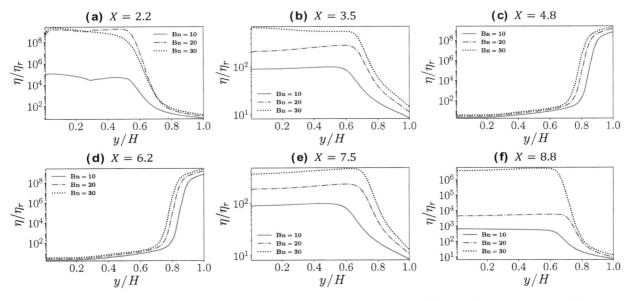

Figure 25. Apparent viscosity profiles along the radial direction at different X positions for different Bingham numbers and $Re = 1.18 \times 10^{-4}(X = x/H)$.

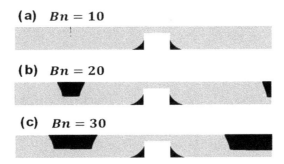

Figure 26. Distribution of the yield surfaces for different Bingham numbers with $Re = 1.18 \times 10^{-4}$. The shaded regions are unyielded.

5. Conclusions

In this paper, we have numerically studied the flow of dense suspension exhibiting yield-stress and shear-thinning effects. The rheological characteristics of such complex fluids need to be understood in engineering process. The dense suspension is modeled as a Herschel-Bulkley fluid and three benchmark geometries are investigated: a straight channel, a channel with a crevice and a pipe with a contraction. According to the numerical simulations, the following conclusions can be drawn:

1. Three representative regularization methods are used to look at the convergence properties on the determination of the yield surface; these methods are (1) the Simple model; (2) the Papanastasiou model and (3) the Bercovier and Engelman model. The yield surface could be described reasonably with a proper choice of the regularization parameter, ε, depending on the regularization method and the flow conditions. However, the maximum value of ε required to represent the yield surface is much lower for the Simple model than for the Papanastasiou and the Bercovier and Engelman models.

2. In the straight channel, for flows with low Reynolds number and high Bingham number, a plug region near the center line where the stress is below the yield stress will form. For shear-thinning fluids, the viscosity parameter n has a similar influence on the velocity profiles as the Bingham number.

3. In the case of the channel with a crevice, even though some vortical structures seem to propagate inside the crevice due to the free shear layer near the interface of the main channel and the crevice, the fluid in the deeper portion of the crevice still has high apparent viscosity because of the existence of the yield stress, forming an unyielded region. Furthermore, near the crevice the flow near the interface of the main channel and the crevice is disturbed a little; this results in a reduction of the plastic viscosity and the apparent viscosity.

4. For the pipe with a contraction, near the neck, the unyielded region reduces significantly due to the enhanced flow disturbance; while the shear rate is nearly zero at the bottom corner of the contraction segment, resulting in a very small yielded region.

5. For the Bingham numbers considered in this work, further increasing the Reynolds number leads to the disappearance of the yield region. The yield phenomenon can still be observed if both the Reynolds number and the Bingham number are increased; this is an important issue and a problem more demanding in computational time; we plan to study this in the near future.

Author Contributions: M.-G.L. developed the framework of the paper and did the numerical simulations. M.M. and W.-T.W. derived the equations. M.-G.L., M.M., F.F. and W.-T.W. prepared the manuscript. All authors have read and agreed to the published version of the manuscript.

Nomenclature

b	body force vector (N)
B	ratio of the yield stress to the viscous stress
Bn	Bingham number
A_1	Related to the symmetric part of velocity gradient (1/s)
div	divergence operator
$grad$	gradient operator
H	reference length (m)
m	material parameter ($m = n - 1$)

n	material parameter
p	pressure (Pa)
Re^*	Generalized Reynolds number
Re	generalized Reynolds number
t	time
T	stress tensor (Pa)
T_y	yield stress (Pa)
T_v	viscous stress (Pa)
v	velocity vector (m/s)
V	dimensionless velocity vector
X, Y	dimensionless Cartesian coordinates
x	position vector (m)
X	dimensionless position vector

Greek symbols

ρ	density of fluid (kg/m3)
η	dynamic viscosity (Pa·s)
η_p	plastic viscosity (Pa·s)
ε	regularization parameter
$\dot{\gamma}$	shear rate (1/s)
τ	dimensionless time
τ_y	yield stress (Pa)

References

1. Ferry, J.D.; Ferry, J.D. *Viscoelastic Properties of Polymers*; John Wiley & Sons: Hoboken, NJ, USA, 1980.
2. Natan, B.; Rahimi, S. The status of gel propellants in year 2000. *Int. J. Energ. Mater. Chem. Propuls.* **2009**, *5*, 172–194. [CrossRef]
3. Bonn, D.; Denn, M.M. Yield stress fluids slowly yield to analysis. *Science* **2009**, *324*, 1401–1402. [CrossRef] [PubMed]
4. Putz, A.M.V.; Burghelea, T.I. The solid–fluid transition in a yield stress shear thinning physical gel. *Rheol. Acta* **2009**, *48*, 673–689. [CrossRef]
5. Barnes, H.A. Thixotropy—A review. *J. Nonnewton. Fluid Mech.* **1999**, *81*, 133–178. [CrossRef]
6. Stickel, J.J.; Powell, R.L. Fluid mechanics and rheology of dense suspensions. *Annu. Rev. Fluid Mech.* **2005**, *37*, 129–149. [CrossRef]
7. Barnes, H.A.; Walters, K. The yield stress myth? *Rheol. Acta* **1985**, *24*, 323–326. [CrossRef]
8. Coussot, P.; Nguyen, Q.D.; Huynh, H.T.; Bonn, D. Avalanche behavior in yield stress fluids. *Phys. Rev. Lett.* **2002**, *88*, 175501. [CrossRef] [PubMed]
9. Zhu, L.; De Kee, D. Slotted-plate device to measure the yield stress of suspensions: Finite element analysis. *Ind. Eng. Chem. Res.* **2002**, *41*, 6375–6382. [CrossRef]
10. Qian, Y.; Kawashima, S. Distinguishing dynamic and static yield stress of fresh cement mortars through thixotropy. *Cem. Concr. Compos.* **2018**, *86*, 288–296. [CrossRef]
11. Malkin, A.; Kulichikhin, V.; Ilyin, S. A modern look on yield stress fluids. *Rheol. Acta* **2017**, *56*, 177–188. [CrossRef]
12. Coussot, P. Yield stress fluid flows: A review of experimental data. *J. Nonnewton. Fluid Mech.* **2014**, *211*, 31–49. [CrossRef]
13. Gomes, A.F.C.; Marinho, J.L.G.; Santos, J.P.L. Numerical simulation of drilling fluid behavior in different depths of an oil well. *Braz. J. Pet. Gas* **2019**, *13*, 309–322. [CrossRef]
14. Saeid, N.H.; Busahmin, B.S. Numerical investigations of drilling mud flow characteristics in vertical well. *An Int. J. (ESTIJ)* **2017**, *6*, 2250–3498.
15. Subramanian, R.; Azar, J.J. Experimental study on friction pressure drop for nonnewtonian drilling fluids in pipe and annular flow. *Soc. Pet. Eng. Int. Oil Gas Conf. Exhib. China 2000 IOGCEC 2000* **2000**. [CrossRef]
16. Ovarlez, G.; Mahaut, F.; Deboeuf, S.; Lenoir, N.; Hormozi, S.; Chateau, X. Flows of suspensions of particles in yield stress fluids. *J. Rheol.* **2015**, *59*, 1449–1486. [CrossRef]

17. Saasen, A.; Ytrehus, J.D. Viscosity models for drilling fluids—Herschel-bulkley parameters and their use. *Energies* **2020**, *13*, 5271. [CrossRef]

18. Soo, S.L. *Multiphase Fluid Dynamics*; Science Press: Beijing, China, 1990.

19. Rajagopal, K.R.; Tao, L. *Mechanics of Mixtures*; World Scientific: Singapore, 1995; Volume 35.

20. Johnson, G.; Massoudi, M.; Rajagopal, K.R. Flow of a fluid infused with solid particles through a pipe. *Int. J. Eng. Sci.* **1991**, *29*, 649–661. [CrossRef]

21. Massoudi, M. A mixture theory formulation for hydraulic or pneumatic transport of solid particles. *Int. J. Eng. Sci.* **2010**, *48*, 1440–1461. [CrossRef]

22. Phillips, R.J.; Armstrong, R.C.; Brown, R.A.; Graham, A.L.; Abbott, J.R. A constitutive equation for concentrated suspensions that accounts for shear-induced particle migration. *Phys. Fluids A* **1992**, *4*, 30–40. [CrossRef]

23. Zhou, Z.; Wu, W.-T.; Massoudi, M. Fully developed flow of a drilling fluid between two rotating cylinders. *Appl. Math. Comput.* **2016**, *281*, 266–277. [CrossRef]

24. Tao, C.; Kutchko, B.G.; Rosenbaum, E.; Wu, W.T.; Massoudi, M. Steady flow of a cement slurry. *Energies* **2019**, *12*, 2604. [CrossRef]

25. Peker, S.M.; Helvaci, S.S. *Solid-liquid Two Phase Flow*; Elsevier: Amsterdam, The Netherlands, 2011.

26. Slattery, J.C. *Advanced Transport Phenomena*; Cambridge University Press: Cambridge, UK, 1999.

27. Li, Y.; Wu, W.-T.; Liu, X.; Massoudi, M. The effects of particle concentration and various fluxes on the flow of a fluid-solid suspension. *Appl. Math. Comput.* **2019**, *358*, 151–160. [CrossRef]

28. Massoudi, M.; Vaidya, A. On some generalizations of the second grade fluid model. *Nonlinear Anal. Real World Appl.* **2008**, *9*, 1169–1183. [CrossRef]

29. Roscoe, R. Suspensions, flow properties of disperse systems. In *Flow Properties of Disperse Systems*; Hermans, J.J., Ed.; North Holland Publishing Company: Amsterdam, The Netherlands, 1953.

30. Tabuteau, H.; Coussot, P.; de Bruyn, J.R. Drag force on a sphere in steady motion through a yield-stress fluid. *J. Rheol.* **2007**, *51*, 125–137. [CrossRef]

31. Coussot, P.; Tocquer, L.; Lanos, C.; Ovarlez, G. Macroscopic vs. local rheology of yield stress fluids. *J. NonNewton Fluid Mech.* **2009**, *158*, 85–90. [CrossRef]

32. Cao, Q.-L.; Massoudi, M.; Liao, W.-H.; Feng, F.; Wu, W.-T. Flow characteristics of water-HPC gel in converging tubes and tapered injectors. *Energies* **2019**, *12*, 1643. [CrossRef]

33. Mas, R.; Magnin, A. Experimental validation of steady shear and dynamic viscosity relation for yield stress fluids. *Rheol. Acta* **1997**, *36*, 49–55. [CrossRef]

34. Denn, M.M.; Bonn, D. Issues in the flow of yield-stress liquids. *Rheol. Acta* **2011**, *50*, 307–315. [CrossRef]

35. White, J.L. A plastic-viscoelastic constitutive equation to represent the rheological behavior of concentrated suspensions of small particles in polymer melts. *J. NonNewton Fluid Mech.* **1979**, *5*, 177–190. [CrossRef]

36. Prager, W. *Introduction to Mechanics of Continua*; Courier Corporation: North Chelmsford, MA, USA, 2004.

37. Putz, A.; Frigaard, I.A.; Martinez, D.M. On the lubrication paradox and the use of regularisation methods for lubrication flows. *J. NonNewton Fluid Mech.* **2009**, *163*, 62–77. [CrossRef]

38. Frigaard, I.A.; Nouar, C. On the usage of viscosity regularisation methods for visco-plastic fluid flow computation. *J. NonNewton Fluid Mech.* **2005**, *127*, 1–26. [CrossRef]

39. Allouche, M.; Frigaard, I.A.; Sona, G. Static wall layers in the displacement of two visco-plastic fluids in a plane channel. *J. Fluid Mech.* **2000**, *424*, 243–277. [CrossRef]

40. Bercovier, M.; Engelman, M. A finite-element method for incompressible non-Newtonian flows. *J. Comput. Phys.* **1980**, *36*, 313–326. [CrossRef]

41. Papanastasiou, T.C. Flows of materials with yield. *J. Rheol.* **1987**, *31*, 385–404. [CrossRef]

42. Von Kampen, J.; Madlener, K.; Ciezki, H.K. Characteristic flow and spray properties of gelled fuels with regard to the impinging jet injector type. In Proceedings of the 42nd AIAA Joint Propulsion Conference, Sacramento, California, USA, 9–12 July 2006; Volume 4, pp. 2639–2650.

43. Abdali, S.S.; Mitsoulis, E.; Markatos, N.C. Entry and exit flows of Bingham fluids. *J. Rheol.* **1992**, *36*, 389–407. [CrossRef]

44. Mitsoulis, E.; Tsamopoulos, J. Numerical simulations of complex yield-stress fluid flows. *Rheol. Acta* **2017**, *56*, 231–258. [CrossRef]

45. Alexandrou, A.N.; McGilvreay, T.M.; Burgos, G. Steady herschel-bulkley fluid flow in three-dimensional expansions. *J. NonNewton Fluid Mech.* **2001**, *100*, 77–96. [CrossRef]
46. Neofytou, P.; Drikakis, D. Non-Newtonian flow instability in a channel with a sudden expansion. *J. NonNewton Fluid Mech.* **2003**, *111*, 127–150. [CrossRef]

3

Numerical Study on Thermal Hydraulic Performance of Supercritical LNG in Zigzag-Type Channel PCHEs

Zhongchao Zhao *, Yimeng Zhou, Xiaolong Ma, Xudong Chen, Shilin Li and Shan Yang

School of Energy and Power, Jiangsu University of Science and Technology, Zhenjiang 212000, China; ymzhou@stu.just.edu.cn (Y.Z.); marlon@stu.just.edu.cn (X.M.); xudongchen@stu.just.edu.cn (X.C.); shilinli@stu.just.edu.cn (S.L.); shanyang33@stu.just.edu.cn (S.Y.)
* Correspondence: zhongchaozhao@just.edu.cn

Abstract: In this paper, we study a promising plate-type heat exchanger, the printed circuit heat exchanger (PCHE), which has high compactness and is suitable for high-pressure conditions as a vaporizer during vaporization. The thermal hydraulic performance of supercritical produce liquefied natural gas (LNG) in the zigzag channel of PCHE is numerically investigated using the SST κ-ω turbulence model. The thermo-physical properties of supercritical LNG from 6.5 MPa to 10MPa were calculated using piecewise-polynomial approximations of the temperature. The effect of the channel bend angle, mass flux and inlet pressure on local convection heat transfer coefficient, and pressure drop are discussed. The heat transfer and pressure loss performance are evaluated using the Nusselt and Euler numbers. Nu/Eu is proposed to evaluate the comprehensive heat transfer performance of PCHE by considering the heat transfer and pressure drop characteristics to find better bend angle and operating conditions. The supercritical LNG has a better heat transfer performance when bend angle is less than 15° with the mass flux ranging from 207.2 kg/(m²·s) to 621.6 kg/(m²·s), which improves at bend angle of 10° and lower compared to 15° at mass flux above 414.4 kg/(m²·s). The heat transfer performance is better at larger mass flux and lower operating pressures.

Keywords: printed circuit heat exchanger; supercritical LNG; zigzag type; heat transfer performance

1. Introduction

Natural gas (NG) is an advantageous energy source for various applications due to its clean nature and its environmental and economic advantages [1]. NG is liquefied to produce liquefied natural gas (LNG) for long-distance transportation and storage, and is regasified before terminal utilization [2]. Therefore, an efficient and reliable LNG vaporizer is a key component in a LNG vaporization system. The common LNG vaporizers, such as intermediate fluid vaporizers, open rack vaporizers (ORVs), super ORVs and submerged combustion vaporizers [3–5] do not satisfy the requirement of high efficiency and compactness in finite volume vaporization processes. Hence, the printed circuit heat exchanger (PCHE), which is a prospective plate-type heat exchanger with high compactness that can operate under high pressure and low temperature, has been investigated by many researchers [6,7].

In recently years, four PCHE channel morphologies have been studied, namely the straight, zigzag, S-shape, and airfoil shapes. Figley et al. [8] conducted numerical simulations to investigate the thermal hydraulic performance of the straight-channel PCHE using helium. The thermal effectiveness and overall heat transfer coefficient were defined and calculated to describe the overall heat transfer performance of PCHE. Kim et al. [9] predicted the thermal performance by developing a mathematical expression of geometric parameters, material properties, and flow conditions to express the effectiveness of cross, parallel, and counterflow PCHEs. Yoon et al. [10] developed a friction factor and Nusselt number relationship of laminar flow in various bend angles for a semi-circular zigzag channel PCHE. Ishizuk et al. [11] studied heat transfer and flow characteristics of zigzag-type PCHE experimentally using supercritical carbon dioxide and obtained the total heat transfer efficiency and pressure drop. Tsuzuki et al. [12] performed numerical analysis of the transition section of a zigzag-type channel PCHE, and encountered the presence of vortices and local circulation flow. Ngo et al. [13,14] conducted three-dimensional simulations of the pressure drop and Nusselt number of PCHEs with S-shaped fins in a supercritical carbon dioxide loop. Kim et al. [15] studied airfoil PCHEs using numerical analysis and indicated that the stream line was smooth and the vortices and countercurrents disappeared in the airfoil channel. Zhao et al. [16] investigated the heat transfer characteristics of an airfoil PCHE numerically using supercritical LNG and optimized the arrangement of airfoil fins.

In spite of the superior pressure drop performance of PCHEs with S-shaped and airfoil fins, compared to straight and zigzag-type PCHEs, discontinuous fins lack durability at high pressure operating conditions, which is attributed to the confined junction area of the fins in the diffuser bonding [17]. This in turn leads to an increase in manufacturing and maintenance costs [18,19]. Therefore, it makes sense to further investigate continuous channels and optimize channel shapes.

Previously, investigations were mainly concerned with the heat transfer performance and pressure drop of zigzag PCHEs, but few combined the heat transfer and pressure drop performance to study the optimization of zigzag PCHEs in terms of bend angle and operating conditions. In addition, supercritical fluids can be used to improve heat transfer performance due to their favorable properties, like high density, low viscosity, and high thermal conductivity compared to conventional fluids [20]. Most studies on the flow and heat transfer characteristics of fluids in PCHE have been conducted using helium and water [21–23], and many researchers also used supercritical water and supercritical carbon dioxide to investigate PCHEs [24,25]. However, studies on the heat transfer performance of supercritical LNG in zigzag PCHE under low temperature and high pressure conditions are rare [16].

In this study, the flow and heat transfer characteristics of supercritical LNG in zigzag-type PCHEs were numerically simulated at an operating pressure of 6.5–10 MPa. The effects of bend angle and mass flux on heat transfer performance were also investigated. The local convection heat transfer coefficient and pressure drop of supercritical LNG under different conditions are discussed. Dimensionless parameters such as the Nusselt number (Nu) and Euler number (Eu) are analyzed to assess the heat transfer and pressure loss performance, Nu/Eu of the bend angles and operating conditions are investigated by considering the performance of heat transfer and the pressure drop of supercritical LNG in PCHEs.

2. Numerical Approach

2.1. Physical Model and Boundary Conditions

In this study, the thermal hydraulic performance of supercritical LNG in zigzag PCHEs is investigated. The cross flow PCHE core model with a full length of 400 mm using supercritical LNG in the cold side and R22 in the hot side is shown in Figure 1a. The supercritical LNG and R22 flow in the semicircular channel with a diameter of 1.5 mm, the solid is composed of steel with thermal conductive coefficient of 16.27 W/(m·K). In this paper, we only study the performance of supercritical LNG in the cold channel. Considering that the cold side of the PCHE contains hundreds of channels, it is unrealistic to consider all the channels, and it is therefore necessary to simplify the cold channels' model. Supercritical LNG flows in parallel in each cold channel, so some assumptions on its flow in the cold side are made. The mass flux is the same in every channel, and there is no temperature difference and heat loss between neighboring channels. The flow of supercritical LNG is steady and uniformly distributed. Based on these assumptions, the cold channel can be simplified to a single model with a geometry of 2 mm × 1.75 mm. The cross-section of the fluid channel is semicircular with a diameter of 1.5 mm (Figure 1b). The bend angles α vary from 0° (which is a straight channel) to 45° (Figure 1c).

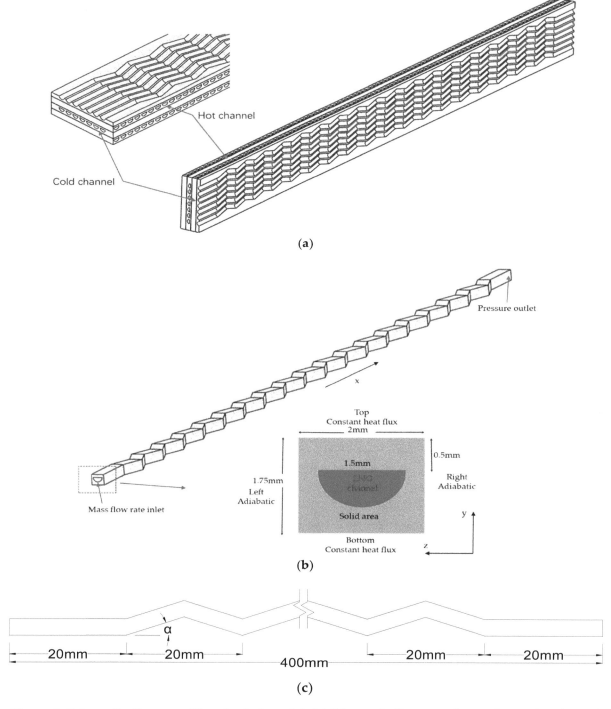

Figure 1. Schematic diagram of the physical model: (**a**) Schematic diagram of cross flow printed circuit heat exchanger (PCHE) model; (**b**) The computational single channel and boundary conditions; (**c**) Top view of zigzag channel.

The adjacent cold channels do not exhibit temperature difference and heat transfer loss; only the supercritical LNG in cold channels absorb heat from top and bottom hot channels. Three types of boundary conditions were applied in the model: fluid inlet, fluid outlet, and wall. The mass flow rate boundary condition was set at the inlet of the supercritical LNG channel whereas at the outlet the pressure-outlet boundary condition was applied (Figure 1b). The left and right walls of the single model are set to adiabatic boundary conditions, and the constant heat flux condition was used at top and bottom walls. The details of the boundary conditions are presented in Table 1.

Table 1. Boundary conditions in detail.

	Inlet		Outlet	Left/Right Walls	Top/Bottom Walls
Pressure (MPa)	Temperature (K)	Mass flux (kg/m²·s)	Pressure outlet	Adiabatic	Constant heat flux (W/m²)
10	121	207.2			7.5×10^4

2.2. Thermo-Physical Properties of Supercritical LNG

In this paper, the operating pressure of LNG considered varies from 6.5 MPa to 10 MPa, which is supercritical pressure. Supercritical LNG has gas-like properties, such as low viscosity, and liquid-like characteristics, like high density and high thermal conductivity. The thermo-physical properties of supercritical LNG, i.e., density, specific heat, thermal conductivity and viscosity, are affected by temperature and pressure. The properties' values were obtained from the NIST Standard Reference Database (REFPROP) [26]. For the numerical simulations, the temperature was changed from 121 K to 385 K. At such a large temperature difference, the properties of supercritical LNG change dramatically, using the average values will cause the inaccurate calculation results in ANSYS Fluent. Therefore, as shown in Figure 2, the thermal properties of supercritical LNG were approximated as piecewise-polynomial functions of temperature. The piecewise-polynomial functions of temperature at 10 MPa is shown in Table 2. The error percentages of various properties using the proposed approximation are shown in Figure 3. The errors were within ±2.5%, which indicates the fitted piecewise-polynomial function approximations are suitable.

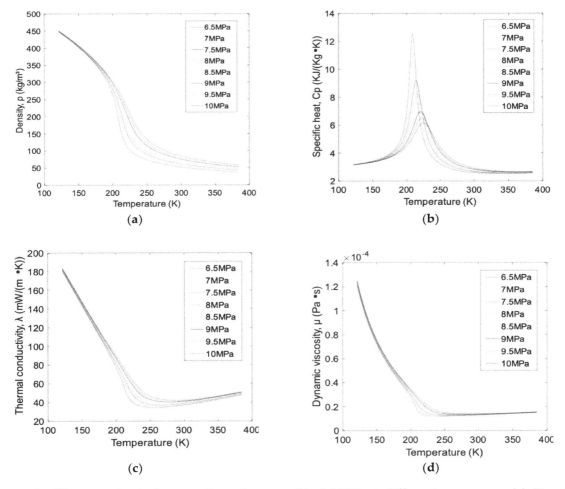

Figure 2. Thermo-physical properties of supercritical LNG at different pressures: (a) Density; (b) Specific heat; (c) Thermal conductivity; (d) Dynamic viscosity.

Table 2. Piecewise-polynomial functions at 10 MPa.

Temperature Range (K)	Density
121–223	$\rho = -1.29548 \times 10^{-6}T^4 + 7.459 \times 10^{-4}T^3 - 0.16537T^2 + 15.12763T - 2.45869$
223–271	$\rho = -5.30962 \times 10^{-4}T^3 + 0.43352T^2 - 119.03258T + 11{,}091.66885$
271–385	$\rho = 1.32923 \times 10^{-7}T^4 - 1.93454 \times 10^{-4}T^3 + 0.10675T^2 - 26.70158T + 2634.06161$
	Specific Heat
121–223	$c_p = 0.00436T^3 - 1.83425T^2 + 263.05475T - 9567.33957$
223–261	$c_p = 0.06448T^3 - 46.88971 \times 10^3T^2 + 11{,}278.86482T - 892{,}240.75366$
261–385	$c_p = -0.00111T^3 + 1.18621T^2 - 423.72673T + 53{,}206.50774$
	Thermal Conductivity
12--235	$\lambda = 1.07039 \times 10^{-8}T^3 - 4.35253 \times 10^{-6}T^2 - 6.44568 \times 10^{-4}T + 0.30675$
235–262	$\lambda = 7.96913 \times 10^{-6}T^2 - 0.00432T + 0.62882$
262–385	$\lambda = -6.69403 \times 10^{-9}T^3 + 7.36829 \times 10^{-6}T^2 - 0.00258 \times 10^{-2}T + 0.33402$
	Viscosity
121–218	$\mu = -7.71822 \times 10^{-11}T^3 + 4.71489 \times 10^{-8}T^2 - 1.01809 \times 10^{-5}T + 8.02631 \times 10^{-4}$
218–254	$\mu = 6.9276 \times 10^{-9}T^2 - 3.52808 \times 10^{-6}T + 4.64105 \times 10^{-4}$
254–385	$\mu = -2.07823 \times 10^{-12}T^3 + 2.1834 \times 10^{-9}T^2 - 7.43585 \times 10^{-7}T + 9.67009 \times 10^{-5}$

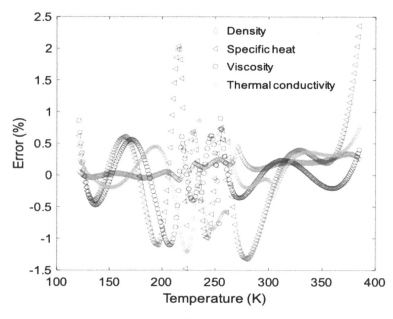

Figure 3. Error curves of linear interpolation functions.

2.3. Numerical Method and Grid Independence

The commercial software FLUENT was used to solve the 3D numerical model. Considering the inlet parameters, the flow corresponded to turbulent flow regimes. Some turbulence models have been studied and used in the literature; these include the κ-ε standard model, the RNG κ-ε model, the shear-stress transport (SST) κ-ω model and the low Reynolds number turbulence model [27,28]. In this study, the SST κ-ω model was used because of its more accurate results on heat transfer of supercritical fluids [29–32].

The governing equations for heat transfer were the continuity, momentum, and energy equations, respectively:

Continuity equation:

$$\frac{\partial}{\partial x_i}(\rho u_i) = 0, \tag{1}$$

where ρ is the density, and u_i is the velocity vector.

Momentum equation:

$$\frac{\partial}{\partial x_i}(\rho u_i u_j) = -\frac{\partial p}{\partial x_i} + \rho g_i + \frac{\partial}{\partial x_j}[(\mu + \mu_t)\frac{\partial u_i}{\partial x_j}], \tag{2}$$

where p is the pressure, μ and μ_t are the molecular and turbulent viscosities, respectively.

Energy equation:

$$\frac{\partial}{\partial x_i}(u_i(\rho E + p)) = \frac{\partial}{\partial x_i}\left(k_{eff}\frac{\partial T}{\partial x_i} + u_i \tau_{ij}\right), \tag{3}$$

where k_{eff} is effective conductivity, $k_{eff} = k + k_t$, and k_t is the turbulent thermal conductivity.

The transport equations are expressed as follows:

$$\frac{D(\rho\kappa)}{Dt} = \frac{\partial}{\partial x_j}\left[(\mu + \sigma_\kappa \mu_t)\frac{\partial \kappa}{\partial x_j}\right] + \tau_{ij}\frac{\partial u_i}{\partial x_j} - \beta^* \rho \omega \kappa \tag{4}$$

$$\frac{D(\rho\omega)}{Dt} = \frac{\partial}{\partial x_j}\left[(\mu + \sigma_{\omega 1} \mu_t)\frac{\partial \omega}{\partial x_j}\right] + \frac{\gamma}{\nu_t}\tau_{ij}\frac{\partial u_i}{\partial x_j} - \beta \rho \omega^2 + 2(1 - F_1)\rho \sigma_{\omega 2}\frac{1}{\omega}\frac{\partial \kappa}{\partial x_j}\frac{\partial \omega}{\partial x_j} \tag{5}$$

$$\omega = \frac{\varepsilon}{\beta^* \kappa}; \nu_t = \frac{a_1 \kappa}{max(a_1 \omega; \Omega F_2)} \tag{6}$$

$$F_1 = \tan h(arg_1^4); F_2 = \tan h(arg_2^2) \tag{7}$$

$$arg_1 = \min\left(max\left(\frac{\sqrt{\kappa}}{0.09\omega y}; \frac{500\nu}{y^2 \omega}\right); \frac{4\rho\sigma_{\omega 2}\kappa}{CD_{\kappa\omega}y^2}\right); CD_{\kappa\omega} = max\left(2\frac{\rho\sigma_{\omega 2}}{\omega}\frac{1}{\omega}\frac{\partial \kappa}{\partial x_j}\frac{\partial \omega}{\partial x_j}, 10^{-20}\right) \tag{8}$$

$$arg_2 = max\left(2\frac{\sqrt{\kappa}}{0.09\omega y}; \frac{500\nu}{y^2 \omega}\right) \tag{9}$$

where ε is the turbulent kinetic energy dissipation rate, Ω is vorticity, and y is the distance from the wall. The constants and damping functions of the SST κ-ω model are shown in Table 3.

Table 3. Constants and functions used in the shear-stress transport (SST) model.

	$\sigma_{\omega 1}$	$\sigma_{\omega 2}$	κ	α_1	β^*
SST	0.5	0.865	0.41	0.31	0.09

The local convective heat transfer coefficient was calculated using Equation (10):

$$h = \frac{q}{T_w - T_b} = \frac{q}{T_w - (T_{out} + T_{in})/2}, \tag{10}$$

where q is the constant heat flux from the top and bottom walls, T_w is the wall temperature and T_b is average temperature of the inlet and the outlet.

Nu was defined as:

$$Nu = \frac{hD_h}{\lambda}; D_h = 4A/l \tag{11}$$

where D_h is the hydraulic diameter and λ is the local thermal conductivity of LNG, A is the cross-sectional area of the semicircular fluid channel and l is the circumference of the semicircular fluid channel section.

The local Fanning friction coefficient (f) was defined in terms of the pressure drop and is expressed by Equation (12):

$$f = \frac{\Delta P_f D_h}{2L\rho_b v_b^2}, \tag{12}$$

$$\Delta P_f = \Delta P - \Delta P_a = \Delta P - \left(\rho_{out}v_{out}^2 - \rho_{in}v_{in}^2\right), \tag{13}$$

where ΔP is the total pressure and was obtained from Fluent directly, ΔP_f and ΔP_a are the frictional and accelerated pressure drops, respectively, L is the channel length, ρ_b and v_b are the bulk density and velocity of LNG, respectively.

The Reynolds number (Re) is given by Equation (14):

$$R_e = \frac{\rho_b v_d D_h}{\mu_b}, \tag{14}$$

The Euler number (Eu) is defined as Equation (15):

$$Eu = \frac{\Delta P}{\rho_b v_b^2/2}, \tag{15}$$

For the solution methods, the SIMPLE algorithm was applied to establish the coupling of velocity and pressure. The momentum, turbulent kinetic energy, turbulent dissipation rate and energy were discretized using the second order upwind scheme. The calculation was considered to converge when the residuals were less than 10^{-6}.

The mesh on the computational domain was generated using GAMBIT. The grid independence was verified to confirm numerical result accuracy. The mesh size of solid, fluid and boundary layer's scale in fluid affect the grid numbers. The influence of the grid numbers on the convective heat transfer coefficient is shown in Table 4. Case 4 has a larger relative error compared to the other cases. The heat transfer coefficient in Cases 1, 2, and 3 is nearly the same. The relative error of the heat transfer coefficient between Cases 1 and 7 is only 0.08%. Therefore, considering the calculation accuracy and time, the 2,988,329 grid nodes (Case 1), showing in Figure 4, was selected in the present work.

Table 4. Boundary layers study.

Case	Scale of Boundary Layer	Rows of Boundary Layer	Cells of Nodes	Heat Transfer Coefficient W/(m²·K)	Relative Error (%)
1	0.01	5	2,988,329	2678.57	3.4%
2	0.01	8	3,589,947	2680.44	3.47%
3	0.003	5	2,974,634	2678.32	3.4%
4	0.03	5	2,697,546	2669.23	3.04%
5			815,644	2590.42	0
6	0.01	5	1,962,788	2636.39	1.8%
7			4,456,851	2680.62	3.48%

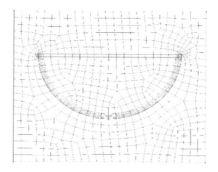

Figure 4. Cross section of computational grids.

2.4. Model Validation

To validate the accuracy and reliability of the model, the simulation results of temperature difference and pressure drop were compared to previous experimental results [30]. The experimental setup is shown in Figure 5. Since LNG is flammable and explosive, the straight-channel cross flow PCHE used supercritical nitrogen as the cold side fluid and R22 as the hot side fluid. The length of the PCHE cold channel was 520 mm, the inlet temperature was 102 K, and the operating pressure varied from 6.5 MPa to 10 MPa. In the simulation, a straight channel model with a length of 520 mm was selected, which was the same as the experimental case. The boundary conditions are shown in Figure 1b. Supercritical nitrogen was used as the working fluid to confirm the correctness the experiment. The inlet pressure changed from 6.5 MPa to 10 MPa and the inlet temperature was 102 K. The comparison of temperature difference and pressure drop between simulation and experimental results is listed in Table 5. The maximum errors of temperature difference and pressure drop are 2.1% and 10.25%, respectively. The simulation pressure drop was less than the experimental, which may be attributed to the following factors: (1) the channel was assumed to be smooth in the numerical study while the PCHE channel of the experiment was rough, (2) the header pressure drop of inlet and the outlet were neglected in the numerical study, but may have been large in the experiment, and (3) the deviation of temperature and pressure transmissions. The simulation results are in accordance with the experiment, illustrating that the simulation model and method were credible.

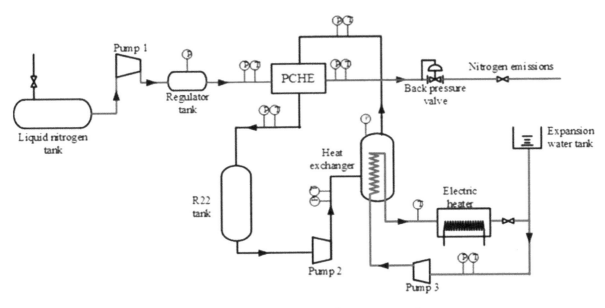

Figure 5. Schematic diagram of experimental setup.

Table 5. Relative error of simulation and experiment results.

Pressure (MPa)	Temperature Difference (K)		Relative Error (%)	Pressure Difference (Pa)		Relative Error (%)
	Experiment	Simulation		Experiment	Simulation	
6.5	178.9	175.1	2.1	16,612.35	15,167.36	10.25
7	180.3	178	1.27	15,636.09	14,521.4	9.95
7.5	182.1	180.4	0.94	14,742.24	14,071.62	7.05
8	182.6	182.3	0.16	13,847.55	13,188.32	6.62
8.5	183.5	184	0.27	13,035.22	12,504.69	6.81
9	185.7	185.5	0.11	12,156.68	11,810.66	4.70
9.5	186.1	185.8	0.16	11,342.33	10,862.6	4.42
10	186.6	186.4	0.11	10,578.6	10,189.45	3.82

3. Results and Discussion

Compared with traditional vaporizers, the heat transfer performance of PCHE is better, but the pressure drop is larger due to the small size of the channel, resulting in an increase of operating costs. A number of studies have shown that a supercritical fluid can enhance heat transfer and reduce pressure drop. The supercritical LNG vaporized by the PCHE is suitable for long distance transport and utilization. In this paper, the flow and heat transfer characteristics of supercritical LNG were studied at 6.5–10MPa, the performance of supercritical LNG at 10 MPa was discussed in detail in the following.

3.1. Effect of Bend Angles of the Zigzag Channel

The heat transfer performance of supercritical LNG is influenced significantly by the bend angle in the zigzag channel of a PCHE. The effect of the bend angles on the local convection heat transfer coefficient and bulk temperature of the LNG along the streamwise direction at a mass flux of 207.2 kg/(m²·s) and operating pressure of 10 MPa are shown in Figure 6. As the bend angle increases, the local convection heat transfer coefficient and bulk temperature increase, which is due to the fact that with the increase of the bend angle, disturbance and turbulence will increase. Moreover, the local heat transfer coefficient increases and then decreases along the streamwise direction, and reaches a peak when the bulk temperature is near the pseudo-critical temperature at bend of 0°–15°. This is because that the thermo-physical properties of supercritical LNG vary drastically at different temperatures, and the specific heat in particular reaches an extremum near the pseudo-critical temperature (Figure 2). However, at bend angles of 25°–45°, the convection heat transfer coefficient is greater at Np = 2–4. When the LNG flows into the channel, its velocity is not large in the Np = 2–4. The flow separation is not dramatic and the velocity of vortices is not much different that of the fluid. As the flow develops, the velocity increases. The velocity difference of vortices and fluid becomes large and the vortices and flow separation increase (as shown in Figure 7b), leading to a decrease in the convection heat transfer coefficient. In addition, the heat transfer coefficient is larger at the inlet due to the entrance effect.

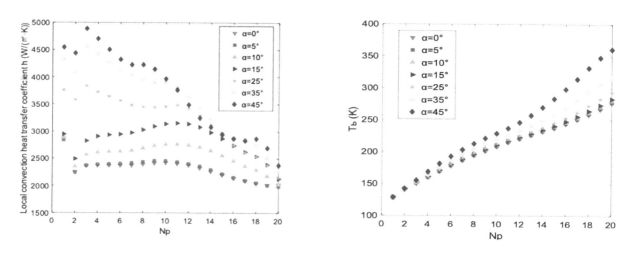

Figure 6. Local convection heat transfer coefficient and bulk temperature at different bend angles along the streamwise direction.

Figure 8 shows the pressure drop and Fanning friction coefficient (f) at different bend angles along the streamwise direction. The pressure drop increases with the bend angles and increases along the streamwise direction. The density and viscosity of the supercritical LNG decrease as the temperature increases (as shown in Figure 2a,d), resulting in a velocity increase and thus in pressure drop. Figure 7 shows the velocity vectors of different cross-sections along the channel and velocity vectors of Np = 10–12 at different bend angles. The velocity increases with the increase of bend angles, which is attributed to the increase of turbulence (Figure 7a). Flow separation and reverse flow appear in larger bend angles (Figure 7b), which increases flow resistance, resulting in an increased pressure drop. The velocity increases along the flow direction, which also increase the pressure drop along the streamwise. The Fanning friction coefficient f increases with the bend angle, which is the same trend observed in pressure drop. However, f decreases along the flow direction; this difference between pressure drop and f is due to the increase of velocity.

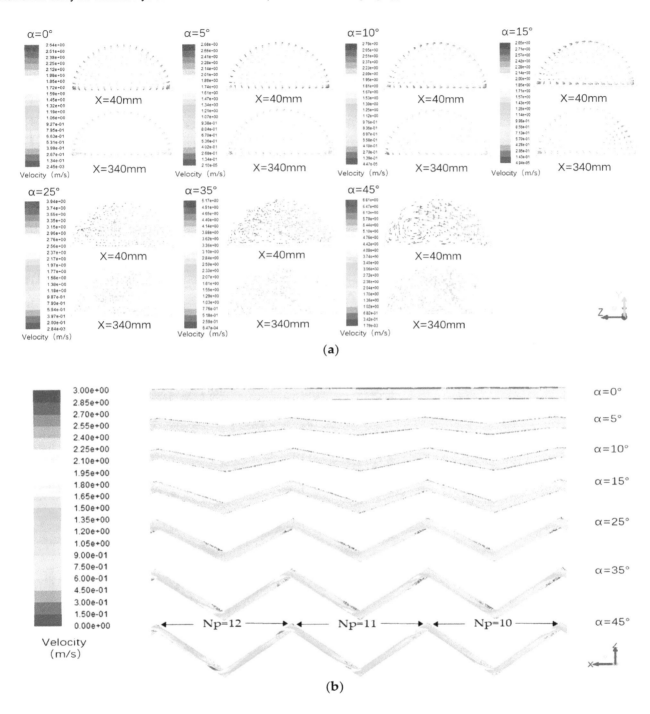

Figure 7. (a) Velocity vectors of different cross-sections along the channel with bend angles of $0°$ to $45°$; **(b)** Velocity vectors of $Np = 10$–12 at bend angles of $0°$ to $45°$.

Figure 9 shows the Nusselt (Nu) and Euler (Eu) numbers along the streamwise direction at different bend angles. It can be seen that Nu increases with the bend angle, and reaches its maximum value near the pseudo-critical temperature, then decreases along the streamwise direction. This is because the thermal conductivity decreases intensely as the temperature rises before pseudo-critical temperature and rises slightly when the temperature surpasses the pseudo-critical value, reaching a minimum near the pseudo-critical temperature, which leads to a maximum for Nu near the pseudo-critical temperature. The Euler number increases as the bend angle rises, which is consistent with the change in pressure drop.

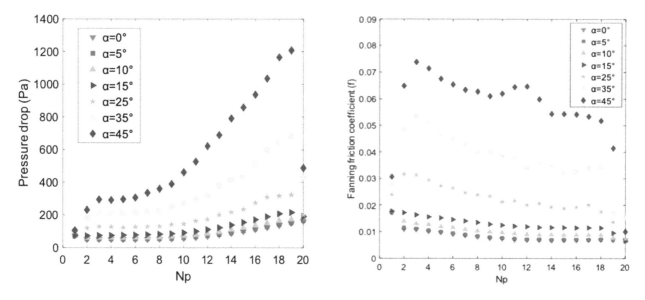

Figure 8. Pressure drop and fanning friction coefficient at different bend angles along the streamwise direction.

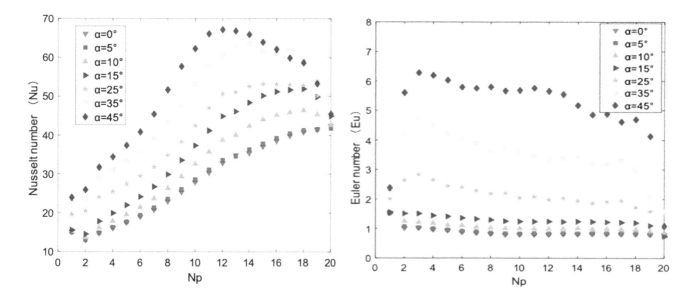

Figure 9. Local Nusselt number (Nu) and Euler (EU) numbers at different bend angles along the streamwise direction.

The objective of improving the heat exchanger's performance can be realized by increasing heat transfer performance and reducing pressure drop. It is therefore essential to comprehensively consider the heat transfer and pressure loss characteristics of supercritical LNG in PCHE. In this study, the ratio of Nusselt to Euler numbers (Nu/Eu) is proposed to evaluate the performance of supercritical LNG in the channel, where a larger ratio indicates better heat transfer performance. Figure 10 shows Nu/Eu at different bend angles. Nu/Eu reaches its peak value at a bend angle of 10°. When the bend angle exceeds 10°, the growth rate of Nu is much less than that of Eu. Nu and Eu at 0° and 5° are nearly the same, while from 5° to 10°, the increase of Nu is much greater than that of Eu.

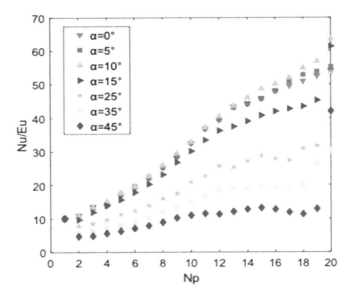

Figure 10. Nu/Eu at different bend angles along the streamwise direction.

3.2. Effect of Mass Flux

The effect of mass flux on flow and heat transfer performance of supercritical LNG was investigated at the bend angle of 10° and an operating pressure of 10 MPa. As shown in Figure 11, the local convective heat transfer coefficient and pressure drop increase significantly as the mass flux increases because of the enhancement of turbulent flow. When the mass flux is increased by 2 times, the local heat transfer coefficient increased 1.4 times, and at the same time, the pressure drop increases 3.3 times. The Nu and Eu are shown in Figure 12. The Nu increases as the mass flux is raised. However, at the last third of the channel, Nu peaks at a mass flux of 301.8 kg/(m²·s), and then decreases at further mass flux increase. This is because as mass flux is increased, the temperature of LNG decreases, its viscosity increases and its velocity decreases. When the heat flux is kept constant, the heat absorbed by the LNG per unit volume is reduced.

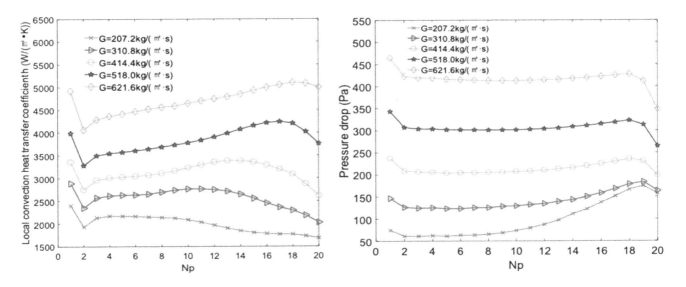

Figure 11. Effect of mass flux on local convection heat transfer coefficient and pressure drop along the streamwise direction.

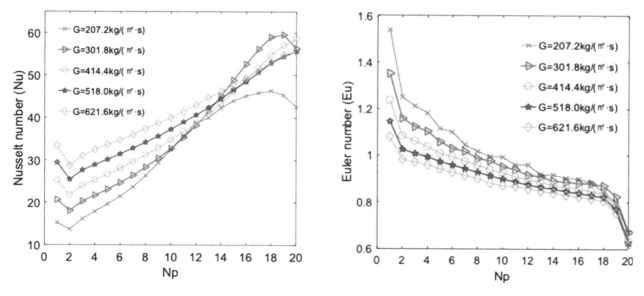

Figure 12. Effect of mass flux on Nu and Eu along the streamwise direction.

Eu increases as the mass flux decreases. The pressure drop and density increase as mass flux rises, but v_b^2 decreased. The increase rate of $\rho_b v_b^2$ is more than that of ΔP, which results in an increase of Eu as mass flux is reduced.

The influence of mass flux on Nu/Eu is shown in Figure 13. The Nu/Eu is increased as the mass flux is raised. However, at the last third of the channel, Nu/Eu peaks at a mass flux of 301.8 kg/(m²·s). The figure suggests that the heat transfer performance of the whole channel is improved as the mass flux increases, but with the development of the fluid flow, the local heat transfer performance is reduced at the last third of the channel, owing to the reduction of heat absorbed capacity by the unit volume fluid.

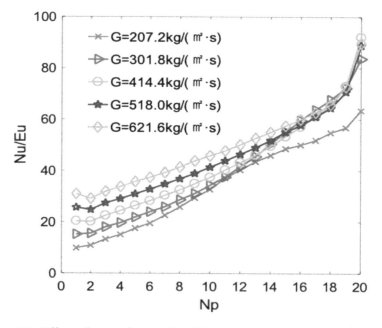

Figure 13. Effect of mass flux on Nu/Eu along the streamwise direction.

Figure 14 shows the impact of mass flux on Nu/Eu at different channel bend angles. The Nu/Eu is significantly reduced when the bend angle exceeds 15°, indicating that the increase of pressure drop is much higher than that of heat transfer performance. Consequently, the comprehensive heat transfer performance is not good when the bend angles exceeds 15°. When the mass flux varies from 207.2 kg/(m²·s) to 621.6 kg/(m²·s), Nu/Eu are higher when the bend angle is less than 15°. However, Nu/Eu at bend angles of 10° and lower are increased compared to 15° at mass fluxes above 414.4 kg/(m²·s). It can be concluded that the supercritical LNG in the PCHE has better heat transfer performance when the bend angle is less than 15° with the mass flux ranging from 207.2 kg/(m²·s) to 621.6 kg/(m²·s), and that it improves at bend angles of 10° and lower compared to 15° at mass fluxes above 414.4 kg/(m²·s).

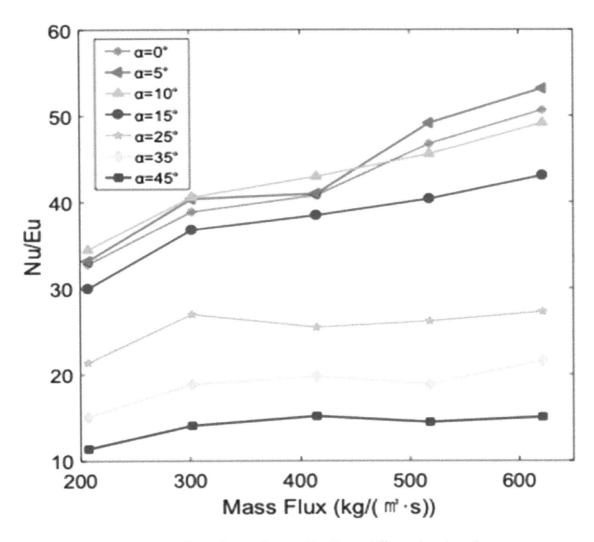

Figure 14. Effect of mass flux on Nu/Eu at different bend angles.

3.3. Effect of Inlet Pressure

At a bend angle of 10° and a mass flux of 207.2 kg/(m²·s), the inlet pressure was varied from 6.5 MPa to 10 MP. The corresponding local heat transfer coefficient and pressure drop values are shown in Figure 15. Before the pseudo-critical temperature, the specific heat and thermal conductivity are slightly affected by the inlet pressure (as shown in Figure 2b,c), so the convection heat transfer coefficient changes slightly with the inlet pressure. After the pseudo-critical temperature, the inlet pressure effect on the local convective heat transfer coefficient is greater because of the specific heat is influenced by inlet pressure more. In addition, the reduction rate of the specific heat rises rapidly as the inlet pressure is reduced. Therefore, the local convection heat transfer coefficient decreases with the decrease of inlet pressure along the streamwise direction. This shows that the specific heat depends on inlet pressure and has a great influence on local convective heat transfer coefficient when temperature exceeds the pseudo-critical temperature.

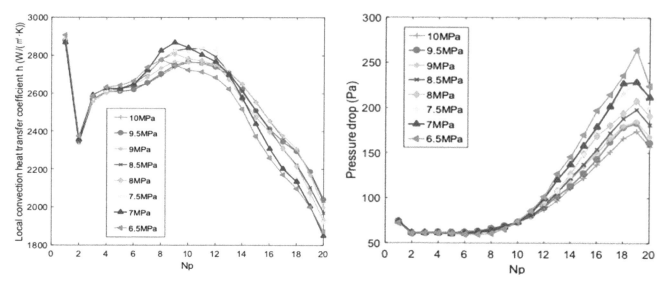

Figure 15. Effect of inlet pressure on local convection heat transfer coefficient and pressure drop along the streamwise direction.

With an increase of inlet pressure, the pressure drop decreases. As inlet pressure rises, the density and dynamic viscosity are larger (Figure 2a,d), which lowers the velocity of supercritical LNG. Hence, the pressure drop is reduced as the inlet pressure is increased. However, up until $Np = 10$, the pressure drop is uninfluenced by the changing inlet pressure. This is because the supercritical LNG has liquid-like properties, so the influence of inlet pressure on density and dynamic viscosity as well as velocity is small (as shown in Figure 2), which leads to pressure drop being only slightly effected by the inlet pressure. After $Np = 10$, the supercritical LNG has gas-like properties, so the influence of inlet pressure on density and dynamic viscosity is greater, leading to the large effect on velocity and causing the pressure drop to change more obviously with the inlet pressure. At the last portion of the channel, the pressure drop reduced, because the last pitch of the channel is straight, which reduces turbulence.

The effect of inlet pressure on Nu and Eu are shown in Figure 16, where both Nu and Eu decrease with as inlet pressure increased. Nu is inversely proportional to thermal conductivity, so as the inlet pressure increases, the thermal conductivity increased, which decreases Nu. Eu is increased as the inlet pressure is reduced, which the same behavior is observed in pressure drop, indicating that the pressure drop performance is larger when the inlet pressure is lower.

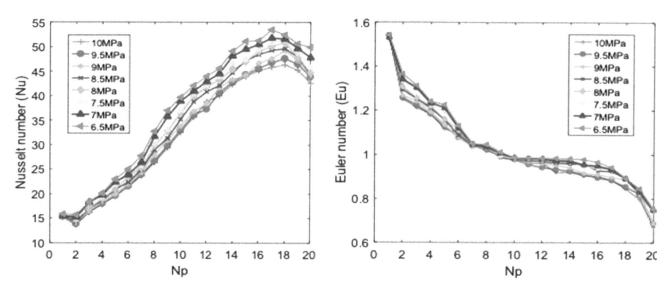

Figure 16. Effect of inlet pressure on Nu and Eu along the streamwise direction.

Although both Nu and Eu increase with a decrease in inlet pressure, Nu/Eu increases, as shown in Figure 17. This is because with the decrease of pressure, the increase rate of Nu is larger than that of Eu. Nu/Eu reaches its maximum at 6.5 MPa, indicating that supercritical LNG will have better heat transfer performance in PCHE at a lower inlet pressure.

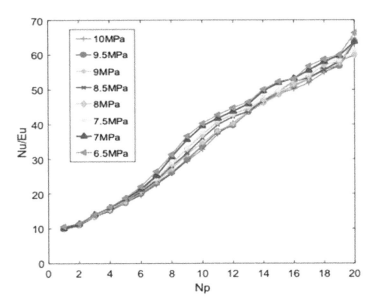

Figure 17. Effect of inlet pressure on Nu/Eu along the streamwise direction.

4. Conclusions

In this study, the flow and heat transfer characteristics of supercritical LNG in zigzag channel of PCHE are numerically investigated at different operating conditions. Some conclusions can be drawn as follows:

(1) The local convection heat transfer coefficient rises and then falls along the streamwise direction, with the peak value appearing at the pseudo-critical temperature. The pressure drop also increases along the streamwise direction.

(2) As the channel bend angle is increased, the local convection heat transfer coefficient and pressure drop rise, and so do the Nu and Euler numbers. The enhancement of heat transfer capability of supercritical LNG is mainly owed to increased turbulence. The increase of pressure drop is mainly due to the rising of velocity and the increase of flow resistance caused by the existence of vortices.

(3) The local convective heat transfer coefficient and pressure drop increase significantly as the mass flux is increased due to the enhancement of turbulent flow. When the mass flux is increased by 2-fold, the local heat transfer coefficient rises by 1.4 times, and the pressure drop increases 3.3 times. The Nu increases as mass flux is increased. However, at the last third of the channel, Nu decreases as the mass flux is raised because of the decreased heat per unit volume absorbed by the LNG. This suggests that when the mass flux is raised, the heat transfer performance of the whole channel is better, but with the development of the fluid's flow, the local heat transfer performance is reduced at the last third of the channel owing to the reduction of heat-absorbed capacity by the unit volume fluid.

(4) The improvement of heat transfer performance with bend angle depends on the mass flux. The supercritical LNG has better heat transfer performance when the bend angle is less than 15° when the mass flux ranges from 207.2 kg/(m²·s) to 621.6 kg/(m²·s), and improves at bend angles of 10° and lower compared to 15° at mass fluxes above 414.4 kg/(m²·s).

(5) Before the pseudo-critical temperature, the local convective heat transfer coefficient changes little with the inlet pressure, while it increases when the temperature surpasses pseudo-critical

point. The pressure drop is reduced as the inlet pressure increases. Nu and EU decrease with increasing inlet pressure, while Nu/Eu reaches a maximum at 6.5 MPa. The results show that supercritical LNG has a better heat transfer performance in zigzag channel of PCHE at lower operating pressures.

Author Contributions: Conceptualization and supervision, Z.Z.; formal analysis and data curation, Y.Z.; validation, methodology and software, X.M., X.C., S.L., S.Y. and Z.Z.; writing—original draft preparation, Y.Z. and Z.Z.

Nomenclature

T	Temperature (K)
P	Pressure (Pa)
L	length of channel (mm)
f	Fanning factor
v	Velocity (m/s)
Re	Reynolds number
h	Convective heat transfer coefficient (W/(m²·K))
Nu	Nusselt number
Eu	Euler number
C_p	Specific heat (KJ/(kg·K))
D_h	hydraulic diameter (m)
G	mass flux (kg/(m²·s))
ΔP	pressure drop (Pa)
ΔP_f	pressure drop due to friction (Pa)
ΔP_a	pressure drop due to acceleration (Pa)
τ	shear stress at the wall (Pa)

Greek symbols

μ	viscosity (Pa·s)
ρ	density (kg/m³)
λ	thermal conductivity (W/(m·K))

Subscript

w	Wall
b	Bulk mean
in	inlet
out	outlet

References

1. Pham, T.N.; Long, N.V.D.; Lee, S.; Lee, M. Enhancement of single mixed refrigerant natural gas liquefaction process through process knowledge inspired optimization and modification. *Appl. Therm. Eng.* **2017**, *110*, 1230–1239. [CrossRef]
2. Pu, L.; Qu, Z.G.; Bai, Y.H.; Qi, D.; Sun, K.; Yi, P. Thermal performance analysis of intermediate fluid vaporizer for liquefied natural gas. *Appl. Therm. Eng.* **2014**, *65*, 564–574. [CrossRef]
3. Xu, S.; Chen, X.; Fan, Z. Thermal design of intermediate fluid vaporizer for subcritical liquefied natural gas. *J. Nat. Gas Sci. Eng.* **2016**, *32*, 10–19. [CrossRef]
4. Pan, J.; Li, R.; Lv, T.; Wu, G.; Deng, Z. Thermal performance calculation and analysis of heat transfer tube in super open rack vaporizer. *Appl. Therm. Eng.* **2016**, *93*, 27–35. [CrossRef]

5. Han, C.-L.; Ren, J.-J.; Wang, Y.-Q.; Dong, W.-P.; Bi, M.-S. Experimental investigation on fluid flow and heat transfer characteristics of a submerged combustion vaporizer. *Appl. Therm. Eng.* **2017**, *113*, 529–536. [CrossRef]
6. Kim, J.H.; Baek, S.; Jeong, S.; Jung, J. Hydraulic performance of a microchannel PCHE. *Appl. Therm. Eng.* **2010**, *30*, 2157–2162. [CrossRef]
7. Chu, W.-X.; Li, X.-H.; Ma, T.; Chen, Y.-T.; Wang, Q.-W. Experimental investigation on SCO2-water heat transfer characteristics in a printed circuit heat exchanger with straight channels. *Int. J. Heat Mass Transf.* **2017**, *113*, 184–194. [CrossRef]
8. Figley, J.; Sun, X.; Mylavarapu, S.K.; Hajek, B. Numerical study on thermal hydraulic performance of a Printed Circuit Heat Exchanger. *Prog. Nucl. Energy* **2013**, *68*, 89–96. [CrossRef]
9. Kim, W.; Baik, Y.-J.; Jeon, S.; Jeon, D.; Byon, C. A mathematical correlation for predicting the thermal performance of cross, parallel, and counterflow PCHEs. *Int. J. Heat Mass Transf.* **2017**, *106*, 1294–13022. [CrossRef]
10. Yoon, S.-Y.; O'Brien, J.; Chen, M.; Sabharwall, P.; Sun, X. Development and validation of Nusselt number and friction factor correlations for laminar flow in semi-circular zigzag channel of printed circuit heat exchanger. *Appl. Therm. Eng.* **2017**, *123*, 1327–1344. [CrossRef]
11. Ishizuka, T.; Kato, Y.; Muto, Y.; Nikitin, K.; Tri Lam, N. Thermal-hydraulic characteristics of a Printed Circuit Heat Exchanger in a supercritical CO_2 loop. *Nucl. React. Therm. Hydraul.* **2006**, *30*, 109–116.
12. Tsuzuki, N.; Kato, Y.; Ishiduka, T. High performance printed circuit heat exchanger. *Appl. Therm. Eng.* **2007**, *27*, 1702–1707. [CrossRef]
13. Ngo, T.L.; Kato, Y.; Nikitin, K.; Tsuzuki, N. New printed circuit heat exchanger with S-shaped fins for hot water supplier. *Exp. Therm. Fluid Sci.* **2006**, *30*, 811–819. [CrossRef]
14. Ngo, T.L.; Kato, Y.; Nikitin, K.; Ishizuka, T. Heat transfer and pressure drop correlations of microchannel heat exchangers with S-shaped and zigzag fins for carbon dioxide cycles. *Exp. Therm. Fluid Sci.* **2007**, *32*, 560–570. [CrossRef]
15. Kim, D.E.; Kim, M.H.; Cha, J.E.; Kim, S.O. Numerical investigation on thermal hydraulic performance of new printed circuit heat exchanger model. *Nucl. Eng. Des.* **2008**, *238*, 3269–3276. [CrossRef]
16. Zhao, Z.; Zhao, K.; Jia, D.; Jiang, P.; Shen, R. Numerical Investigation on the Flow and Heat Transfer Characteristics of Supercritical Liquefied Natural Gas in an Airfoil Fin Printed Circuit Heat Exchanger. *Energies* **2017**, *10*, 1828. [CrossRef]
17. Lee, S.Y.; Park, B.G.; Chung, J.T. Numerical studies on thermal hydraulic performance of zigzag-type printed circuit heat exchanger with inserted straight channels. *Appl. Therm. Eng.* **2017**, *123*, 1434–1443. [CrossRef]
18. Yoon, S.Y.; No, H.C.; Kang, G.B. Assessment of straight, zigzag, S-shape, and airfoil PCHEs for intermediate heat exchangers of HTGRs and SFRs. *Nucl. Eng. Des.* **2014**, *270*, 334–343. [CrossRef]
19. Kim, I.H.; No, H.C. Physical model development and optimal design of PCHE for intermediate heat exchangers in HTGRs. *Nucl. Eng. Des.* **2012**, *243*, 243–250. [CrossRef]
20. Huang, D.; Wu, Z.; Sunden, B.; Li, W. A brief review on convection heat transfer of fluids at supercritical pressures in tubes and the recent progress. *Appl. Energy* **2016**, *162*, 494–505. [CrossRef]
21. Kim, I.H.; No, H.C. Thermal–hydraulic physical models for a Printed Circuit Heat Exchanger covering He, He-CO_2 mixture, and water fluids using experimental data and CFD. *Exp. Therm. Fluid Sci.* **2013**, *48*, 213–221. [CrossRef]
22. Yu, X.; Yang, X.; Wang, J. Heat Transfer and Pressure Loss of Immediate Heat Exchanger. *Inst. Nucl. New Energy Technol.* **2009**, *43*, 256–259.
23. Sung, J.; Lee, J.Y. Effect of tangled channels on the heat transfer in a printed circuit heat exchanger. *Int. J. Heat Mass Transf.* **2017**, *115*, 647–656. [CrossRef]
24. Nikitin, K.; Kato, Y.; Ngo, L. Printed circuit heat exchanger thermal-hydraulic performance in supercritical CO_2 experimental loop. *Int. J. Refrig.* **2006**, *29*, 807–814. [CrossRef]
25. Jeon, S.; Baik, Y.-J.; Byon, C.; Kim, W. Thermal performance of heterogeneous PCHE for supercritical CO_2 energy cycle. *Int. J. Heat Mass Transf.* **2016**, *102*, 867–876. [CrossRef]
26. Higashi, Y. NIST Thermodynamic and Transport Properties of Refrigerants and Refrigerant Matures (REFPROP). *Netsu Bussei* **2000**, *14*, 1575–1577.
27. Kwon, J.G.; Kim, T.H.; Park, H.J.; Cha, J.E.; Kim, M.H. Optimization of airfoil-type PCHE for the recuperate of small scale Brayton cycle by cost-based objective function. *Nucl. Eng. Des.* **2016**, *298*, 192–200. [CrossRef]

28. Han, C.L.; Ren, J.J.; Dong, W.-P.; Bi, M.-S. Numerical investigation of supercritical LNG convective heat transfer in a horizontal serpentine tube. *Cryogenics* **2016**, *78*, 1–13. [CrossRef]

29. Xu, X.; Ma, T.; Li, L.; Zeng, M.; Chen, Y.; Huang, Y.; Wang, Q. Optimization of fin arrangement and channel configuration in an airfoil fin PCHE for supercritical CO_2 cycle. *Appl. Therm. Eng.* **2014**, *70*, 867–875. [CrossRef]

30. Zhao, Z.; Zhang, X.; Zhao, K.; Jiang, P.; Chen, Y. Numerical investigation on heat transfer and flow characteristics of supercritical nitrogen in a straight channel of printed circuit heat exchanger. *Appl. Therm. Eng.* **2017**, *126*, 717–729. [CrossRef]

31. Yang, J.G.; Wu, H. Explicit Coupled Solution of Two-equation k-ω SST Turbulence Model and Its Application in Turbomachinery Flow Simulation. *Acta Aeronaut. ET Astronaut. Sinaica* **2014**, *35*, 116–124.

32. Ren, Y.; Liu, H.L.; Shu, M.H. Improvement of SST k-ω SST Turbulence Model and Numerical Simulation in Centrifugal Pump. *Trans. Chin. Soc. Agric. Mach.* **2014**, *43*, 123–128.

4

Unsteady Simulation of a Full-Scale CANDU-6 Moderator with OpenFOAM

Hyoung Tae Kim [1], Se-Myong Chang [2,*] and Young Woo Son [2]

[1] Thermal Hydraulic and Severe Accident Research Division, Korea Atomic Energy Research Institute, 989-111 Daedeok-daero, Yuseong-gu, Daejeon 34057, Korea; kht@kaeri.re.kr
[2] School of Mechanical Convergence Systems Engineering, Kunsan National University, 558 Daehak-ro, Gunsan, Jeonbuk 54150, Korea; ywson@kunsan.ac.kr
*   Correspondence: smchang@kunsan.ac.kr

Abstract: Three-dimensional moderator flow in the calandria tank of CANDU-6 pressurized heavy water reactor (PHWR) is computed with Open Field Operation and Manipulation (OpenFOAM), an open-source computational fluid dynamics (CFD) code. In this study, numerical analysis is performed on the real geometry model including 380 fuel rods in the calandria tank with the heat-source distribution to remove uncertainty of the previous analysis models simplified by the porous media approach. Realizable k-ε turbulence model is applied, and the buoyancy due to temperature variation is considered by Boussinesq approximation for the incompressible single-phase Navier-Stokes equations. The calculation results show that the flow is highly unsteady in the moderator. The computational flow visualization shows a circulation of flow driven by buoyancy and asymmetric oscillation at the pseudo-steady state. There is no region where the local temperature rises continuously due to slow circulating flow and its convection heat transfer.

Keywords: CANDU-6; PHWR; moderator; turbulence; OpenFOAM

1. Introduction

CANadian Deuterium Uranium (CANDU) reactors have been introduced in Korea since the late 1980s, and four units of CANDU-6 reactors were constructed in the Wolsong areas [1]. The horizontal fuel channels in a CANDU-6 reactor (a pressurized heavy water reactor, PHWR) are submerged in the heavy water (D_2O) pool which is contained by a cylindrical tank called calandria. One of the important design features of the CANDU-6 reactor is the use of moderator as a heat sink during some postulated accidents such as a large-break Loss of Coolant Accident (LOCA). Therefore, it is one of the major concerns in the CANDU safety analyses to estimate the local subcooling margin of the moderator inside the calandria tank.

Previous experimental studies [2] showed that the film boiling on the outside surface of fuel channels would be unlikely to occur if the local moderator subcooling is sufficient. Therefore, an accurate prediction of the moderator temperature distribution in the calandria tank is needed to confirm the channel integrity [3]. To predict the local temperature of the calandria tank, numerous experimental and numerical studies have been performed so far. Huget et al. [4,5] conducted two-dimensional moderator circulation tests at the STERN Laboratories Inc. (STERN Lab.), and they validated a specific code, MODerator TURbulent Circulation (MODTURC) [4] and its advanced version, Co-Located Advance Solution (MODTURC_CLAS) [5] against the experimental results [6] of the velocity and the temperature distributions.

Temperature distribution in the moderator is highly affected by flow patterns and circulation characteristics which themselves are generated as a result of interactions between the inertia forces (produced by inlet jets) and buoyancy forces (resulting from heat addition) in the calandria (see Figure 1). Given the differences in the moderator heat load, flow rate and inlet nozzle distribution and design, each CANDU reactor has a different flow pattern and temperature distribution during normal operation. Therefore, Korea Atomic Energy Research Institute (KAERI) installed a 1/4 scaled Moderator Circulation Test (MCT) facility [7] that is representative of CANDU-6 reactors with 380 fuel channels. These test results [8] showed that the moderator circulation flow has a mixed flow patterns with combination of inertial forces and buoyancy forces under the CANDU-6 operation conditions. Furthermore, the flow oscillation and unsteady flow behavior were observed, which were not reported in the previous studies [4,6].

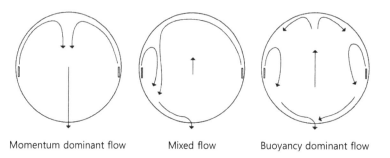

Momentum dominant flow   Mixed flow   Buoyancy dominant flow

Figure 1. Flow pattern inside a calandria by balance of buoyancy and momentum forces.

There have been numerous computational efforts to estimate the thermal hydraulics in the calandria tank using CFD codes. Hadaller et al. [9] obtained a tube bank pressure drop model for tube bundle region of the calandria tank and implemented it into the MODTURC_CLAS code. Yoon et al. [10] used a commercial code, CFX to develop a CFD model with a porous media approach for the core region. However, it is known that porous media modeling provides only average values of flow velocities and temperatures in the moderator and do not give any information about 3-D local flow variables near tube solid walls, which are necessary to implement accurate heat transfer calculations. Recently, porous media modeling in the tube bank region of core using economic computing resources are replaced by the full geometric model of calandria tubes requiring high computing resources. Sarchami et al. [11] used another FLUENT code to model all the calandria tubes as they are without any approximation for the core region. They could show the nature of moderator temperature fluctuations by dynamic flow behavior with completion between the upward moving buoyancy driven flows and the downward moving momentum driven flows. Teyssedou et al. [12] conducted FLUENT code simulation of moderator flow around calandria tubes of CANDU-6 and showed that the standard k-e model is appropriate for turbulence model to perform this kind of simulation. Application of FLUENT and CFX code is successfully performed for the reduced-scale CFD models for various thermal hydraulics problems in nuclear engineering also by the authors [13,14].

In this study, Open Field Operation and Manipulation (OpenFOAM) [15], an open-source CFD solver, is used to simulate the three-dimensional flows improving the computational efficiency by parallel computing which does need no proprietary license. The feasibility on the computation of 3-D flow has been tested and validated by the comparison with other codes by the authors [16], but the models are just focused on the pressure drop in a straight channel. In this paper, the full capacity of OpenFOAM CFD is tested for a turbulent unsteady flow as observed in the 1/4 scale of test [8] to resolve the 3-D structure of circulation flow in the moderator system of a real-scale CANDU-6 reactor.

We have studied the suitable grid levels and the validation of pressure drops with the comparison with various commercial codes such as ANSYS-CFX and COMSOL (COMputer SOLution) Multiphysics as well as experimental data using OpenFOAM [16]. However, the full simulation of CANDU-6 is not yet attempted because of its high complexity in three dimensions. From the dimensional analysis, the complex scale effects between prototype CANDU-6 and model MCT should be considered [7]. Therefore, the full-scale simulation is expected to show the overall flow physics with proper predictions of subcooling margin in this research. In the system codes, the difference of temperature or pressure between inlet and outlet is given as a lumped input parameter. For example, a system analysis code for PHWR, CATHENA [17] can consider a pipe network for tube bundles of fuel channels, but a three-dimensional numerical model is made for the present study to understand sophisticated flow physics such as turbulence diffusion and mixing, convective heat transfer, buoyancy forces, etc. in a moderator pool.

2. Simulation Method

2.1. Open Source Code

OpenFOAM has been developed by Henry Weller and Hrvoje Jasak in Imperial College. The source code has been opened to the public since 2004. This code is operated on the Linux-based O/S such as Ubuntu, so the copyright is absolutely free for every CFD program developer. This code is originated from the object-oriented programming (OOP) concept based on C++ program language. Solvers and libraries are defined as C++ classes. With the post processor ParaView, the graphical visualization becomes possible with a command paraFoam [15]. In this study, OpenFOAM version 2.3.1 (The OpenCFD Ltd., London, UK) is used. The numerical calculations are conducted with two OpenFOAM standard solver, "buoyantBoussinesqSimpleFoam" and "buoyantBoussinesqPimpleFoam".

The governing equations of the solvers are incompressible continuity equation, the Navier-Stokes equations and energy equation for the heat transfer where the buoyant force is related in the source term in the momentum equation with Boussinesq approximation. The realizable k-ε turbulence model is also applied for the low Reynolds number turbulent flow in the moderator. Computation is performed with two stages to save the settling time for the pseudo-steady state: "buoyantBoussinesqSimpleFoam" is for steady flow using Semi-Implicit Method for Pressure-Linked Equations (SIMPLE) algorithm, while "buoyantBoussinesqPimpleFoam" is for unsteady flow using PISO, Pressure Implicit with Splitting of Operator, and SIMPLE (PIMPLE) algorithm because the latter one is known to be better for the time-accurate computation [15].

2.2. Governing Equations and Discretization

The hydraulic governing equations based on the single-phase incompressible flow are written in the vector form:

$$\nabla \cdot \mathbf{V} = 0 \tag{1}$$

$$\rho \left\{ \frac{\partial \mathbf{V}}{\partial t} + (\mathbf{V} \cdot \nabla) \mathbf{V} \right\} = -\nabla p + \rho \mathbf{g} + (\mu + \mu_t) \nabla^2 \mathbf{V} + \mathbf{f}_V \tag{2}$$

where $\mathbf{V}$, ρ, p are velocity vector, density, and pressure while the constants μ and $\mathbf{f}_V$ are dynamic viscosity and body force per unit volume. Equation (1) is the continuity equation for incompressible flow, and the Navier-Stokes momentum equation, Equation (2) is decoupled from energy equation in the source term, or buoyancy force of Boussinesq approximation, $\mathbf{f}_V \approx -\rho \mathbf{g} \beta (T - T_0)$, where β is the thermal expansion in the unit of $1/K$, and $T - T_0$ is the difference of temperature from the reference condition.

A realizable k-ε model, which is better for rotational flow, is used for the simulation of turbulent flow. This model includes two additional equations in a tensor form:

$$\rho \left\{ \frac{\partial k}{\partial t} + (\mathbf{V} \cdot \nabla) k \right\} = \frac{\partial}{\partial x_j} \left\{ \left(\mu + \frac{\mu_t}{\sigma_k} \right) \frac{\partial k}{\partial x_j} \right\} + P_k + P_b - \rho \varepsilon - Y_M + S_k \tag{3}$$

$$\rho \left\{ \frac{\partial \varepsilon}{\partial t} + (\mathbf{V} \cdot \nabla) \varepsilon \right\} = \frac{\partial}{\partial x_j} \left\{ \left(\mu + \frac{\mu_t}{\sigma_\varepsilon} \right) \frac{\partial \varepsilon}{\partial x_j} \right\} + \rho \, \Phi S \varepsilon - \frac{\rho C_{2\varepsilon} \varepsilon^2}{k + \sqrt{\nu \varepsilon}} + C_{1\varepsilon} \frac{\varepsilon}{k} C_{3\varepsilon} P_b + S \varepsilon \tag{4}$$

where $S = \sqrt{2S_{ij}S_{ij}}$ is the modulus of mean rate-of-strain tensor, and $S_{ij} = \frac{1}{2}\left(\frac{\partial u_j}{\partial x_i} + \frac{\partial u_i}{\partial x_j}\right)$. In Equations (3) and (4), the turbulent eddy viscosity is defined as:

$$\mu_t = \rho C_\mu \frac{k^2}{\varepsilon} \tag{5}$$

where Equation (5) is substitute to Equation (2) for the consideration of turbulence. C_μ is not a constant like that of standard k-ε model but a function of S, ω_k (angular velocity in the reference frame), ε_{ijk} (dissipation tensor), and $\overline{\Omega}_{ij}$ (mean rate-of-rotation tensor), and Φ in Equation (4) is specified as

$$\Phi = \max\left[0.43, \frac{kS}{kS + 5\varepsilon}\right] = \begin{cases} 0.43 & if\ kS/(kS + 5\varepsilon) \leq 0.43 \\ kS/(kS + 5\varepsilon) & else \end{cases} \tag{6}$$

Other coefficients in Equations (3) and (4) are listed as $C_{1\varepsilon} = 1.44$, $C_{2\varepsilon} = 1.9$, $C_{3\varepsilon} = -0.03$, $\sigma_k = 1.0$, $\sigma_\varepsilon = 1.2$. The energy equation to get the temperature field for the computation of $\mathbf{f}_V$ in Equation (2) is

$$\rho C_p \left\{ \frac{\partial T}{\partial t} + (\mathbf{V} \cdot \nabla)T \right\} = \left(\lambda + \frac{\mu_t}{\Pr_t}C_p\right)\nabla^2 T + Q_s \tag{7}$$

where T is temperature; C_p is heat capacity; λ is thermal conductivity; $\Pr_t$ is turbulent Prandtl number, assumed as a constant of 0.85 for all the range of fluid, and Q_s is volumetric heat source, which should be specified in next section, Equation (8).

The convection terms of the governing equations are discretized with second order upwind scheme and diffusion terms are calculated with second order centered difference scheme. Turbulence equations and heat transfer equation were discretized with first order upwind scheme.

In the computation using OpenFOAM, SIMPLE algorithm, a kind of finite volume method (FVM) is applied for the iteration until the steady state for Equations (1) and (2). In this method, the pressure gradient term in Equation (2) is isolated, and sub-iterations should be performed between predictor and corrector [14]. The PIMPLE method is used for unsteady time marching, which is specified as no under-relaxation and multiple corrector steps in the calculation of momentum. PIMPLE is far accurate in time and applied to the unsteady computation instead of SIMPLE.

2.3. Boundary and Initial Conditions

The essential boundary conditions in this problem are listed as follows:

- Velocities: no-slip conditions at walls, and the mass flow rate is specified on the inlet, fixed to 127.4 kg/s per each inlet nozzle, or 1019 kg/s in total for the present problem. The inlet turbulent intensity is fixed as 5%, which can make the additional uncertainty for the turbulent flow linked with the full system;
- Pressure: zero pressure gradient conditions at walls and inlet, which should be valid under the assumption that the thickness of boundary layer is very thin. The outlet pressure is fixed by the moderator system;
- Temperature: the inlet temperature is fixed to 47.3 °C.

Total thermal power exerted to the whole system is 100 MW, which should be processed as the source term, Q_s in Equation (7) where the factor 1.089 (of course, the volume blockage of tubes is considered). The equivalent temperature, or the energy dived by density and heat capacity, should be considered in the energy equation of OpenFOAM where the temperature should be specified instead of power. The power distribution is defined as $Q_s(r,z) = Q_s f_r(r)f_z(z)$, and the shape functions are, in the dimensionless form [18,19],

$$f_r(r) = 0.94588 - 0.01989r + 0.0995r^2 - 0.03888r^3 - 0.00256r^4\ (0.0 \leq r[1/m] \leq 3.8)$$
$$f_z(z) = 1.0 - 0.1111z^2\ (-3.0 \leq z[1/m] \leq 3.0) \tag{8}$$

where Equation (8) is obtained from group distributions of fuel bundles measured from the plant data in a Wolsong PHWR [10], and the correlation is regressed with a fourth-order least-square curve.

The initial temperature of the whole computational domain is 47.3 °C, and the flow is assumed stationary in the beginning of computation. Actually, the CANDU-6 moderator is liked with the system network, but we did an independent simulation for the moderator only. The properties of the fluid (D_2O) for simulations are summarized in Table 1.

Table 1. Material properties of the heavy water.

Definition (Symbol)	Symbol	Value	Unit
Density	ρ	1085	kg/m^3
Thermal expansion	β	5×10^{-4}	K^{-1}
Dynamic viscosity	μ	5.5×10^{-4}	kg/(m·s)
Heat Capacity	C_p	4207	J/(kg·K)
Thermal conductivity	λ	0.659	W/(m·K)

2.4. Grid Generation

The prototype of CANDU-6 is such as Figure 2. The 380 circular rods called calandria tubes are allocated symmetry from the central line of tank; the inlet holes are four along each side part, i.e., eight in total with feeding nozzles consisting of four radial diffusers; and there are two outlet exits at the bottom. This prototype has an asymmetric shape for the cross section along the longitudinal direction because the outlet vent hole is tilted from the vertical midline.

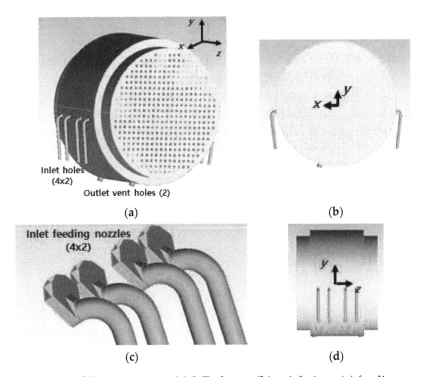

Figure 2. 3-D modeling of the prototype: (**a**) 3-D shape; (**b**) axial view; (**c**) feeding nozzles; (**d**) side view.

Figure 3 shows the three-dimensional unstructured grids at the view of lateral and longitudinal direction. The total grids are 6,740,446 consisting of 5,112,270 for the hexahedral, 13,112 for pyramids, and 1,615,064 for the tetrahedral. They are concentrated at the wall boundary with 15 stretched layers to increase the accuracy in turbulent boundary layers. The computation is done with a message passing interface (MPI) parallel machine where 24 processors are used. Each computational result is stored at multiple folders to assemble them in the post processor.

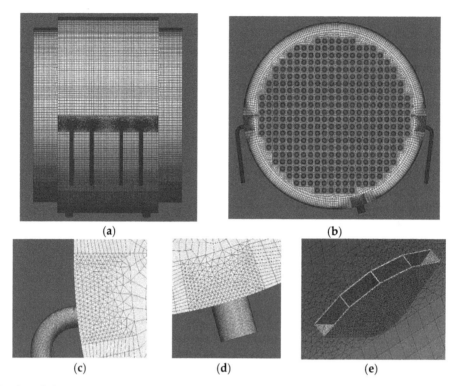

Figure 3. Grids of the prototype: (**a**) side longitudinal; (**b**) planform sectional; (**c**) inlet plumbing; (**d**) outlet exit; and (**d**) feeding nozzle.

3. Result and Discussion

The numerical method is verified and validated in the previous research by the authors [14,16]. The pressure drop with comparison of STERN laboratory experiment shows an error within 16.3% from the experimental data [16] (see Figure 4). The pressure drop is measured for isolated four-row bank of aligned cylinders of 33.02 mm diameter and 71.4 mm spacing. The pressure sensors are in the distance of sixteen blocks of cylinders along the central axis. Three sets of experiments are used for this comparison, specified with the Reynolds number based on the tube diameter, $Re_d = \rho V d / \mu$. Among various codes such as ANSYS-CFX and COMSOL, the open source code OpenFOAM displayed similar or better level of coincidence for all kinds of turbulence models, and k-ε model was the best result. The modeling of two-dimensional heat flow can predict the temperature with a maximum local error of 3.5 °C, which can be a reduced model of CANDU-6 moderator [14].

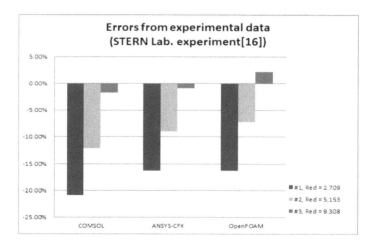

Figure 4. Comparison of numerical results from various codes for the experiment of STERN laboratory.

3.1. Quasi-Steady State

The solution is not converged to a steady state with segregated solvers such as SIMPLE or PIMPLE, but instead it fluctuates with oscillation [2]. In the earlier stage, steady solution is obtained with SIMPLE algorithm. After the computation is stabilized, in the later stage, the solution is time-marched to get the unsteady one. Figure 5 is the temperature at two outlets and the origin of (a) quasi-steady, or the center of calandria, before 12,000 time steps and (b) unsteady procedure to 850 s of physical time. The temperature of two outlets are slightly different from each other because of the asymmetry from flow instability. At the center, the time-averaged temperature is about 85 °C.

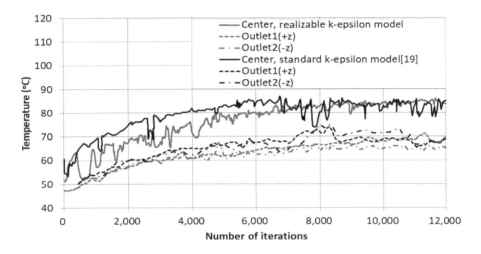

(a) Earlier stage, or steady solution.

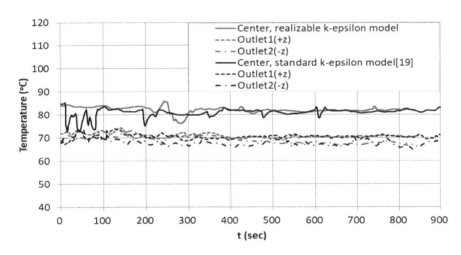

(b) Later stage, or unsteady solution.

Figure 5. Temperature at the center and two outlets.

3.2. Turbulent Model and Scale

In Figure 5, the present computational results are compared with those from the standard k-ε model [19]. With the realizable k-ε turbulent model, the convergence is slower than that with standard one in Figure 5a but shows overall better stable temperature in the pseudo-steady stage in Figure 5b. The inlet turbulence intensity is fixed to 5% [16].

To show how the grid system in Figure 4 can capture turbulent physics in a proper scale, the normalized wall distance, y^+ is plotted in Figure 6, which is ranged widely. At the outer wall, the maximum y^+ exceeds 100 where the fast waterjet sweeps injected from the nozzles. The y^+ is distributed from 10 to 60 at the cylinder walls. However, with the use of the wall function, the value of $y^+ < 80$ should be enough in the most of computational domain of the present problem since the value less than 30 can often make the turbulent wake flow unstable even though that at the tube wall boundary must be maintained near unity [18].

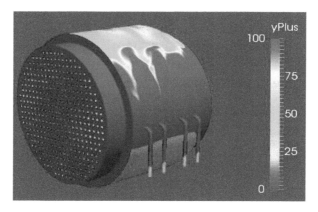

Figure 6. Distribution of y^+ at 840 s.

3.3. Velocity and Temperature Fields in the Unsteady Solution

After the quasi-steady state after 12,000 iterations, the time is reset to zero, and the fields of velocity and temperature are visualized in Figures 7–10 from 615 to 840 s.

Figures 7 and 8 are plotted at the sectional plane $z = 0$ (x-y plane), and the change of velocity and temperature are observed in the series of figures, respectively. In Figure 7, the cooling water from nozzles, initially to the upper direction or the positive y-axis, in both sides meets at a stagnation point, denoted with S in the upper right-hand side, the same tilt direction of vent hole. Please note that it is not symmetric. The flow field seems to be periodic for 225 s time difference. However, the temperature field in Figure 8 presents much more turbulent diffusion, so it becomes very difficult to find the obvious regularity. The period is not resolved from the figures, but the similar flow patterns are repeated with time passing: the cooling waterjet falls from the stagnation point, soaked into gaps of cylinders until the outflow at the vent hole. The heated water maintains balance of temperature at the upper region of tube bundles because of the buoyancy in the momentum transfer. The flow velocity is very slow less than 1 m/s in most of the domain, and no local region is found for the rapid increase of temperature thanks to the mixing of diffusive turbulent flow.

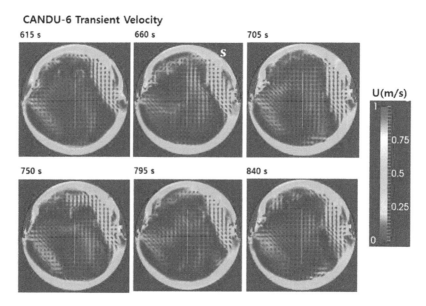

Figure 7. Velocity distribution at $z = 0$; 615–840 s.

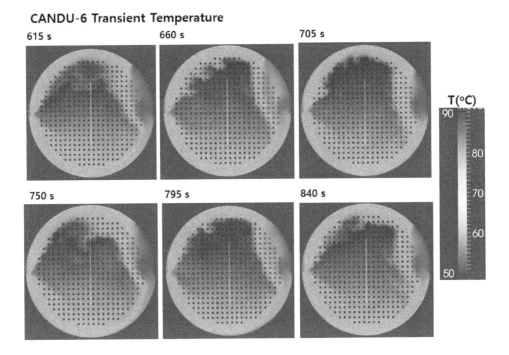

Figure 8. Temperature distribution at $z = 0$; 615–840 s.

Figures 9 and 10 are plotted at the sectional plane $x = 0$ (longitudinal), and the flow is not simple, too. In Figure 9, the flow velocity is so slow, but the marks of calandria tubes are dimly visible like stripes as they decelerate the circulation flow from the no-slip boundary condition. The maximum temperature stays about 89 °C in Figure 10, and cannot be found the region of successive increase of temperature.

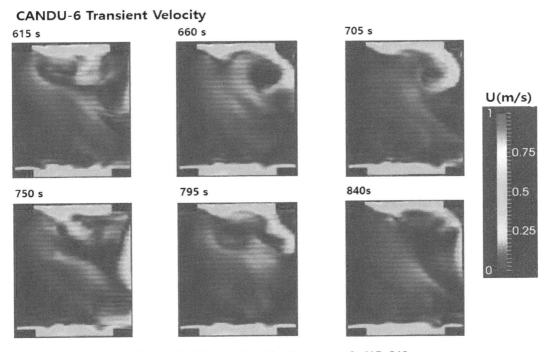

Figure 9. Velocity distribution at $x = 0$; 615–840 s.

CANDU-6 Transient Temperature

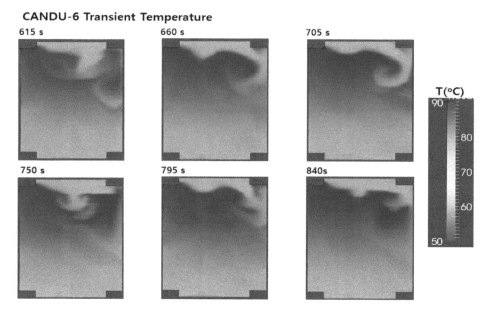

Figure 10. Temperature distribution at $x = 0$; 615–840 s.

The velocity distribution in Figures 7 and 9 show obviously that the flow circulation penetrates the interval of circular rods decelerating the flow with a pressure drop. The diffused flow makes the temperature increase at the upper central region in Figure 8 because flow resistance takes the worse cooling efficient. In the temperature field view of Figure 8, the largest turbulent eddies can be discerned at the interface of different temperature at the upper half plane such as the mushroom shape. They merge and separate continuously, developing a highly complex turbulent structure, so the high temperature difference of about 20 °C is dramatically visualized in both Figures 8 and 10.

In Figure 11, the mean inlet velocity at the nozzle is approximately 2 m/s, and speed at the central section is slower than the side one where a nozzle consists of four sections because the expansion ratio is greater. This fact compensates for the inlet jet flow to maintain a uniform flow along the curve of outer wall, approximately.

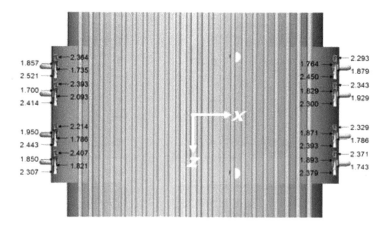

Figure 11. Mean velocities at the inlet nozzle, final time of the simulation (top view, unit: m/s).

3.4. Validation of Numerical Data

In Figure 12, the vertical axis at the center is plotted on the temperature for the last one of Figures 8 and 10 at 840 s. As we had no measures data for the prototype CANDU-6, the temperature distribution is normalized with the reduced-scale model test [14], and compared with other methods of computation as well as a set of coarse but experimental data: the numerical data from ANSYS-CFX, and

MODTURC-CLAS [5,6], etc. Although position and temperature are normalized, the circulating flow derived from buoyancy force reaches the equilibrium of maximum temperature at $0.25 < y/D < 0.3$.

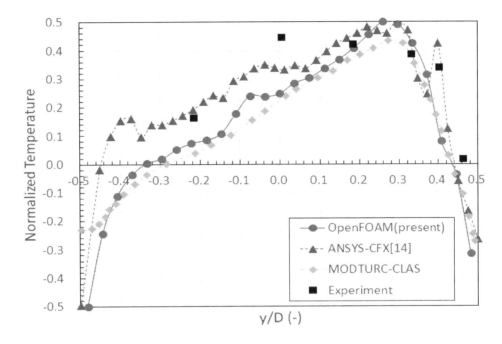

Figure 12. Validation of numerical data with other codes and experiment for the normalized temperature distribution along the vertical axis.

4. Conclusions

A prototype of CANDU-6 reactor is numerically analyzed around a three-dimensional moderator flow in calandria tank with OpenFOAM, an open-source CFD code. The three-dimensional shape including 380 rods in the calandria tank is precisely modeled without porous approximation to avoid parasite errors. The buoyancy term in the incompressible Navier-Stokes equation is considered with Boussinesq approximation of the temperature variation. Turbulence effect is reflected to energy equation as well as momentum equation with the realizable k-ε model.

The computational result shows that there should be no steady solution about the circulation flow, and therefore the unsteady simulation is achieved after getting a quasi-steady with oscillation of flow properties. The flow field is not converged to a steady solution. Instead, it oscillates in the regime of quasi-steady state. After 12,000 iterations from initial condition to the quasi-steady state, the unsteady simulation within 840 s shows no evidence of exact periodic oscillation for physical properties. The observation for 225 s, an approximate period of flow pattern, presents a complex structure of turbulent mixing despite uncertainties originated from the high intensity of turbulence. There are no regions where the temperature rises more than 90 degrees Celsius due to very slow transferring flow. Most of computational region marks the velocity less than 1 m/s. As the inlet nozzle flow going down from the stagnation point, it is highly diffused with the pressure drop due to the calandria tubes. Turbulent eddies were found in the temperature field, continuously developing to merge or separate at the interface of hot and cool fluid. The dimensionless wall distance of the first grid from wall, y^+ was checked as less than 80 in the most of computational domain but should be reduced with finer grids free of wall functions, especially for the outer wall of calandria tank.

Overall, this research presents that the use of open-source software is also very feasible for the application of analysis on the moderator system of PHWR such as CANDU-6. Compared with other commercial codes, the equivalent computation could be obtained from cheaper price and free copyright. However, the use of CFD alone provides a limited perspective. In practice, the CFD boundary condition should be supported by system analysis for possible transient phenomena.

Author Contributions: Conceptualization, H.T.K. and S.-M.C.; Methodology, S.-M.C.; Software, S.-M.C.; Validation, H.T.K., S.-M.C. and Y.W.S.; Formal Analysis, Y.W.S.; Investigation, H.T.K.; Resources, H.T.K.; Data Curation, H.T.K. and S.-M.C.; Writing—Original Draft Preparation, H.T.K.; Writing—Review & Editing, S.-M.C.; Visualization, S.-M.C.; Supervision, S.-M.C.; Project Administration, S.-M.C.; Funding Acquisition, H.T.K.

References

1. Wolsong Units. *2/3/4 Final Safety Analysis Report*; Korea Electric Power Corporation: Naju, Korea, 1995; Chapter 15.
2. Gillespie, G.E. An Experimental Investigation of Heat Transfer from a Reactor Fuel Channel: To Surrounding Water. In Proceedings of the 2nd Annual Conf. Canadian Nuclear Society, Ottawa, ON, Canada, 10 June 1981.
3. Fan, H.Z.; Aboud, R.; Neal, P.; Nitheanandan, T. Enhancement of the Moderator Subcooling Margin using Glass-peened Calandria Tubes in CANDU Reactors. In Proceedings of the 30th Annual Conference of the Canadian Nuclear Society, Calgary, AB, Canada, 31 May–3 June 2009.
4. Huget, R.G.; Szymanski, J.; Midvidy, W. Status of Physical and Numerical Modelling of CANDU Moderator Circulation. In Proceedings of the 10th Annual Conference of the Canadian Nuclear Society, Ottawa, ON, Canada, 4–7 June 1989.
5. Huget, R.G.; Szymanski, J.; Galpin, P.F.; Midvidy, W.I. MODTURC-CLAS: An Efficient Code for Analyses of Moderator Circulation in CANDU Reactors. In Proceedings of the 3rd International Conference on Simulation Methods in Nuclear Engineering, Montreal, QC, Canada, 18–20 April 1990.
6. Khartabil, H.F.; Inch, W.W.; Szymanski, J.; Novog, D.; Tavasoli, V.; Mackinnon, J. Three-dimensional moderator circulation experimental program for validation of CFD code MODTURC_CLAS. In Proceedings of the 21th CNS Nuclear Simulation Symposium, Ottawa, ON, Canada, 24–26 September 2000.
7. Kim, H.T.; Rhee, B.W. Scaled-down moderator circulation test facility at Korea Atomic Energy Research Institute. *Sci. Technol. Nucl. Install.* **2016**, *2016*, 5903602. [CrossRef]
8. Im, S.; Kim, H.T.; Rhee, B.W.; Sung, H.J. PIV measurement of the flow patterns in a CANDU-6 model. *Ann. Nucl. Eng.* **2016**, *98*, 1–11. [CrossRef]
9. Hadaller, G.I.; Fortman, R.A.; Szymanski, J.; Midvidy, W.I.; Train, D.J. Frictional Pressure Drop for Staggered and In Line Tube Bank with Large Pitch to Diameter Ratio. In Proceedings of the 17th Annual Conference of the Canadian Nuclear Society, Fredericton, NB, Canada, 9–12 June 1996.
10. Yoon, C. Development of a CFD Model for the CANDU-6 Moderator Analysis Using a Coupled Solver. *Ann. Nucl. Eng.* **2007**, *35*, 1041–1049. [CrossRef]
11. Sarchami, A.; Ashgriz, N.; Kwee, M. Three Dimensional Numerical Simulation of a Full Scale CANDU Reactor Moderator to Study Temperature Fluctuations. *Int. J. Eng. Phys. Sci.* **2012**, *6*, 275–281. [CrossRef]
12. Teyssedou, A.; Necciari, R.; Reggio, M.; Zadeh, F.M.; Étienne, S. Moderator Flow Simulation around Calandria Tubes of CANDU-6 Nuclear Reactors. *Eng. Appl. Comput. Fluid Mech.* **2014**, *8*, 178–192. [CrossRef]
13. Gim, G.H.; Chang, S.M.; Lee, S.; Jang, G. Fluid-Structure Interaction in a U-Tube with Surface Roughness and Pressure Drop. *Nucl. Eng. Technol.* **2014**, *46*, 633–640. [CrossRef]
14. Kim, H.T.; Chang, S.M. Computational Fluid Dynamics Analysis of the Canadian Deuterium Uranium Moderator Tests at the Stern Laboratories Inc. *Nucl. Eng. Technol.* **2015**, *47*, 284–292. [CrossRef]
15. OpenFOAM User Guide; CFD Direct Ltd.: 2019. Available online: https://openfoam.com/documentation/user-guide/ (accessed on 20 January 2019).
16. Kim, H.; Chang, S.M.; Shin, J.H.; Kim, Y.G. The Feasibility of Multidimensional CFD Applied to Calandria System in the Moderator of CANDU-6 PHWR Using Commercial and Open-Source Codes. *Sci. Technol. Nucl. Install.* **2016**, *2016*, 3194839. [CrossRef]
17. Hanna, B.N. CATHENA: A thermalhydraulic code for CANDU analysis. *Nucl. Eng. Des.* **1998**, *180*, 113–131. [CrossRef]
18. Seo, Y.S.; Chang, S.M.; Yeom, G.S. *CFD Analysis on the Validation Experiment with MCT 1/4 Model*; KAERI Report; Mirae Engineering Co.: Jeonju, Jeonbuk, Korea, 2015.
19. Kim, H.T.; Chang, S.M. OpenFOAM Analysis of CANDU-6 Moderator Flow. In Proceedings of the Transactions of the Korean Nuclear Society Autumn Meeting, Gyeongju, Korea, 29–30 October 2015.

5

Spherical Shaped ($Ag - Fe_3O_4/H_2O$) Hybrid Nanofluid Flow Squeezed between Two Riga Plates with Nonlinear Thermal Radiation and Chemical Reaction Effects

Naveed Ahmed [1][iD], Fitnat Saba [1], Umar Khan [2], Ilyas Khan [3,*][iD], Tawfeeq Abdullah Alkanhal [4], Imran Faisal [5] and Syed Tauseef Mohyud-Din [1]

[1] Department of Mathematics, Faculty of Sciences, HITEC University, Taxila Cantt 47080, Pakistan; naveed.ahmed@hitecuni.edu.pk (N.A.); fitnat_saba89@gmail.com (F.S.); syedtauseefs@hitecuni.edu.pk (S.T.M.-D.)

[2] Department of Mathematics and Statistics, Hazara University, Mansehra 21300, Pakistan; umar_jadoon@hu.edu.pk

[3] Faculty of Mathematics and Statistics, Ton Duc Thang University, Ho Chi Minh City 736464, Vietnam

[4] Department of Mechatronics and System Engineering, College of Engineering, Majmaah University, Majmaah 11952, Kingdom of Saudi Arabia; t.alkanhal@mu.edu.sa

[5] Department of Mathematics, Taibah University, Universities Road, P.O. Box 344 Medina, Kingdom of Saudi Arabia; mfaisal@taibahu.edu.sa

* Correspondence: ilyaskhan@tdt.edu.vn

Abstract: The main concern is to explore an electro-magneto hydrodynamic (EMHD) squeezing flow of ($Ag - Fe_3O_4/H_2O$) hybrid nanofluid between stretchable parallel Riga plates. The benefits of the use of hybrid nanofluids, and the parameters associated to it, have been analyzed mathematically. This particular problem has a lot of importance in several branches of engineering and industry. Heat and mass transfer along with nonlinear thermal radiation and chemical reaction effects have also been incorporated while carrying out the study. An appropriate selection of dimensionless variables have enabled us to develop a mathematical model for the present flow situation. The resulting mathematical method have been solved by a numerical scheme named as the method of moment. The accuracy of the scheme has been ensured by comparing the present result to some already existing results of the same problem, but for a limited case. To back our results further we have also obtained the solution by anther recipe known as the Runge-Kutta-Fehlberg method combined with the shooting technique. The error analysis in a tabulated form have also been presented to validate the acquired results. Furthermore, with the graphical assistance, the variation in the behavior of the velocity, temperature and concentration profile have been inspected under the action of various ingrained parameters. The expressions for skin friction coefficient, local Nusselt number and local Sherwood number, in case of ($Ag - Fe_3O_4/H_2O$) hybrid nanofluid, have been derived and the influence of various parameters have also been discussed.

Keywords: ($Ag - Fe_3O_4/H_2O$) hybrid nanofluid; nonlinear thermal radiation; heat transfer; chemical reaction; mass transfer; method of moment; numerical results

1. Introduction

An unprecedented and staggering development in the field of microfluidics, microelectronics, optical devices, chemical synthesis, transportation, high power engines and microsystems, including mechanical and electrical components, transforms the underpinnings of human life. These expansions further demand efficient cooling techniques, in order to manage the thermal performance, reliability and long-term operational devices. The primitive thermal management techniques (like cooling through liquids) seem to be deficient, in order to meet the challenges of thermal efficiency. Later on, this issue has been resolved by dispersing nano-meter sized structures, within the host fluid, which certainly influences its thermo-mechanical properties. In this regard, Choi [1,2] was considered as the pioneer, who gave this concept and calls it 'Nanofluid'. Many researchers have proposed various theoretical models for thermal conductivity, by following his footsteps. Maxwell [3] worked on a model for the thermal conductivity which is suitable only for the spherical shaped nanoparticles. Further studies in this area lead us to a variety of models, containing the impact of, particle–particle interactions (i.e., Bruggeman model, 1935) [4], particles shapes (i.e., Hamilton and crosser model 1962) [5] and particles distribution (i.e., Suzuki et al. 1969) [6]. Furthermore, researchers have found a number of articles in the literature that covers the different aspects of the nanofluid. Some of them can be found in the references [7–12].

In recent past years, a new class of nanofluids, entitled "Hybrid nanofluid", have come into existence that bears high thermal conductivity as compared to that of mono nanofluid. They have brought a revolution in various heat transfer applications like nuclear system cooling, generator cooling, electronic cooling, automobile radiators, coolant in machining, lubrication, welding, solar heating, thermal storage, heating and cooling in buildings, biomedical, drug reduction, refrigeration, and defense etc. In the case of a regular nanofluid, the critical issue is either they possess a good thermal conductive network or display a better rheological properties. The nanocomposites (single handedly) do not possess all the possible features which are required for a certain application. Therefore, by an appropriate selection of two or more nanoparticles, hybrid nanofluid can lead us to a homogeneous mixture, which possesses all physicochemical properties of various substances that can hardly be found in an individual substance [13,14].

The distinctive features of hybrid nanofluid have gained the attention of worldwide researchers and therefore a number of research articles have been published over the past few years. By employing a new material design concept, Niihara [15] discussed that the mechanical and thermal properties of the host fluid can be greatly enhanced, by the inclusion of nanocomposites.. Jana et al. [16] examined the thermal efficiency of the host fluid, by incorporating single and hybrid nanoparticles. Suresh et al. [17] takes into account a two-step method in order to synthesize water-based $(AI_2O_3 - Cu)$ hybrid nanofluid. Their experimental results reveal an improvement in the viscosity and thermal properties of the prepared hybrid nanofluid. In their next study [18], the effects of $(AI_2O_3 - Cu)$ hybrid nanofluid on the rate of heat transfer have been investigated. Momin [19], in 2013, conducted an experiment to study the impact of mixed convection on the laminar flow of hybrid nanofluid inside an inclined tube. By employing a numerical scheme, Devi and Devi [20] investigated the influence of magneto hydrodynamic flow of H_2O based $(Cu - AI_2O_3)$ hybrid nanofluid, over a porous dilating surface. With the aid of entropy generation, the magneto hydrodynamic flow of water based $(Cu - AI_2O_3)$ hybrid nanofluid, inside a permeable channel, has been discussed by Das et al. [21]. Chamkha et al. [22], numerically analyzed, the time dependent conjugate natural convection of water based hybrid nanofluid, within a semicircular cavity The Blasius flow of hybrid nanofluid with water, taken as a base fluid over a convectively heated surface, has been examined by Olatundun and Makinde [23]. Besides, in [24], Hayat and Nadeem incorporated the silver (Ag) and copper oxide (CuO) as nanoparticles within the water, to enhance the rate of heat transfer, over the linearly stretching surface.

These days, researchers have been attracted, to analyze the squeezing flows in various geometries. Due to their significance, they have been involved in many practical and industrial situations, like biomechanics, food processing, and chemical and mechanical engineering. They have also been utilized, in order to examine the formation of lubrication, polymer processing, automotive engines, bearings, injection, gear, appliances etc. These flow phenomena have been observed in different hydro dynamical machines and devices, where the normal velocities are enforced by the moving walls of the channel. Stefan [25] was the pioneer behind this concept. Later on, Shahmohamadi et al. [26]

employed an analytical technique, to examine the time-dependent axisymmetric flow of a squeezed nature. Recently, the effects of squeezing flow on nanofluid, confined between parallel plates, have been investigated by M. Sheikholeslami et al. [27]. They also utilized the Adomian's decomposition method to find the solution of the respective flow model. Khan et al. [28] have taken into account, the viscous dissipation effects along with slip condition, to analyze the two-dimensional squeezing flow of copper-water based nanofluid. For solution methodology, they have employed a variation of the parameters method. In 2017, the squeezing effects on the magneto hydrodynamic flow of Casson fluid (inside a channel) have been thoroughly inspected by Ahmed et al. [29]. They have modelled the respective flow problem and then solved it both numerically (Runge-Kutta scheme of fourth order) and analytically (Variation of parameters method).

Gallites and Lilausis [30] came up with the idea of an electromagnetic actuator device, in order to set up the crossed magnetic and electric fields, that appropriately provoked the wall's parallel Lorentz forces. The purpose of that device was to control the flow characteristics, which usually have a span wise arrangement of alternating and invariable magnets that specifically mounted a plane surface. The device, sometimes indicated as Riga plate [31], provided an aid to reduce the pressure drag, as well as the friction of submarines, that can be achieved by reducing the turbulence production and a boundary layer separation. A number of research articles have been published, in order to explore the distinctive features of the laminar flow of a fluid due to Riga plate. By assuming the least electrical conductivity effects, Pantokratoras and Magyari [32] investigated the flow behavior along with free convection. 1n 2011, Pantokratoras [33] reported the performance of Blasius flow, enforced by the Riga plate. He also encounterd the Sakiadis flow in his study. Later on, Magyari and Pantokratoras [34] took into account the Blasius flow of the liquid, which at the same time is electrically conducting, induced by Riga surface. The electro magneto hydrodynamic flow of nanofluid, induced by Riga plate along with the slip consequences, have been examined by Ayub et al. [35]. In 2017, Hayat et al. [36], discussed the squeezing flow of a fluid between two parallel Riga plates, together with convective heat transfer. The thermal radiative effects accompanied by chemical reaction, were also a part of their study. Moreover, Hayat et al. [37] investigated the electro magneto squeezing flow of carbon nanotube's suspended nanofluid between two parallel rotatory Riga plates along with viscous dissipation effects. They have considered the melting heat transfer condition, which basically revealed that the heat conducting process to the solid surface, involved the combine effects of both sensible and melting heat, which significantly enhances the temperature of the solid surface to its melting temperature.

The thermal radiation is a significant mode of heat transfer [38,39], which seems to be dominant, in order to transfer the net amount of heat, even in the existence of free or forced convection. The transfer of heat via radiation have been significantly found in many engineering and industrial applications, including airplanes, space vehicles, satellites, and atomic-force plant. In this context, many researchers have comprehensively discussed the radiative heat transfer phenomena. Some of the most relevant have been found in [40–44].

The literature survey revealed the fact that no single step has been taken in order to analyze the salient features of $(Ag − Fe_3O_4/H_2O)$ hybrid nanofluid, between two parallel Riga plates. This article encounters the influential behavior of the viscid flow of $(Ag − Fe_3O_4/H_2O)$ hybrid nanofluid between two parallel Riga plates, where the lower plate experiences a stretching velocity, while the upper plate enforces a squeezing flow. The transfer of heat and mass along with nonlinear thermal radiative and chemical reaction effects would also be a part of this study. By employing the suitable similarity transforms, a mathematical model for the present flow situation have been accomplished. Method of moment along with Runge-Kutta-Fehlberg method have been considered to find the solution of the model. Tables have been provided which presents the validity of the acquired results. Furthermore, the graphical aid has been provided, to demonstrate the influence of various ingrained entities, on the velocity and temperature along with concentration profiles. The expressions related to the coefficient of skin friction, local Nusselt number and local Sherwood number have also been developed and discussed with the help of graphs.

2. Formulation of the Governing Equations

Two parallel Riga plates have been under consideration, among which an electro-magneto hydrodynamic (EMHD) flow of $(Ag - Fe_3O_4/H_2O)$ hybrid nanofluid has been flowing. The flow is also time dependent and incompressible. Cartesian coordinates have been chosen in such away, that the $\breve{x}$−axis coincides with the horizontal direction, whereas the $\breve{\mathscr{Y}}$−axis is placed normal to it. The lower plate positioned at $\breve{\mathscr{Y}} = 0$, experiences a stretching velocity $\mathscr{U}_w(\breve{x}) = a\breve{x}/(1 - \lambda\breve{t})$. Besides, the upper Riga plate, owing the place at $\breve{\mathscr{Y}} = \mathscr{b}(\breve{t}) = \frac{-\lambda}{2}\left(a/v_f(1 - \lambda\breve{t})\right)^{-0.5}$. It is further assumed that the flow of $(Ag - Fe_3O_4/H_2O)$ hybrid nanofluid is a squeezing flow, having the velocity $\breve{v}_b = d\mathscr{b}/d\breve{t}$. Moreover, the nonlinear thermal radiation and chemical reaction effects are also considered. Figure 1 displays the configuration of the flow model.

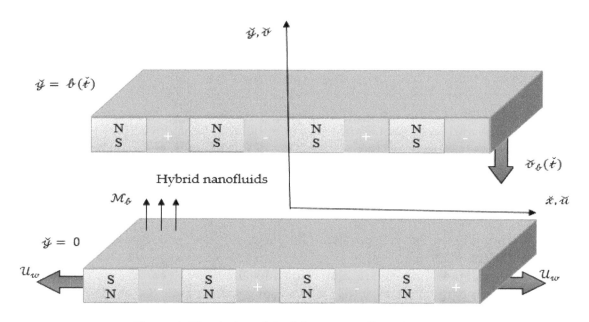

Figure 1. Physical model of the present flow situation.

The Navier-Stokes equations, suitable for the present flow situation, are given as [36]:

$$\frac{\partial \breve{v}}{\partial \breve{\mathscr{Y}}} + \frac{\partial \breve{u}}{\partial \breve{x}} = 0, \tag{1}$$

$$\frac{\partial \breve{p}}{\partial \breve{x}} + \breve{\rho}_{hnf}\left(\frac{\partial \breve{u}}{\partial \breve{t}} + \frac{\partial \breve{u}}{\partial \breve{x}}\breve{u} + \frac{\partial \breve{u}}{\partial \breve{\mathscr{Y}}}\breve{v}\right) = \mu_{hnf}\left(\frac{\partial^2 \breve{u}}{\partial \breve{\mathscr{Y}}^2} + \frac{\partial^2 \breve{u}}{\partial \breve{x}^2}\right) + \frac{Exp\left(-\pi\breve{\mathscr{Y}}/\ell\right)}{8(\pi\breve{j}_0\mathscr{M}_0)^{-1}}, \tag{2}$$

$$\frac{\partial \breve{p}}{\partial \breve{\mathscr{Y}}} + \breve{\rho}_{hnf}\left(\frac{\partial \breve{v}}{\partial \breve{t}} + \frac{\partial \breve{v}}{\partial \breve{x}}\breve{u} + \frac{\partial \breve{v}}{\partial \breve{\mathscr{Y}}}\breve{v}\right) = \mu_{hnf}\left(\frac{\partial^2 \breve{v}}{\partial \breve{\mathscr{Y}}^2} + \frac{\partial^2 \breve{v}}{\partial \breve{x}^2}\right), \tag{3}$$

$$\frac{\partial \breve{T}}{\partial \breve{t}} + \frac{\partial \breve{T}}{\partial \breve{x}}\breve{u} + \frac{\partial \breve{T}}{\partial \breve{\mathscr{Y}}}\breve{v} = \frac{k_{hnf}}{(\rho C_p)_{hnf}}\left(\frac{\partial^2 \breve{T}}{\partial \breve{\mathscr{Y}}^2} + \frac{\partial^2 \breve{T}}{\partial \breve{x}^2}\right) - \frac{1}{(\breve{\rho}C_p)_{hnf}}\left(\frac{\partial \breve{q}_r}{\partial \breve{\mathscr{Y}}} + \frac{\partial \breve{q}_r}{\partial \breve{x}}\right), \tag{4}$$

$$\frac{\partial \breve{\mathfrak{C}}}{\partial \breve{t}} + \frac{\partial \breve{\mathfrak{C}}}{\partial \breve{x}}\breve{u} + \frac{\partial \breve{\mathfrak{C}}}{\partial \breve{\mathscr{Y}}}\breve{v} = \mathfrak{D}_{hnf}\left(\frac{\partial^2 \breve{\mathfrak{C}}}{\partial \breve{\mathscr{Y}}^2} + \frac{\partial^2 \breve{\mathfrak{C}}}{\partial \breve{x}^2}\right) - c_1\left(\breve{\mathfrak{C}} - \breve{\mathfrak{C}}_b\right), \tag{5}$$

where, $\breve{u}$, signifies the horizontal component of velocity, while the vertical one is symbolized by $\breve{v}$. The dimensional pressure, temperature and concentration, are respectively shown by $\breve{p}$, $\breve{T}$ and $\breve{\mathfrak{C}}$. Furthermore, ℓ denotes the width between magnets and electrodes. $\mathscr{M}_0(Tesla)$ represents the magnetization of the permanent magnets, while, $\check{\jmath}_0(m^{-2}A)$ is the applied current density in the electrodes. The first order coefficient for a chemical reaction, is presented by c_1. In addition, $\check{q}_r$ symbolizes the rate of heat flux. The expression for the thermal radiative term has been successively proposed by Rosseland [38], which is given as:

$$\check{q}_r = -\frac{16\check{\sigma}\tilde{T}^3}{3\mathbb{k}}\frac{\partial\tilde{T}}{\partial\tilde{y}},$$ (6)

where, the coefficient for mean absorption is given by $\mathbb{k}$, while $\check{\sigma}$ stands for Stefan-Boltzmann constant. Therefore, after incorporating the Equation (6) in Equation (5), the energy equation can be generalized as follows:

$$\frac{\partial\tilde{T}}{\partial\check{t}} + \frac{\partial\tilde{T}}{\partial\check{x}}\check{u} + \frac{\partial\tilde{T}}{\partial\check{y}}\check{v} = \frac{k_{hnf}}{(\check{\rho}C_p)_{hnf}}\left(\frac{\partial^2\tilde{T}}{\partial\check{y}^2} + \frac{\partial^2\tilde{T}}{\partial\check{x}^2}\right) + \frac{1}{(\check{\rho}C_p)_{hnf}}\frac{16\check{\sigma}}{3\mathbb{k}}\frac{\partial}{\partial\check{y}}\left(\tilde{T}^3\frac{\partial\tilde{T}}{\partial\check{y}}\right).$$ (7)

The auxiliary conditions, specifying the current flow situation, are given as:

$$\check{u} = \mathscr{U}_w(\check{x}) = a\check{x}/(1-\lambda\check{t}), \check{v} = 0, \left(\tilde{T} - \tilde{T}_0\right) = 0, \left(\tilde{\mathfrak{C}} - \tilde{\mathfrak{C}}_0\right) = 0 \quad \text{at} \quad \check{y} = 0,$$ (8)

$$\check{v} = \frac{d\theta}{d\check{t}} = \frac{-\lambda}{2}\left(\frac{a(1-\lambda\check{t})}{v_f}\right)^{-0.5}, \check{u} = 0, \left(\tilde{T} - \tilde{T}_\theta\right) = 0, \left(\tilde{\mathfrak{C}} - \tilde{\mathfrak{C}}_\theta\right) = 0 \quad \text{at} \quad \check{y} = \theta(\check{t}),$$ (9)

where, $\tilde{T}_0$ and $\tilde{T}_\theta$ simultaneously, indicates the temperatures of the plates situated at $\check{y} = 0$ and $\check{y} = \theta(\check{t})$. The concentration of nanoparticles at the bottom plate is denoted by $\tilde{\mathfrak{C}}_0$, while, $\tilde{\mathfrak{C}}_\theta$ is the nanoparticles concentration at the top wall. Moreover, the rate, with which the lower surface is being stretched is a, while, λ represents the constant characteristics parameter.

In the aforementioned equations, $v_f = \mu_f/\check{\rho}_f$ denotes the effective kinematic viscosity. Furthermore, μ_{hnf} and μ_f simultaneously represents the effective dynamic viscosities of the hybrid nanofluid and mono nanofluid that significantly influence the flow behavior of the host fluid. Brinkman [45], in 1952, proposed a model for the effective dynamic viscosity $\left(\mu_{nf}\right)$ of a mono nanofluid which is given below:

$$\mu_{nf} = \frac{\mu_f}{(1-\varphi)^{5/2}},$$ (10)

where, φ denotes the nanoparticle volume fraction. Thus, in the case of hybrid nanofluid, the effective dynamic viscosity $\left(\mu_{hnf}\right)$ is defined as [22]:

$$\mu_{hnf} = \frac{\mu_f}{(1-\varphi_h)^{5/2}},$$ (11)

where, $\varphi_h = \varphi_1 + \varphi_2$ (in case of hybrid nanofluid) is a net volume fraction of distinct nanoparticles.

The effective density $\left(\check{\rho}_{nf}\right)$ presented by Pak and Cho [46] and the heat capacity $(\check{\rho}C_p)_{nf}$ [47] of mono nanofluid, can be respectively given by:

$$\check{\rho}_{nf} = \check{\rho}_f + \varphi\left(\check{\rho}_p - \check{\rho}_f\right),$$ (12)

$$(\check{\rho}C_p)_{nf} = (\check{\rho}C_p)_f + \varphi\left((\check{\rho}C_p)_p - (\check{\rho}C_p)_f\right).$$ (13)

By following the rules of mixture principle, the effective density $\left(\check{\rho}_{hnf}\right)$ [22,48] and heat capacity $\left(\check{\rho}C_p\right)_{hnf}$ [22,48] of hybrid nanofluid, can be estimated via Equations (14) and (15).

$$\check{\rho}_{hnf} = \varphi_1\check{\rho}_{p1} + \varphi_2\check{\rho}_{p2} + (1-\varphi_h)\check{\rho}_f, \tag{14}$$

$$\left(\check{\rho}C_p\right)_{hnf} = \varphi_1\left(\check{\rho}C_p\right)_{p1} + \varphi_2\left(\check{\rho}C_p\right)_{p2} + (1-\varphi_h)\left(\check{\rho}C_p\right)_f. \tag{15}$$

The thermal conductivity $\left(\mathcal{k}_{nf}\right)$ is the fundamental property, defining the heat transfer characteristics of the mono nanofluid. Maxwell suggested a correlation [3], for the mono nanofluid, by considering the spherical shaped nanoparticles, whose mathematical expression is given by:

$$\mathcal{k}_{nf} = \mathcal{k}_f\frac{\mathcal{k}_p(1+2\varphi) + 2\mathcal{k}_f(1-\varphi)}{\mathcal{k}_p(1-\varphi) + \mathcal{k}_f(2+\varphi)}. \tag{16}$$

In the case of hybrid nanofluid, the thermal conductivity ratio can be accomplished by modifying the Maxwell correlation [22] as:

$$\frac{\mathcal{k}_{hnf}}{\mathcal{k}_f} = \frac{\frac{\varphi_1\mathcal{k}_{p1}+\varphi_2\mathcal{k}_{p2}}{\varphi_h} + 2\mathcal{k}_f + 2\left(\varphi_1\mathcal{k}_{p1} + \varphi_2\mathcal{k}_{p2}\right) - 2\varphi_h\mathcal{k}_f}{\frac{\varphi_1\mathcal{k}_{p1}+\varphi_2\mathcal{k}_{p2}}{\varphi_h} + 2\mathcal{k}_f - \left(\varphi_1\mathcal{k}_{p1} + \varphi_2\mathcal{k}_{p2}\right) + \varphi_h\mathcal{k}_f}. \tag{17}$$

In 1935, another correlation, for spherical nanoparticles, has been introduced by Bruggeman [4], which usually considers the impact of nano clusters on the thermal conductivity. By mixture principle, this model can be extended for the estimation of thermal conductivity ratio of the hybrid nanofluid and is given by:

$$\frac{\mathcal{k}_{hnf}}{\mathcal{k}_f} = \frac{1}{4}\left[(3\varphi_h - 1)\left(\frac{\frac{\varphi_1\mathcal{k}_{p1}+\varphi_2\mathcal{k}_{p2}}{\varphi_h}}{\mathcal{k}_f}\right) + (2 - 3\varphi_h) + (\Delta)^{1/2}\right], \tag{18}$$

where,

$$\Delta = \left[\begin{array}{l}(3\varphi_h - 1)^2\left(\frac{\frac{\varphi_1\mathcal{k}_{p1}+\varphi_2\mathcal{k}_{p2}}{\varphi_h}}{\mathcal{k}_f}\right)^2 + (2 - 3\varphi_h)^2 + \\ 2\left(2 + 9\varphi_h - 9\varphi_h^2\right)\left(\frac{\frac{\varphi_1\mathcal{k}_{p1}+\varphi_2\mathcal{k}_{p2}}{\varphi_h}}{\mathcal{k}_f}\right)\end{array}\right]. \tag{19}$$

The molecular diffusivity [22,49–51], of the species concentration, for mono nanofluid and hybrid nanofluid are simultaneously defined as:

$$\mathcal{D}_{nf} = (1-\varphi)\mathcal{D}_f, \tag{20}$$

$$\mathcal{D}_{hnf} = (1-\varphi_h)\mathcal{D}_f. \tag{21}$$

In all the above expressions, φ_1 and φ_2 simultaneously, represents the volume concentration of magnetite (Fe_3O_4) and silver (Ag) nanoparticles in hybrid nanofluids. The viscosity, density and specific heat of host fluid are respectively denoted by μ_f, $\check{\rho}_f$ and $(C_p)_f$. At constant pressure, $(C_p)_{p1}$ and $(C_p)_{p2}$ respectively, denotes the specific heat of magnetite and silver nanoparticles. The densities, of magnetite and silver nanoparticles, are specified by $\check{\rho}_{p1}$ and $\check{\rho}_{p2}$ respectively. $\mathcal{k}_f$ and $\mathcal{D}_f$ represents the thermal conductivity and mass diffusivity of the water (H_2O). The thermal conductivities of magnetite and silver nanocomposites, are respectively symbolized by $\mathcal{k}_{p1}$ and $\mathcal{k}_{p2}$.

The prescribed form of similarity transforms, which deals with the process of conversion of Equations (1)–(3) and (7) into a nonlinear set of ordinary differential equations (ODE), are given as:

$$\Psi = \left(\frac{(1-\lambda \check{t})}{a v_f}\right)^{-0.5} \check{x} \check{F}(\chi), \quad \chi = \check{\mathscr{Y}}(\theta(\check{t}))^{-1}, \quad \check{u} = \frac{\partial}{\partial \check{\mathscr{Y}}}(\Psi) = \mathscr{U}_w \check{F}'(\chi),$$

$$\hat{v} = -\frac{\partial}{\partial \check{x}}(\Psi) = -\left(\frac{(1-\lambda \check{t})}{a v_f}\right)^{-0.5} \check{F}(\chi), \quad \check{\mathscr{T}}(\chi) = \frac{\check{T} - \check{T}_\theta}{\check{T}_0 - \check{T}_\theta}, \quad \check{\mathscr{C}}(\chi) = \frac{\check{\mathfrak{C}} - \check{\mathfrak{C}}_\theta}{\check{\mathfrak{C}}_0 - \check{\mathfrak{C}}_\theta}. \tag{22}$$

where, the superscript ′ stands for $d/d\chi$. Thus, by opting Brinkman (11) and Bruggeman (18) models, the dimensionless mode of a system of nonlinear ordinary differential equations, for $(Ag - Fe_3O_4/H_2O)$ hybrid nanofluid, along with radiation and chemical reaction parameters has been accomplished that can be written as:

$$\check{F}^{iv} + \check{Y}_1 \left[\check{F}'\check{F}'' - \check{F}\check{F}''' - \frac{\gamma}{2}\left(3\check{F}'' + \chi\check{F}'''\right)\right] - (1-\varphi_h)^{5/2} \mathscr{M}_\theta \mathscr{P} e^{-\mathscr{P}\chi} = 0, \tag{23}$$

$$\left(\left(\frac{\mathscr{k}_{hnf}}{\mathscr{k}_f} + Rd((1-\check{\mathscr{T}}) + \check{\mathscr{T}}\theta_w)^3\right)\check{\mathscr{T}}\prime\right)' + Pr\check{Y}_2\left(\check{F} - \frac{\gamma}{2}\chi\right)\check{\mathscr{T}}' = 0. \tag{24}$$

$$\check{\Theta}'' + \frac{Sc}{(1-\varphi_h)}\left(\check{F} - \frac{\gamma}{2}\chi\right)\check{\Theta}' - \frac{Sc}{(1-\varphi_h)}c_\mathscr{R}\check{\Theta} = 0. \tag{25}$$

where, $\check{F}$, $\check{\mathscr{T}}$ and $\check{\Theta}$, all are the dependent functions of dimensionless variable χ. Furthermore, the dimensionless auxiliary conditions, supporting the present flow situation, are therefore suggested as:

$$\check{F}(0) = 0, \quad \check{F}(1) = \frac{\gamma}{2}, \quad \check{F}\prime(0) - 1 = 0, \quad \check{F}\prime(1) = 0, \tag{26}$$

$$\check{\mathscr{T}}(0) - 1 = 0, \quad \check{\mathscr{T}}(1) = 0, \tag{27}$$

$$\check{\Theta}(0) - 1 = 0, \quad \check{\Theta}(1) = 0. \tag{28}$$

In the above-mentioned system of Equations (23) and (25), $\gamma = \lambda/a$ represents a dimensionless squeeze number, while, $\mathscr{M}_\theta = \pi \mathscr{j}_0 \mathscr{M}_0 \check{x} / 8 \check{\rho}_f \mathscr{U}_w^2$ is the modified Hartman number and $\mathscr{P} = \pi\theta(\check{t})/\ell$ is the dimensionless parameter. Moreover, the radiation parameter is denoted by $Rd = 16\check{\sigma}\check{T}_\theta^3/3\mathbb{k}\,\mathscr{k}_f$. Prandtl number is symbolized by $Pr = \left(\mathscr{k}_f/(\check{\rho}C_p)_f v_f\right)^{-1}$. Besides, $Sc = v_f/\mathfrak{D}_f$ signifies, the Schmidt number. The chemical reaction is indicated by $c_\mathscr{R} = c_1(1-\lambda\check{t})/a$.

Moreover, the constants $\check{Y}_1$ and $\check{Y}_2$, embroiled in the governing dimensionless model, can be mathematically stated as:

$$\left.\begin{aligned}\check{Y}_1 &= \frac{v_f}{v_{hnf}} = \frac{\left(1 - \varphi_h + \varphi_1\frac{\check{\rho}_{p1}}{\check{\rho}_f} + \varphi_2\frac{\check{\rho}_{p2}}{\check{\rho}_f}\right)}{(1-\varphi_h)^{-5/2}}, \\ \check{Y}_2 &= \frac{(\check{\rho}C_p)_{hnf}}{(\check{\rho}C_p)_f} = 1 - \varphi_h + \varphi_1\frac{(\check{\rho}C_p)_{p1}}{(\check{\rho}C_p)_f} + \varphi_2\frac{(\check{\rho}C_p)_{p2}}{(\check{\rho}C_p)_f}.\end{aligned}\right\} \tag{29}$$

The coefficient of skin friction, local heat transferal rate (i.e., local Nusselt number) and local Sherwood number, for the present flow situation, opt the following dimensionless expressions:

$$\hat{\mathscr{C}}_{f\check{x}} = \frac{\tau_w}{\check{\rho}_{hnf}\mathscr{U}_w^2}, \quad Nu_{\check{x}} = \frac{\theta\mathscr{k}_f^{-1}}{\left(\check{T}_0 - \check{T}_\theta\right)}(\check{q}_w + \check{q}_r) \quad \text{and} \quad Sh_{\check{x}} = \frac{\theta\mathfrak{D}_f^{-1}}{\left(\check{\mathfrak{C}}_0 - \check{\mathfrak{C}}_\theta\right)}\check{q}_m, \tag{30}$$

where, τ_w indicates the shear stress, while, the heat and mass fluxes, at both of the walls, are simultaneously signifies by $\check{q}_w$ and $\check{q}_m$. They are respectively defined as:

$$
\tau_w = \mu_{hnf}\left(\frac{\partial \check{u}}{\partial \check{y}}\right)_{\check{y} = \{\begin{smallmatrix} 0 \\ \mathscr{B}(\check{t}) \end{smallmatrix}}, \quad \check{q}_w = -\mathscr{k}_{hnf}\left(\frac{\partial \check{T}}{\partial \check{y}}\right)_{\check{y} = \{\begin{smallmatrix} 0 \\ \mathscr{B}(\check{t}) \end{smallmatrix}}, \quad \check{q}_m = -\mathscr{D}_{hnf}\left(\frac{\partial \check{\mathscr{C}}}{\partial \check{y}}\right)_{\check{y} = \{\begin{smallmatrix} 0 \\ \mathscr{B}(\check{t}) \end{smallmatrix}} \tag{31}
$$

Subsequently, by incorporating Equations (6) and (31) into Equation (30), we finally achieved the dimensionless forms of skin friction, the Nusselt number, and the Sherwood number, both at the top and bottom walls, which can be expressed as:

$$
Re_{\check{x}}^{0.5}\hat{\mathscr{C}}_{lower} = \frac{1}{\Upsilon_1}\check{F}''(0), \qquad\qquad Re_{\check{x}}^{0.5}\hat{\mathscr{C}}_{upper} = \frac{1}{\Upsilon_1}\check{F}''(1), \tag{32}
$$

$$
\begin{aligned}
(1 - \lambda\check{t})^{0.5}Re_{\check{x}}^{-0.5}Nu_{lower} &= -\left(\frac{\mathscr{k}_{hnf}}{\mathscr{k}_f} + Rd(\theta_w)^3\right)\check{\mathscr{T}}'(0), \\
(1 - \lambda\check{t})^{0.5}Re_{\check{x}}^{-0.5}Nu_{upper} &= -\left(\frac{\mathscr{k}_{hnf}}{\mathscr{k}_f} + Rd\right)\check{\mathscr{T}}'(1),
\end{aligned} \tag{33}
$$

and

$$
Re_{\check{x}}^{-0.5}Sh_{lower} = -(1 - \varphi_h)\check{\Theta}'(0), \qquad\qquad Re_{\check{x}}^{-0.5}Sh_{upper} = -(1 - \varphi_h)\check{\Theta}'(1), \tag{34}
$$

where, $Re_{\check{x}} = \check{x}\mathscr{U}_w/v_f$ denotes the local Reynolds number.

3. Solution Procedure

Method of moments (MM), one of the sub-class of the method of weighted residual (MWR), has been considered, in order to tackle the system of differential equations coupled with boundary conditions. From an accuracy point of view, a comparison has also been made, between the results achieved by Method of moments (MM) and Runge-Kutta-Fehlberg method (RKF). For this purpose, a mathematical software Maple 16 has been used.

Method of Moments

Let $\mathscr{D}$, an arbitrary differential operator, acting upon $\check{F}(\chi)$ generate a function $\mathscr{g}(\chi)$, which is given as:

$$
\mathscr{D}\left(\check{F}(\chi)\right) = \mathscr{g}(\chi). \tag{35}
$$

In order to approximate the solution of the above-mentioned problem, a trial solution has been defined, which is in the form of a linear combination of base function. These basis functions, also hold the property of linearly independence. Mathematically, it can be expressed as:

$$
\check{F}(\chi) \cong \tilde{F}(\chi) = \psi_0 + \sum_{i=1}^{n} c_i \psi_i, \tag{36}
$$

where, the essential boundary conditions are usually incorporated in ψ_0. By substituting back Equation (36) in Equation (35), one can acquired an exact solution, in the form of the trial solution that satisfies the given problem (35), which is an extremely rare situation. More often, it does not satisfies the given problem and therefore, left an expression that represents the error or the residual as under:

$$
\widetilde{\mathscr{R}}(\chi) = \mathscr{D}\left(\check{F}(\chi)\right) - \mathscr{g}(\chi) \neq 0. \tag{37}
$$

The proper selection of weights enabled us to construct weighted residual error. The values of unknown constants c_i's have been accomplished after the minimization procedure, that is:

$$
\int_{\chi} \widetilde{\mathscr{R}}(\chi)\,\mathscr{W}_i(\chi)d\chi = 0, \qquad i = 1, 2, \ldots, n. \tag{38}
$$

The above equation generate a system of algebraic equations, whose solution finally lead us to determine the unknown constants c_i's, and thus, a numerical solution has been obtained after plugging them back into the trial solution.

It is pertinent to mention that the weight functions involved in the method of moments (MM), are defined as:

$$\mathscr{W}_i(\chi) = \frac{\partial}{\partial c_j} c_j \chi^j, \qquad j = 0, 1, \ldots, n - 1. \tag{39}$$

For the present flow problem, the system of the trial solution, under consideration, are defined as:

$$\check{F}(\chi) = \frac{4}{5}\chi^3 - \frac{17}{10}\chi^2 + \chi + \sum_{i=1}^{5} c_i \chi(\chi - 1)^i, \tag{40}$$

$$\check{\mathscr{T}}(\chi) = 1 - \chi + \sum_{i=1}^{5} d_i \chi(\chi - 1)^i, \tag{41}$$

$$\check{\Theta}(\chi) = 1 - \chi + \sum_{i=1}^{5} e_i \chi(\chi - 1)^i. \tag{42}$$

By following the procedure as suggested above, the numerical solution has been achieved by substituting the above set of trial solutions into the governing dimensionless system of equations, which are nonlinear in nature. Thus, by assigning some specific values to the parameters, the approximate solution for the velocity and temperature along with concentration profiles are as under:

$$\check{F}(\chi) \cong \tilde{F}(\chi) = 0.0387970586596651373\chi^6 - 0.135831460629211753\chi^5 +$$
$$0.150915945544620123\chi^4 + 0.750688356622255526\chi^3 - \tag{43}$$
$$1.70467856655171612\chi^2 + 1.00010866635438700\chi,$$

$$\check{\mathscr{T}}(\chi) \cong \tilde{\mathscr{T}}(\chi) = -0.0866389688520037421\chi^6 + 0.500946012785362327\chi^5 -$$
$$1.06954770759623097\chi^4 + 0.911080929453323640\chi^3 - \tag{44}$$
$$0.0448789189043188319\chi^2 - 1.21096134688613266\chi + 1.0,$$

$$\check{\Theta}(\chi) \cong \tilde{\Theta}(\chi) = -0.00281815275323297988\chi^6 +$$
$$0.0301378227343711141\chi^5 - 0.0834665005070162475\chi^4 +$$
$$0.0635487269722848497\chi^3 + 0.0509906671544004958\chi^2 - \tag{45}$$
$$1.05839256360080736\chi + 1.0.$$

The above solutions are obtained for certain values of parameters, which are given as:

$$\gamma = c_{\mathscr{R}} = Rd = 0.2, \quad Sc = 0.5, \quad \theta_w = 1.1, \quad \mathscr{M}_\beta = 1.5, \quad \mathscr{P} = 10, \quad \varphi_1 = \varphi_2 = 0.01. \tag{46}$$

Table 1 displays some important thermal and physical properties of carrier fluid (H_2O) [52] and the nanoparticles. These values play a key role in order to obtain the above solutions.

Table 1. Thermo-mechanical properties of H_2O, Fe_3O_4 and Ag nanoparticles [51–53].

	$H_2O(f)$	$Fe_3O_4\ (\varphi_1)$	$Ag\ (\varphi_2)$
$\check{\rho}\ (kg\ m^{-3})$	997.1	5180	10,500
$C_p\ (J\ kg^{-1}K^{-1})$	4179	670	235
$\hbar\ (Wm^{-1}K^{-1})$	0.613	9.7	429
Pr	6.2	—	—

The subsequent Tables 2–4 respectively, provide a comparison between the results obtained via MM and RKF, for velocity, temperature, and concentration profiles. The values, as suggested above, remains the same for ingrained parameters. From these tables, one can clearly visualize the validity of the acquired results. Furthermore, for the tabulated values, the significant digit is set to 4.

Table 2. Comparison of the results obtained for $\ddot{}(\chi)$ for $(Ag - Fe^3O^4/H^2O)$ hybrid nanofluid with $(\varphi_1 = 0.01)$.

χ	NM	MM	Abs Error
0.0	0	0	0
0.1	0.08361971055	0.0837285414	0.0001.088308511
0.2	0.1377601897	0.1380405799	0.0002803901712
0.3	0.1674522669	0.1678007463	0.0003484793851
0.4	0.1776981193	0.1779703975	0.0002722782048
0.5	0.1734032258	0.1735144536	0.000111227772
0.6	0.1593783977	0.1593361686	0.00000422290531
0.7	0.1403592885	0.1402398353	0.0001194532992
0.8	0.121027384	0.1209214235	0.0001059605161
0.9	0.1060297995	0.1059871531	0.0000042646395
1.0	0.1	0.1	0.0000000000000

Table 3. Comparison of the results obtained for $\breve{\mathscr{T}}(\chi)$ for $(Ag - Fe_3O_4/H_2O)$ hybrid nanofluid with $(\varphi_1 = 0.01)$.

χ	NM	MM	Abs Error
0.0	1	1	0.0000000000000
0.1	0.8793088955	0.8792641251	0.0000044770436
0.2	0.7618050723	0.7617447028	0.0000060369490
0.3	0.6498328728	0.6497624809	0.0000070391950
0.4	0.5442201803	0.5441384063	0.0000081773904
0.5	0.4447153684	0.4446388103	0.0000076558102
0.6	0.3504084469	0.3503582131	0.0000050233744
0.7	0.2600611457	0.2600397494	0.0000213962680
0.8	0.1723422761	0.1723332129	0.0000090632458
0.9	0.08599850137	0.08599072095	0.0000007780414
1.0	0	0	0.0000000000000

Table 4. Comparison of the results obtained for $\breve{\Theta}(\chi)$ for $(Ag - Fe_3O_4/H_2O)$ hybrid nanofluid with $(\varphi_1 = 0.01)$.

χ	NM	MM	Abs err
0.0	1	1	0.0000000000000
0.1	0.8947297917	0.8947261509	0.0000003640704
0.2	0.7907519656	0.7907454211	0.0000006544490
0.3	0.688190526	0.6881823084	0.0000008217550
0.4	0.5870371838	0.5870289256	0.0000008258260
0.5	0.4871828347	0.4871760929	0.0000006741763
0.6	0.3884468482	0.3884424019	0.0000004446265
0.7	0.2906036897	0.29060125	0.0000002439726
0.8	0.1934071528	0.1934058456	0.0000001307163
0.9	0.09661287932	0.09661218516	0.0000000694153
1.0	0	0.0000000000000	0.0000000000000

The reliability of the obtained results have been further checked by reproducing the results for Skin friction coefficient, which were previously presented by Hayat et al. [36]. The results were obtained for the regular fluid ($\varphi_1 = \varphi_2 = 0$). Table 5 has been prepared to check the validity of the obtained results. It has been observed that the results obtained via Method of Moments are in good agreement with the previously existing results. Moreover, Method of Moments offers less computational complexity as compared to Homotopy analysis method. From the table, it has also been detected that the skin friction coefficient displays a decline with the increasing squeezing parameter (γ). However, a reversed behavior has been observed for increasing values of Modified Hartmann number $\mathcal{M}_b$.

Table 5. Comparison of the results obtained for Skin friction coefficient with ($\varphi_1 = \varphi_2 = 0$).

$\mathcal{M}_b$	γ	NM [36]	HAM [36]	MM
1.5	0.5	0.467511	0.467511	0.467511
1.0	-	0.452395	0.452395	0.452395
0.0	-	0.422159	0.422159	0.422159
1.5	0.3	1.08543	1.08543	1.08543
-	0.1	1.69635	1.69634	1.69634

4. Results and Discussions

The goal is to graphically elucidate the influential behavior of velocity, temperature and concentration profiles, due to the various ingrained entities. A pictorial view, from Figures 2–20, has been presented for the above-mentioned purpose. Figures 2–4 displays the performance of velocity profile, under the action of the squeezing parameter, Modified Hartmann number and solid volume fraction. The variations in velocity component $\breve{F}'(\chi)$, due to squeezing parameter γ, have been depicted in Figure 2a. For $\gamma > 0$, i.e., when the upper plate moves in the downward direction, the fluid nearby the upper wall experiences a force, which in turn enhances the fluid velocity in that region. As γ increases sufficiently, the velocity component $\breve{F}'(\chi)$ also increases and gradually depreciates the reversal behavior of the flow. The velocity component $\breve{F}(\chi)$ also experiences an increment in the region, adjacent to the upper wall, which is mainly due to the squeezing behavior of the upper plate and this phenomena has been clearly observed through Figure 2b. Figure 3 demonstrates the impact of Modified Hartmann number $\mathcal{M}_b$ on the axial and normal components of the velocity distribution. Since the magnetic field experiences an exponential decline, therefore velocity component $\breve{F}'(\chi)$ seems to be increased in the lower region of the channel. The fact behind is that the application of magnetic field generates the Lorentz forces, which in turn opposes the fluid flow. But in the present situation, the magnetic field decreases, so the Lorentz forces decreases and consequently, an increment in velocity has been perceived in the region close to the lower Riga plate. Besides, in the upper half, the velocity displays an opposite behavior as compared to the lower half of the channel, which may be due to the downward squeezing motion of the upper plate. Figure 3b exhibits an increment in the normal component of velocity $\breve{F}(\chi)$ with the increasing Modified Hartmann number, which is primarily be due to the decreasing effects of Lorentz force. It can be detected from Figure 4a that the axial velocity decreases in the lower region with the increasing nanoparticles concentration, while an opposite behavior has been perceived in the upper portion of the channel. The reason behind is that the nanoparticle's concentration resists the fluid to move and therefore decreases the fluid velocity. Figure 4b depicts a decline in the normal component of the velocity $\breve{F}(\chi)$ with increasing nanoparticle's concentration, which opposes the fluid motion. Moreover, the inset pictures reveal the fact that the velocity for the (Fe_3O_4/H_2O) nanofluid mostly attains the higher values as compared to the ($Ag - Fe_3O_4/H_2O$) hybrid nanofluid.

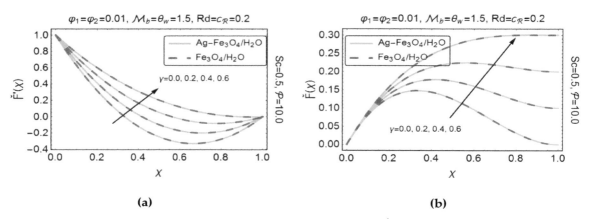

Figure 2. Impact of particular values of γ on **(a)** $\breve{F}'(\chi)$ and **(b)** $\breve{F}(\chi)$.

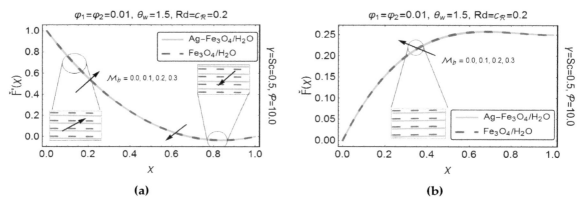

Figure 3. Impact of particular values of $\mathcal{M}_b$ on **(a)** $\breve{F}'(\chi)$ and **(b)** $\breve{F}(\chi)$.

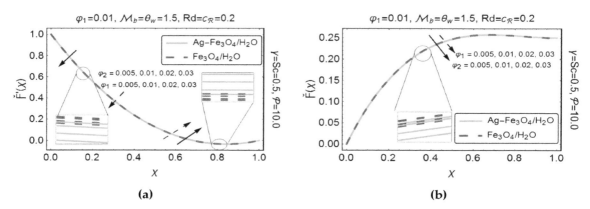

Figure 4. Impact of particular values of φ_1 and φ_2 on **(a)** $\breve{F}'(\chi)$ and **(b)** $\breve{F}(\chi)$.

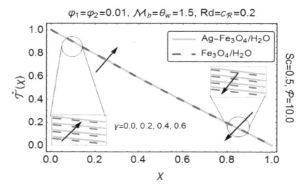

Figure 5. Impact of particular values of γ on $\breve{\mathcal{T}}(\chi)$.

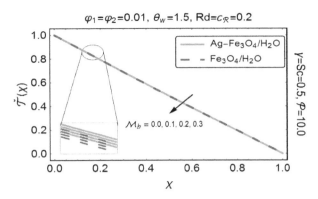

Figure 6. Impact of particular values of $\mathcal{M}_\beta$ on $\breve{\mathcal{T}}(\chi)$.

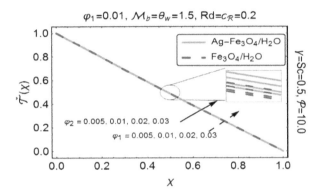

Figure 7. Impact of particular values of φ_1 and φ_2 on $\breve{\mathcal{T}}(\chi)$.

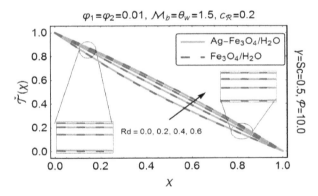

Figure 8. Impact of particular values of Rd on $\breve{\mathcal{T}}(\chi)$.

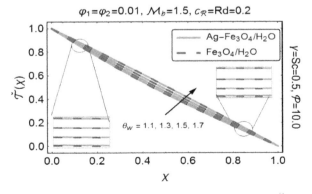

Figure 9. Impact of particular values of θ_w on $\breve{\mathcal{T}}(\chi)$.

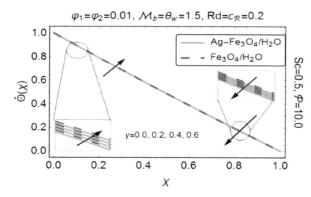

Figure 10. Impact of particular values of γ on $\breve{\Theta}(\chi)$.

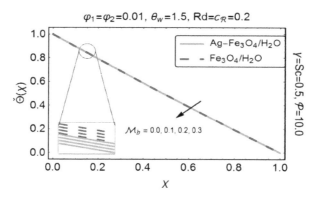

Figure 11. Impact of particular values of $\mathcal{M}_\beta$ on $\breve{\Theta}(\chi)$.

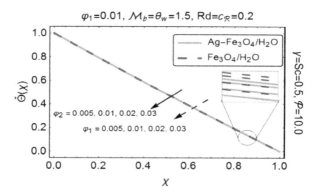

Figure 12. Impact of particular values of φ_1 and φ_2 on $\breve{\Theta}(\chi)$.

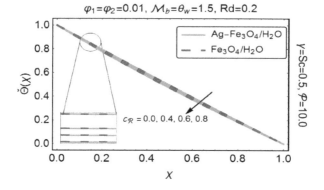

Figure 13. Impact of particular values of $c_{\mathscr{R}}$ on $\breve{\Theta}(\chi)$.

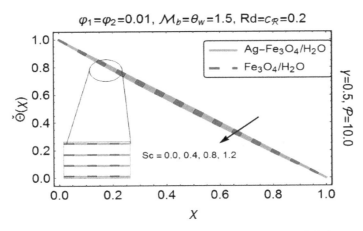

Figure 14. Impact of particular values of Sc on $\breve{\Theta}(\chi)$.

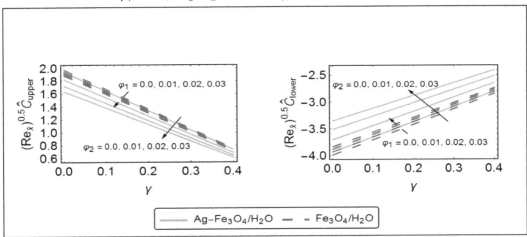

Figure 15. Coefficient of skin friction drag for particular values of φ_1 and φ_2.

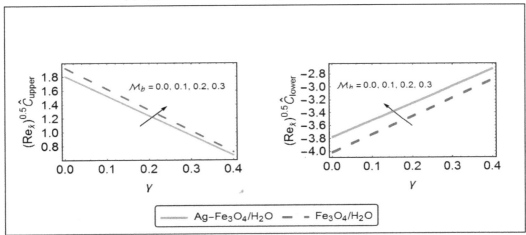

Figure 16. Coefficient of skin friction drag for particular values of $\mathcal{M}_b$

φ_1=0.01, $\mathcal{M}_b$=θ_w=1.5, Rd=$c_\mathcal{R}$=0.2, Sc=0.5, $\mathcal{P}$=10.0

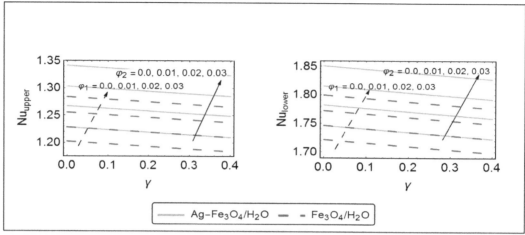

Figure 17. Local Nusselt number for particular values of φ_1 and φ_2.

φ_1=φ_2=0.01, γ=Sc=0.5, $\mathcal{M}_b$=1.5, $c_\mathcal{R}$=0.2, $\mathcal{P}$=10.0

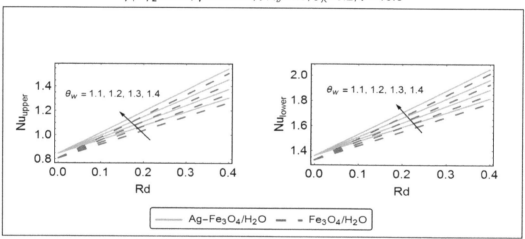

Figure 18. Local Nusselt number for particular values of θ_w.

φ_1=0.01, $\mathcal{M}_b$=θ_w=1.5, Rd=$c_\mathcal{R}$=0.2, Sc=0.5, $\mathcal{P}$=10.0

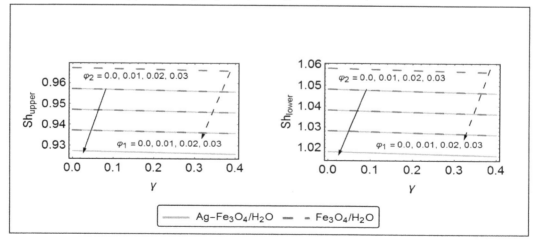

Figure 19. Sherwood number for particular values of φ_1 and φ_2.

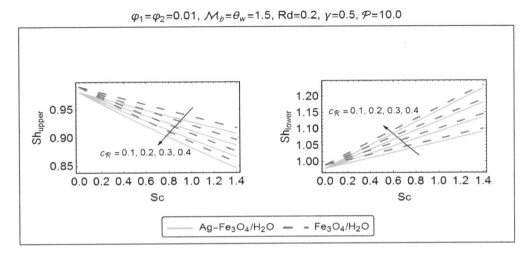

Figure 20. Sherwood number for particular values of $c_{\mathscr{R}}$.

The upcoming figures give a pictorial description of the variations in temperature distribution, for various embedded parameters. Figure 5 displays the impact of squeezing parameter γ on temperature profile. When the upper plate squeezed down, i.e., $\gamma > 0$, it exerts a force on the nearby fluid and enhances its velocity, but since the temperature seems to be dominant at the lower wall therefore the fluid in the adjacent region experiences the higher temperature values as compared to the region nearby the channel's upper wall. To demonstrate the impact of Modified Hartmann number $\mathscr{M}_\beta$ on temperature distribution, Figure 6 has been plotted. It has been found that temperature reveals lower values, as $\mathscr{M}_\beta$ increases. As the impact of Lorentz force on velocity profile produce a friction on the flow, which mainly be responsible to produce more heat energy. In the present flow situation, since the magnetic field exponentially decreases, so the Lorentz force decreases which in turn generates less friction force and consequently, decreases the heat energy and therefore decreases the fluid's temperature as well as the thermal boundary layer thickness. From Figure 7, one can clearly observe an increment in temperature profile with increasing nanoparticles concentration. The fact behind is that, the inclusion of nanoparticles with different volume fractions augments the thermal properties of the host fluid and therefore increases its temperature. It has also been observed that the temperature of $(Ag - Fe_3O_4/H_2O)$ hybrid nanofluid shows its supremacy over the (Fe_3O_4/H_2O) nanofluid, which definitely be due to the rising values of the thermal conductivity for $(Ag - Fe_3O_4/H_2O)$ hybrid nanofluid.

Figure 8 has been sketched, to highlight the temperature behavior under the influence of radiation parameter Rd. An upsurge has been encountered in temperature, for increasing Rd. The fact behind is that, the increasing Rd corresponds to the decrement in mean absorption coefficient, which in turn raises the fluid temperature. The temperature also depicts a rising behavior with increasing θ_w (see Figure 9). The increasing θ_w implies that the temperature differences between the lower and upper walls significantly rises and subsequently, an increment in temperature has been recorded.

The next set of figures provide us an aid, to visualize the deviations, in concentration profile, caused by various embedded parameters. Figure 10 demonstrates the influence of squeezing parameter γ on the concentration profile. When the upper plate moves vertically downward, i.e., $\gamma > 0$, it suppresses the adjacent fluid layers and enhances its velocity, but since the concentration shows its supremacy at the lower wall, therefore the concentration profile shows its dominancy in the region, close to the lower wall, as compared to the region adjacent to the upper wall. To demonstrate the impact of Modified Hartmann number $\mathscr{M}_\beta$ on concentration profile, Figure 11 has been painted. As explained earlier that the Lorentz forces, in present flow situation, experience a decline, which as a result generate less friction force and therefore, decrease the concentration profile along with concentration boundary layer thickness. From Figure 12, one can clearly detect a decline in concentration profile, as nanoparticle fraction increases. Moreover, it has been noticed that the concentration profile for $(Ag - Fe_3O_4/H_2O)$ hybrid nanofluid possesses lower values as compared to the (Fe_3O_4/H_2O) nanofluid.

Figure 13 portrays the influence of chemical reaction parameter $c_{\mathscr{R}}$ on concentration profile. A clear decline has been perceived in the concentration of species with the growing values of chemical reaction parameter $c_{\mathscr{R}}$. Since the chemical reaction, in the present flow analysis, is due to the consumption of the chemicals, therefore, the concentration profile experiences a decline with the increasing values of $c_{\mathscr{R}}$. The variations in concentration profile, under the action of Schmidt number Sc, has been presented in Figure 14. It has been observed that the increasing values of Schmidt number Sc causes a decline in the concentration of the species. Since the Schmidt number is the ratio of momentum diffusivity to mass diffusivity. Therefore, the increment in Schmidt number consequently implies a decline in mass diffusivity, which in turn decreases the concentration profile.

Figures 15 and 16 display the impact of various ingrained parameters on the skin friction coefficient, both at the upper and lower Riga surfaces. It has been detected from Figure 15 that increasing the nanoparticles concentration certainly enhances the coefficient of skin friction drag at the lower Riga plate. However, at the upper plate, an opposite behavior has been clearly visible. As far as squeezing parameter γ is concerned, the skin friction coefficient exhibits an increasing behavior, in the region adjacent to the lower plate. However, a decline has been perceived at the upper wall. From Figure 16, one can clearly observes an increment in skin friction coefficient, with the increasing $\mathscr{M}_{\beta}$, both at the upper and lower Riga plates. Moreover, the skin friction coefficient for $(Ag - Fe_3O_4/H_2O)$ hybrid nanofluid possesses higher values, at the bottom of the channel, as compared to the (Fe_3O_4/H_2O) nanofluid.

Figures 17 and 18 have been plotted, to assess the consequences of various embedded entities on the local rate of heat transfer i.e., Nusselt number. From Figure 17, one can clearly detect an increment in heat transfer, with increasing nanoparticle volume fraction, at both of the plates. Since the nanoparticle's inclusion, in the base fluid, is responsible for rising its temperature, therefore an augmentation in the heat transfer rate is quite obvious. By varying the squeezing number γ horizontally, the local Nusselt number at the upper as well as on the lower plates, indicate a decreasing behavior. Figure 18 depicts the variations in heat transfer rate, with growing values of radiation parameter Rd and temperature difference parameter θ_w. Since both the parameters (Rd and θ_w) significantly amplifies the temperature of the fluid, therefore, they play a key role in enhancing the local Nusselt number, both at the upper and lower Riga plates. Besides, it has been observed from both the figures that the $(Ag - Fe_3O_4/H_2O)$ hybrid nanofluid shows its supremacy in transferring the heat, both at the upper and lower Riga plates.

Figures 19 and 20 depict the variations in the rate of mass transfer, i.e., Sherwood number under the action of various involved parameters. Figure 19 reveals a decline in the Sherwood number with increasing nanoparticle concentration, both at the upper and lower Riga plates. Since the increasing nanoparticle's volume fraction certainly opposes the fluid motion, therefore a decrement in Sherwood number is quite obvious. The rate, with which mass flows, also shows a decreasing behavior when the squeezing parameter γ increases horizontally. From Figure 20, one can observes a clear enhancement in the rate of mass flow in the region nearby the lower plate, when the chemical reaction parameter $c_{\mathscr{R}}$ increases curve wise and Schmidt number Sc varies along the horizontal axis. On the other hand, a reverse behavior has been perceived at the upper plate. Moreover, (Fe_3O_4/H_2O) nanofluid remains dominant in transferring the mass, both at the upper and lower Riga plates.

Hybrid nanofluids, being advanced version of nanofluids, considerably influences the thermo-mechanical properties of the working fluid, particularly the thermal conductivity. For the said purpose, Tables 6 and 7 have been designed, to see the deviations in thermo-mechanical properties of the $(Ag - Fe_3O_4/H_2O)$ hybrid nanofluid and (Fe_3O_4/H_2O) nanofluid. It has been detected that the density of $(Ag - Fe_3O_4/H_2O)$ hybrid nanofluid depicts an increment, as compared to (Fe_3O_4/H_2O) nanofluid. While, the specific heat clearly experiences a decline with the increasing nanoparticles fraction. As far as thermal conductivity is concerned, $(Ag - Fe_3O_4/H_2O)$ hybrid nanofluid shows a dominant behavior against the (Fe_3O_4/H_2O) nanofluid. Besides, the Bruggeman model (18), for thermal conductivity shows its proficiency over the Maxwell's model (17). The reason is that, the Bruggeman model is more focused on the maximum interactions between randomly dispersed particles. It usually involves the spherical shaped particles, with no limitation on the particles concentration. On the other hand, the Maxwell's model depends on the nanoparticle's volume fraction and the thermal conductivity of the base fluid and the spherical shaped particles.

Table 6. Variation in thermo-physical properties of $(Ag - Fe_3O_4/H_2O)$ hybrid nanofluid with $(\varphi_1 = 0.01)$.

φ_2	$\tilde{\rho}$	$\tilde{\rho}C_p \times 10^6$	k_{hnf} (14)	k_{hnf} (13)
0.00	1038.929	4.159918091	0.628640157	0.628422974
0.02	1228.987	4.125930473	0.67317743	0.6695051201
0.04	1419.045	4.091942855	0.72042306	0.7092486827
0.06	1609.103	4.057955237	0.77474778	0.7506805172
0.08	1799.161	4.023967619	0.83787931	0.7939214931
0.10	1989.219	3.989980001	0.91213707	0.8390950667

Table 7. Variation in Thermo-physical properties of (Fe_3O_4/H_2O) nanofluid with $(\varphi_2 = 0)$.

φ_1	$\tilde{\rho}$	$\tilde{\rho}C_p \times 10^6$	k_{nf} (14)	k_{nf} (13)
0.00	997.1	4.1668809	0.613	0.613
0.02	1080.758	4.152955282	0.644998753	0.6441068291
0.04	1164.416	4.139029664	0.680047429	0.6762841143
0.06	1248.074	4.125104046	0.718527614	0.7095880769
0.08	1331.732	4.111178428	0.760870799	0.7440789486
0.10	1415.39	4.09725281	0.80756218	0.779821329

5. Conclusions

This article discloses the salient features of nonlinear thermal radiation, in the squeezing flow of $(Ag - Fe_3O_4/H_2O)$ hybrid nanofluid, between two Riga plates along with a chemical reaction. Method of moment has been employed for the solution point of view. The obtained results are then compared with the numerical results (obtained via Runge-Kutta-Fehlberg algorithm). Both the methods depict an excellent agreement between the results.

Further investigations are as follows:

- Velocity profile seems to be an increasing function of both squeezing parameter γ and modified Hartmann number $\mathcal{M}_\beta$.

- A decrement in the velocity behavior has been perceived, with increasing nanoparticle concentration.

- The velocity profile for $(Ag - Fe_3O_4/H_2O)$ hybrid nanofluid mostly remains on the lower side.

- The amplification in temperature has been recorded for increasing squeezed number γ and nanoparticle concentration, while a reversed behavior has been noticed for increasing modified Hartman number $\mathcal{M}_\beta$.

- The temperature behaves in an increasing manner with the rising Rd and θ_w. Besides, the temperature profile possesses a dominant behavior for $(Ag - Fe_3O_4/H_2O)$ hybrid nanofluid.

- The concentration profile demonstrates a decreasing behavior, with increasing modified Hartman number $\mathcal{M}_\beta$ and nanoparticle volume fraction.

- The increment in chemical reaction parameter $c_\mathcal{R}$ and Schmidt number Sc depicts a clear decline in the concentration profile.

- Skin friction coefficient for $(Ag - Fe_3O_4/H_2O)$ hybrid nanofluid displays an increasing behavior, in the region adjacent to the lower Riga plate, against the varying squeezing parameter γ, modified Hartman number $\mathcal{M}_\beta$ and the nanoparticle concentration.

- The local heat transfer rate, for $(Ag - Fe_3O_4/H_2O)$ hybrid nanofluid, shows its proficiency for varying nanoparticle concentration, radiation parameter Rd and temperature difference parameter θ_w and this phenomena has been detected at both the plates.

- The augmentation of Schmidt number Sc and chemical reaction parameter $c_{\mathcal{R}}$ enhances the Sherwood number, at the lower plate, while a reversed phenomenon has been observed at the upper plate.

Author Contributions: All the authors equally contributed to the paper. Final draft has been read and approved by all the authors.

Acknowledgments: We are thankful to the anonymous reviewers for their valuable comments which really improved the quality of presented work.

Nomenclature

ℓ	Width between magnets and electrodes
$\mathcal{M}_0$	Magnetization of the permanent magnets, $Tesla$
j_0	Applied current density in the electrodes, $m^{-2}A$
c_1	First order chemical reaction coefficient
$\mathcal{D}$	Molecular diffusivity
k	Thermal conductivity, W/mK
$\check{p}$	Pressure
C_p	Specific heat at constant pressure, $J/kg.K$
$\check{u}$	Axial velocity component, m/s
$\check{v}$	Normal velocity component, m/s
$\Bbbk$	Coefficient for mean absorption
$\mathcal{M}_\beta$	Modified Hartman number
a	Rate of stretching
$c_{\mathcal{R}}$	Chemical reaction parameter
Pr	Prandtl number
Rd	Radiation parameter
Nu	Nusselt number
Sc	Schmidt number
Re	Reynolds number
Sh	Sherwood number
Ag	Silver nanoparticles
H_2O	Water
Fe_3O_4	Magnetite nanoparticles
EMHD	Electro-magneto hydrodynamic

Greek Symbols

φ	Solid volume fraction
μ	Dynamic viscosity, $N.s/m^2$
$\check{\rho}$	Density, kg/m^3
v	Kinematic viscosity, m^2/s
χ	Similarity variable
$\check{\sigma}$	Stefan-Boltzmann constant
λ	Constant characteristics parameter
γ	Dimensionless squeeze number
θ_w	Temperature difference parameter
$\check{\rho}C_p$	Heat capacitance

Subscripts

hnf	Hybrid Nanofluid
nf	Nanofluid
f	Base fluid
$p1$	Solid nanoparticles of Fe_3O_4
$p2$	Solid nanoparticles of Ag

References

1. Choi, S.U.S. Enhancing thermal conductivity of fluids with nanoparticles. In *Developments and Applications of Non-Newtonian Flows*; Siginer, D.A., Wang, H.P., Eds.; ASME: New York, NY, USA, 1995; Volume 231, pp. 99–105.

2. Choi, S.U.S.; Zhang, Z.G.; Yu, W.; Lockwood, F.E.; Grulke, E.A. Anomalous thermal conductivity enhancement in nanotube suspensions. *Appl. Phys. Lett.* **2001**, *79*, 2252–2254. [CrossRef]

3. Maxwell, J.C. *A Treatise on Electricity and Magnetism*, 3rd ed.; Clarendon; Oxford University Press: Oxford, UK, 1904.

4. Bruggeman, D.A.G. Berechnzcrrg verschCedcner physikalducher Eonstanten von heterogenew Yuhstan.xen 1. *Ann. Phys.* **1935**, *416*, 636–664. [CrossRef]

5. Hamilton, R.L.; Crosser, O.K. Thermal conductivity of heterogeneous two-component systems. *Ind. Eng. Chem. Fundam.* **1962**, *1*, 187–191. [CrossRef]

6. Suzuki, A.; Ho, N.F.H.; Higuchi, W.I. Predictions of the particle size distribution changes in emulsions and suspensions by digital computation. *J. Colloid Interface Sci.* **1969**, *29*, 552–564. [CrossRef]

7. Buongiorno, J. Convective transport in nanofluids. *J. Heat Transf.* **2006**, *128*, 240–250. [CrossRef]

8. Xue, Q.Z. Model for thermal conductivity of carbon nanotube-based composites. *Phys. B Condens. Matter.* **2005**, *368*, 302–307. [CrossRef]

9. Iijima, S. Helical microtubules of graphitic carbon. *Nature* **1991**, *354*, 56–58. [CrossRef]

10. Timofeeva, E.V.; Gavrilov, A.N.; McCloskey, J.M.; Tolmachev, Y.V.; Sprunt, S.; Lopatina, L.M.; Selinger, J.V. Thermal conductivity and particle agglomeration in alumina nanofluids: Experiment and theory. *Phys. Rev. E* **2007**, *76*, 061203. [CrossRef]

11. Masoumi, N.; Sohrabi, N.; Behzadmehr, A. A new model for calculating the effective viscosity of nanofluids. *J. Phys. D Appl. Phys.* **2009**, *42*, 055501. [CrossRef]

12. Thurgood, P.; Baratchi, S.; Szydzik, C.; Mitchell, A.; Khoshmanesh, K. Porous PDMS structures for the storage and release of aqueous solutions into fluidic environments. *Lab Chip* **2017**, *17*, 2517–2527. [CrossRef]

13. Sarkar, J.; Ghosh, P.; Adil, A. A review on hybrid nanofluids: Recent research, development and applications. *Renew. Sustain. Energy Rev.* **2015**, *43*, 164–177. [CrossRef]

14. Ranga Babu, J.A.; Kumar, K.K.; Srinivasa Rao, S. State-of-art review on hybrid nanofluids. *Renew. Sustain. Energy Rev.* **2017**, *77*, 551–565. [CrossRef]

15. Niihara, K. New Design Concept of Structural Ceramics. *J. Ceram. Soc. Jpn.* **1991**, *99*, 974–982. [CrossRef]

16. Jana, S.; Salehi-Khojin, A.; Zhong, W.H. Enhancement of fluid thermal conductivity by the addition of single and hybrid nano-additives. *Thermochim. Acta* **2007**, *462*, 45–55. [CrossRef]

17. Suresh, S.; Venkitaraj, K.P.; Selvakumar, P.; Chandrasekar, M. Synthesis of Al_2O_3-Cu/water hybrid nanofluids using two step method and its thermo physical properties. *Colloids Surf. A Physicochem. Eng. Asp.* **2011**, *388*, 41–48. [CrossRef]

18. Suresh, S.; Venkitaraj, K.P.; Selvakumar, P.; Chandrasekar, M. Effect of Al_2O_3–Cu/water hybrid nanofluid in heat transfer. *Exp. Therm. Fluid Sci.* **2012**, *38*, 54–60. [CrossRef]

19. Momin, G.G. Experimental investigation of mixed convection with water-Al_2O_3 & hybrid nanofluid in inclined tube for laminar flow. *Int. J. Sci. Technol. Res.* **2013**, *2*, 195–202.

20. Devi, S.P.A.; Devi, S.S.U. Numerical investigation of hydromagnetic hybrid Cu-Al_2O_3/water nanofluid flow over a permeable stretching sheet with suction. *Int. J. Nonlinear Sci. Numer.* **2016**, *17*, 249–257. [CrossRef]

21. Das, S.; Jana, R.N.; Makinde, O.D. MHD Flow of Cu-Al_2O_3/Water Hybrid Nanofluid in Porous Channel: Analysis of Entropy Generation. *Defect Diffus. Forum* **2017**, *377*, 42–61. [CrossRef]

22. Chamkha, A.J.; Miroshnichenko, I.V.; Sheremet, M.A. Numerical analysis of unsteady conjugate natural convection of hybrid water-based nanofluid in a semi-circular cavity. *J. Therm. Sci. Eng. Appl.* **2017**, *9*, 1–9. [CrossRef]

23. Olatundun, A.T.; Makinde, O.D. Analysis of Blasius flow of hybrid nanofluids over a convectively heated surface. *Defect Diffus. Forum* **2017**, *377*, 29–41. [CrossRef]

24. Hayat, T.; Nadeem, S. Heat transfer enhancement with Ag–CuO/water hybrid nanofluid. *Results Phys.* **2017**, *7*, 2317–2324. [CrossRef]

25. Stefan, M.J. Versuch Uber die Scheinbare Adhasion, sitzungsber. *Abt. II, Osterr. Akad. Wiss., MathNaturwiss.kl.* **1874**, *69*, 713–721.

26. Shahmohamadi, H.; Rashidi, M.M.; Dinarvand, S. Analytic approximate solutions for unsteady two-dimensional and axisymmetric squeezing flows between parallel plates. *Math. Probl. Eng.* **2008**, *2008*, 1–13. [CrossRef]

27. Sheikholeslami, M.; Ganji, D.D.; Ashorynejad, H.R. Investigation of squeezing unsteady nanofluid flow using ADM. *Powder Technol.* **2013**, *239*, 259–265. [CrossRef]

28. Khan, U.; Ahmed, N.; Asadullah, M.; Mohyud-Din, S.T. Effects of viscous dissipation and slip velocity on two-dimensional and axisymmetric squeezing flow of Cu-water and Cu-kerosene nanofluids. *Propul. Power Res.* **2015**, *4*, 40–49. [CrossRef]

29. Ahmed, N.; Khan, U.; Khan, S.I.; Bano, S.; Mohyud-Din, S.T. Effects on magnetic field in squeezing flow of a Casson fluid between parallel plates. *J. King Saud Univ. Sci.* **2017**, *29*, 119–125. [CrossRef]

30. Gailitis, A.; Lielausis, O. On a possibility to reduce the hydrodynamic resistance of a plate in an electrolyte. *Appl. Magnetohydrodyn.* **1961**, *12*, 143–146.

31. Avilov, V.V. *Electric and Magnetic Fields for the Riga Plate*; Technical Report; FRZ: Rossendorf, Germany, 1998.

32. Pantokratoras, A.; Magyari, E. EMHD free-convection boundary-layer flow from a Riga-plate. *J. Eng. Math.* **2009**, *64*, 303–315. [CrossRef]

33. Pantokratoras, A. The Blasius and Sakiadis flow along a Riga-plate. *Prog. Comput. Fluid Dyn. Int. J.* **2011**, *11*, 329–333. [CrossRef]

34. Magyari, E.; Pantokratoras, A. Aiding and opposing mixed convection flows over the Riga-plate. *Commun. Nonlinear Sci. Numer. Simul.* **2011**, *16*, 3158–3167. [CrossRef]

35. Ayub, M.; Abbas, T.; Bhatti, M.M. Inspiration of slip effects on electromagnetohydrodynamics (EMHD) nanofluid flow through a horizontal Riga plate. *Eur. Phys. J. Plus* **2016**, *131*, 193. [CrossRef]

36. Hayat, T.; Khan, M.; Imtiaz, M.; Alsaedi, A. Squeezing flow past a Riga plate with chemical reaction and convective conditions. *J. Mol. Liq.* **2017**, *225*, 569–576. [CrossRef]

37. Hayat, T.; Khan, M.; Khan, M.I.; Alsaedi, A.; Ayub, M. Electromagneto squeezing rotational flow of Carbon (C)-Water (H2O) kerosene oil nanofluid past a Riga plate: A numerical study. *PLoS ONE* **2017**, *12*, e0180976. [CrossRef] [PubMed]

38. Rosseland, S. *Astrophysik und Atom-Theoretische Grundlagen*; Springer-Verlag: Berlin, Germany, 1931.

39. Magyari, E.; Pantokratoras, A. Note on the effect of thermal radiation in the linearized Rosseland approximation on the heat transfer characteristics of various boundary layer flows. *Int. J. Heat Mass Transf.* **2011**, *38*, 554–556. [CrossRef]

40. Rashidi, M.M.; Mohimanian pour, S.A.; Abbasbandy, S. Analytic approximate solutions for heat transfer of a micropolar fluid through a porous medium with radiation. *Commun. Nonlinear Sci. Numer. Simul.* **2011**, *16*, 1874–1889. [CrossRef]

41. Noor, N.F.M.; Abbasbandy, S.; Hashim, I. Heat and mass transfer of thermophoretic MHD flow over an inclined radiate isothermal permeable surface in the presence of heat source/sink. *Int. J. Heat Mass Transf.* **2012**, *55*, 2122–2128. [CrossRef]

42. Mohyud-Din, S.T.; Khan, S.I. Nonlinear radiation effects on squeezing flow of a Casson fluid between parallel disks. *Aerosp. Sci. Technol.* **2016**, *48*, 186–192. [CrossRef]

43. Khan, U.; Ahmed, N.; Mohyud-Din, S.T.; Bin-Mohsin, B. Nonlinear radiation effects on MHD flow of nanofluid over a nonlinearly stretching/shrinking wedge. *Neural Comput. Appl.* **2017**, *28*, 2041–2050. [CrossRef]

44. Saba, F.; Ahmed, N.; Hussain, S.; Khan, U.; Mohyud-Din, S.T.; Darus, M. Thermal Analysis of Nanofluid Flow over a Curved Stretching Surface Suspended by Carbon Nanotubes with Internal Heat Generation. *Appl. Sci.* **2018**, *8*, 395. [CrossRef]

45. Brinkman, H.C. The viscosity of concentrated suspensions and solutions. *J. Chem. Phys.* **1952**, *20*, 571. [CrossRef]

46. Pak, B.C.; Cho, Y.I. Hydrodynamic and heat transfer study of dispersed fluids with submicron metallic oxide particles. *Exp. Heat Transf.* **1998**, *11*, 151–170. [CrossRef]

47. Xuan, Y.; Roetzel, W. Conceptions for heat transfer correlation of nanofluids. *Int. J. Heat Mass Transf.* **2000**, *43*, 3701–3707. [CrossRef]

48. Ho, C.J.; Huang, J.B.; Tsai, P.S.; Yang, Y.M. Preparation and properties of hybrid water-based suspension of Al2O3 nanoparticles and MEPCM particles as functional forced convection fluid. *Int. J. Heat Mass Transf.* **2010**, *37*, 490–494. [CrossRef]

49. Mamut, E. Characterization of heat and mass transfer properties of nanofluids. *Rom. J. Phys.* **2006**, *51*, 5–12.

50. Singh, P.; Kumar, M. Mass transfer in MHD flow of alumina water nanofluid over a flat plate under slip conditions. *Alexandria Eng. J.* **2015**, *54*, 383–387. [CrossRef]

51. Reddy, N.; Murugesan, K. Numerical Investigations on the Advantage of Nanofluids under DDMC in a Lid-Driven Cavity. *Heat Tran. Asian Res.* **2017**, *46*, 1065–1086. [CrossRef]

52. Bergman, T.L.; Lavine, A.S.; Incropera, F.P.; Dewitt, D.P. *Fundamentals of Heat and Mass Transfer*, 4th ed.; John Wiley & Sons: New York, NY, USA, 2002.

53. Sheikholeslami, M.; Ganji, D.D. Free convection of Fe_3O_4-water nanofluid under the influence of an external magnetic source. *J. Mol. Liq.* **2017**, *229*, 530–540. [CrossRef]

6

Unsteadiness of Tip Leakage Flow in the Detached-Eddy Simulation on a Transonic Rotor with Vortex Breakdown Phenomenon

Xiangyu Su , Xiaodong Ren⑩, Xuesong Li *⑩ and Chunwei Gu

Institute of Gas Turbine, Department of Energy and Power Engineering, Tsinghua University, Beijing 100084, China; su-xy17@mails.tsinghua.edu.cn (X.S.); rxd@mail.tsinghua.edu.cn (X.R.); gcw@mail.tsinghua.edu.cn (C.G.)
* Correspondence: xs-li@mail.tsinghua.edu.cn

Abstract: Tip leakage vortex (TLV) in a transonic compressor rotor was investigated numerically using detached-eddy simulation (DES) method at different working conditions. Strong unsteadiness was found at the tip region, causing a considerable fluctuation in total pressure distribution and flow angle distribution above 80% span. The unsteadiness at near choke point and peak efficiency point is not obvious. DES method can resolve more detailed flow patterns than RANS (Reynolds-averaged Navier–Stokes) results, and detailed structures of the tip leakage flow were captured. A spiral-type breakdown structure of the TLV was successfully observed at the near stall point when the TLV passed through the bow shock. The breakdown of TLV contributed to the unsteadiness and the blockage effect at the tip region.

Keywords: tip leakage flow; detached-eddy simulation; vortex breakdown; transonic compressor

1. Introduction

Driven by the pressure gradient inside the clearance of the rotor and casing, tip leakage is an unavoidable phenomenon in the field of turbomachinery, which scholars have studied for a long time in both compressible [1–4] and incompressible [5–8] fields. As for axial compressors, tip leakage flow plays an even more significant role due to its close relationship with loss and stall characteristics [9,10], which are highly valued in the design or analyzing processes. Aiming at reducing the impact of tip leakage flow, plenty of flow control methods (such as air injection [11], bleeding [12], casing treatment [13] and plasma actuation [14]) have been studied in recent years.

As for low-speed compressors, flow structures as well as unsteadiness of tip leakage vortex (TLV) have been widely studied and certain achievements were obtained through numerical efforts and experiments. As the flow coefficient decreases, the interface between the TLV and the incoming flow moves upstream [15], and the trajectory of which will be aligned with the leading edge when the rotor finally encounters a spike-type stall inception [16]. The criterion for spike-initiated numerical stall that leading-edge spillage and trailing-edge backflow are both essential was proved effective in low-speed compressor experiments [17]. Leading-edge spillage was later found to be an accompanying phenomenon, whose fundamental cause is probably the tornado-like vortex, resulted from the interaction between TLV and leading-edge separation [18]. In rotors with a large gap, attention was also paid to the effects of double-leakage tip clearance flow, which generates a vortex rope and subsequent extra mixing loss in the adjacent blade passage [19]. With the increase of stage loading, the importance of tip leakage flow in high-speed or transonic compressor has been increasingly emphasized. There are indeed similarities in the basic structures and mechanisms of the TLV in low-speed compressors and transonic ones. Nevertheless, conclusions for tip leakage flow in low-speed compressors are not entirely suitable for transonic ones, due to further larger pressure gradient and the existence of shock wave. When the TLV crosses the shock, it interacts both with the shock wave and with the pressure-side secondary flows generating a leakage-interaction-region of low speed, high entropy and high turbulence [2]. The interaction results in extra complexity and less stability in the tip flow field, which is considered a hot spot. Strong self-induced unsteadiness was found in the TLV with a characteristic frequency near 60% BPF (blade passing frequency) [4,20,21] and the oscillation of passage shock was revealed as well [22]. Moreover, the casing boundary layer along with the blade surface boundary layer may participate in the interacting process [23], especially at near stall point where a shock-induced separation inside the boundary layer is likely to happen in most cases. Under these circumstances, the TLV and its interaction can make a great impact on the overall performance and eventually lead to a spike-type stall inception in transonic compressors [24].

Despite the considerable efforts made by scholars in turbomachinery community, the complete mechanism of TLV and its influence on the tip flow field are still not fully understood, especially in transonic compressors. Previous investigations have shown that the interaction between the shock and TLV contributes a lot to the unsteadiness in the tip flow field, but failed to reveal this interacting process in detail. On the other hand, Reynolds-averaged Navier–Stokes (RANS) method is routinely adopted in most simulations among previous studies; however, traditional turbulence models have native defects in predicting the unsteady and vortical flows such as the TLV [25]. As a result, DES (detached-eddy simulation) method is thought to be an alternative in capturing separated or vortical flow with bearable cost [26]. Up to now, many scholars [18,27–30] have applied DES methods to the turbomachinery field and achieved satisfactory results.

In this paper, we carried out DES investigations of a transonic compressor rotor, focused on the structure of the TLV, the interaction with shock wave as well as the unsteady characteristics. Due to the limits of computing resources, a single-row DES calculation was adopted. This paper is organized as follows: the compressor and the numerical method chosen in the present study are demonstrated in Section 2, with the validation results shown in Section 3.1. Section 3.2 mainly deals with the unsteadiness related to the TLV. Detailed structures of the TLV at different working conditions are shown in Section 3.3, focused on the vortex breakdown phenomenon at near stall point. Section 4 mainly deals with the mechanism of leakage vortex breakdown. Finally, a short conclusion is drawn in Section 5.

2. Methodology

2.1. Testing Case

The compressor investigated in the present study is an in-house 1.5-stage transonic axial compressor with 22 rotor blades and a tip clearance of 0.82% chord length, which is modeled from the first stage of an F-class gas turbine. Its schematic structure and design parameters are shown in Figure 1 and Table 1, respectively.

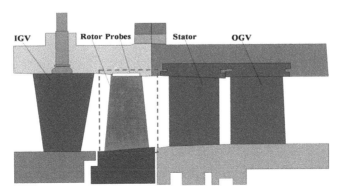

Figure 1. Schematic structure of the 1.5-stage compressor.

Table 1. Design parameters of the compressor.

Parameter	Value	Unit
Rotor blade number	22	-
Rotating speed	24,840	rpm
Mass flow	12	kg/s
Total pressure ratio	1.3	-
Tip Mach number	1.25	-
Tip clearance	0.82%	chord length

2.2. Mesh for DES Calculation

The computational domain is the rotor part (indicated by red dashed line in Figure 1) with inlet and outlet boundaries extended for one-axial chord and two-axial chord, respectively. An unstructured hex-dominant mesh is employed in the DES calculation. As shown in Figure 2, a relatively coarse grid with a refined tip region is adopted to make a trade-off between the flow field resolution and the calculating resources required. To capture the detailed structure of the TLV, the local grid scale at the tip region needs to be at least an order of magnitude smaller than the tip clearance scale in three dimensions, which would be a great challenge if we use a conventional structured mesh. The final mesh is an hybrid grid containing hexahedrons, tetrahedrons, and prisms, with a total element number of 12.3 million, 10 nodes applied in boundary layers to ensure $y^+ < 1$ and 27 nodes applied in the tip clearance region with $\Delta x^+, \Delta y^+, \Delta z^+ < 20$.

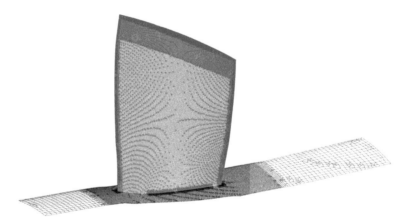

Figure 2. Overview of the detached-eddy simulation (DES) Computational Grid.

2.3. Solver Theory and Calculation Settings

A commercial solver package, FLUENT (14.0, ANSYS, Inc., Canonsburg, PA, USA), was used in the present work, which is a three-dimensional, time-accurate code with implicit second-order scheme, long applied to the field of axial compressors and tip leakage flow [31–37]. The compressible forms of the Reynolds-averaged Navier–Stokes equations were solved in the fluid domain with gravity and volumetric heat source neglected:

$$\frac{\partial \bar{\rho}}{\partial t} + \frac{\partial}{\partial x_j}(\bar{\rho}\tilde{u}_j) = 0 \tag{1}$$

$$\frac{\partial}{\partial t}(\bar{\rho}\tilde{u}_i) + \frac{\partial}{\partial x_j}(\bar{\rho}\tilde{u}_i\tilde{u}_j) = -\frac{\partial \bar{p}}{\partial x_i} + \frac{\partial}{\partial x_j}(\tau_{ij} + \tau'_{ij}) \tag{2}$$

$$\frac{\partial}{\partial t}(\bar{\rho}\tilde{E}) + \frac{\partial}{\partial x_j}[(\bar{\rho}\tilde{E} + \bar{p})\tilde{u}_j] = \frac{\partial}{\partial x_j}[(\tau_{ij} + \tau'_{ij})\tilde{u}_i] - \frac{\partial}{\partial x_j}[(\lambda + \lambda')\frac{\partial \tilde{T}}{\partial x_j}] \tag{3}$$

where τ_{ij} is the viscous stress tensor, τ'_{ij} is the Reynolds stress tensor, and $\lambda' = -C_p(\mu_t/Pr_t)$ is the turbulent thermal conductivity.

The ideal air was chosen as the fluid material, which follows the equation of state:

$$\bar{p} = \bar{\rho}\frac{R_m}{M}\tilde{T} \tag{4}$$

The properties of air are: molecular weight $M = 28.966$ g/mol, specific heat capacity at the constant pressure $C_p = 1004.4$ kJ/(kg $\cdot$ K), thermal conductivity $\lambda = 0.0261$ W/(m $\cdot$ K), and dynamic viscosity μ is determined by the Sutherland's formula [38].

The constitutive equations of Newtonian fluid were adopted to model the viscous stress term:

$$\tau_{ij} = 2\mu\bar{S}_{ij} - \frac{2}{3}\mu\bar{S}_{kk}\delta_{ij} \tag{5}$$

where $\bar{S}_{ij} = [(\partial\bar{u}_i/\partial x_j + \partial\bar{u}_j/\partial x_i)/2]$ is the deformation rate tensor. The Reynolds stress term was modeled using the Boussinesq hypothesis, as follows:

$$\tau'_{ij} = 2\mu_t\bar{S}_{ij} - \frac{2}{3}(\rho k + \mu_t\bar{S}_{kk})\delta_{ij} \tag{6}$$

For the detached-eddy simulation, the DES97 model, first developed by Spalart et al. [39], was adopted in the present study with a default DES coefficient C_{DES} of 0.65 [40]. In the DES97 model, the near-wall distance d in the original S-A turbulence model [41] has been replaced by the DES length scale $\tilde{d}$, as shown in Equation (7):

$$\frac{D\tilde{v}}{Dt} = c_{b1}\tilde{S}\tilde{v} + \frac{1}{\sigma}[\nabla \cdot ((v + \tilde{v})\nabla\tilde{v}) + c_{b2}(\nabla\tilde{v})^2] - c_{w1}f_w(\frac{\tilde{v}}{\tilde{d}})^2 \tag{7}$$

where $\tilde{v}$ is the working viscosity and v is the kinematic viscosity. $\tilde{S}$ is a function of another scalar $\bar{S}$ which can be chosen from the vorticity magnitude or the deformation rate [41]. c_{b1}, c_{b2}, c_{w1}, σ and f_w are coefficients of the S-A turbulence model, whose definitions can all be found in [41]. The DES length scale $\tilde{d}$ is defined as follows:

$$\tilde{d} \equiv \min(d, C_{DES}\Delta) \tag{8}$$

where Δ is the local grid scale and C_{DES} is a coefficient in this model. According to Equation (8), DES length scale $\tilde{d}$ will recover to the near-wall distance d when $d \ll \Delta$. This always happens inside a boundary layer so that the original S-A turbulence model is activated. Nevertheless, in the mainstream with high Reynolds number, the production term will balance with the destruction term [39] so that Equation (7) becomes:

$$v_t = (\frac{c_{b1}C^2_{DES}}{c_{w1}f_w})\Delta^2\bar{S} \propto \Delta^2\bar{S} \tag{9}$$

If $\bar{S}$ is defined as the deformation rate and we choose C_{DES} properly, Equation (9) can be the Smagorinsky-Lilly model [42] for LES:

$$v_{SGS} = 2C_s\Delta^2\bar{S} \tag{10}$$

where $\bar{S}$ is the deformation rate. It is worth noting that the RANS equation and the LES equation are formally identical at some time. Taking Equation (2) for example, once a suitable turbulence model is introduced into the momentum equations, the equation itself will no longer carry any information concerning their derivation (averaging). This is true if we always adopt eddy viscosity models in RANS or LES. The tensor τ'_{ij} can be the Reynolds stress for RANS when we consider the superscript "~" as "time-averaging". Whereas, τ'_{ij} can also be the sub-grid-scale stress for LES when the superscript was treated as "spatial-filtering". In general, the DES length scale $\tilde{d}$ acts as a switch for RANS and LES. Therefore, the DES97 model can use LES method in the mainstream and activating RANS method (with S-A turbulence model) inside the boundary layer.

We conduct DES calculations at three working conditions, namely near choke point (NC), peak efficiency point (PE) and near stall point (NS), as shown in Table 2. The physical time of each time step is 1×10^{-6} s, which is small enough to include at least 110 time steps in one blade passing period. Mass flow rate and static pressure were monitored during the calculation to ensure a good convergence. Pressure monitors are located on the tip region of the rotor, with eight points on the casing and one on the blade tip, as shown in Figure 3. As for boundary conditions, adiabatic nonslip-wall conditions were adopted for all solid walls. Radial distributions of total pressure, total temperature, and flow angles were given at the inlet using UDF (user-defined function) files. Static pressure distribution was specified at the outlet.

Table 2. Calculation settings.

Parameter	Setting
Computational domain	One R1 passage
Rotating speed	22,000 rpm, 24,840 rpm
Working condition	NC, PE, NS
Solver	FLUENT
Model	DES97
Time step	1×10^{-6} s
Inner iteration	15

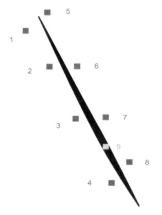

Figure 3. Monitor points in DES calculation (red: on casing, blue: on blade tip).

2.4. Grid Independence Study

The grid independence study was based on RANS calculations with 7 sets of grids at PE, shown in Table 3. The current mesh for DES is the NO.7 grid, which is confirmed to be grid-independent according to Figure 4. We may conclude that the current mesh for DES can provide us a grid-independent result in RANS region. However, the LES region is naturally grid-dependent because the cut-off scale is related to local grid scale. A finer mesh always means a better resolution of the flow field. So an appropriate mesh for LES should meet the requirements of $\Delta x^{+}, \Delta y^{+}, \Delta z^{+}$, which has already been checked in Section 2.2.

Table 3. Grids in the grid independence study.

Mesh No.	1	2	3	4	5	6	7
Elements/million	0.19	0.47	0.80	1.16	2.87	5.83	12.30

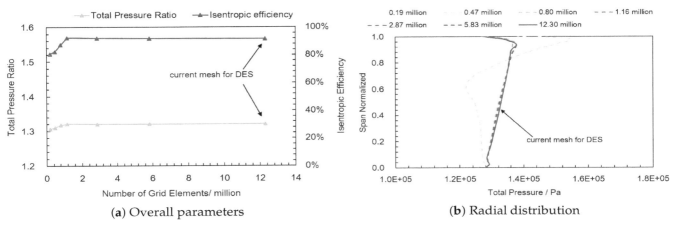

(a) Overall parameters   (b) Radial distribution

Figure 4. Grid independence study.

3. DES Results

3.1. Validation

DES results are compared with the experimental data as well as multistage RANS results to ensure a reasonable prediction of the overall flow field, as shown in Figure 5. Please note that the point NS1 and point NS2 were both at near stall conditions. These two points were at same working conditions except for the outlet static pressure. We slightly raised the static pressure (1kPa on average) at the outlet boundary for NS1 then we got NS2, aimed at further approaching the stall limit. Numerical calculations have a good prediction for the performance trend of the compressor at different rotating speeds, especially for the total pressure ratio. As for DES results, the maximum deviations of averaged mass flow rate and total pressure ratio at three working conditions are 0.25% and 0.94%, respectively. Other flow details were compared with corresponding RANS results. Figures 6 and 7 shows radial distributions of total pressure ratio and relative flow angle. Parameter distributions of the DES results were consistent with RANS results, indicating the predictions of averaged flow field are not worse than those of RANS. In addition, Figure 8 shows the comparison of relative Mach contour at 99.3% span (slightly below the blade tip) at near stall point at design speed. These two results near the top region have no conflicts in the shock location or the leakage flow behavior, indicating tip leakage flow was correctly captured in DES calculations. In general, DES results are relatively reliable in the present simulation and can be used for following analysis of the tip leakage flow. Besides, there is no experimental data at the design speed. So we conducted the DES calculation at 22,000 rpm instead and compared it to the experimental data aiming to validate the numerical method.

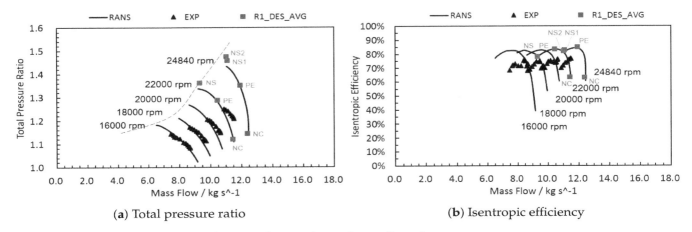

(a) Total pressure ratio   (b) Isentropic efficiency

Figure 5. Comparison of overall performance.

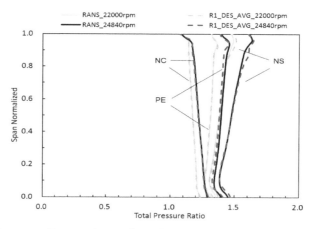

Figure 6. Comparison of total pressure ratio distribution.

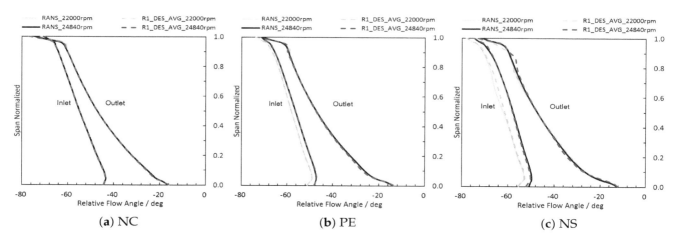

(a) NC   (b) PE   (c) NS

Figure 7. Comparison of relative flow angle distribution. NC: near choke point; PE: peak efficiency point; NS: near stall point.

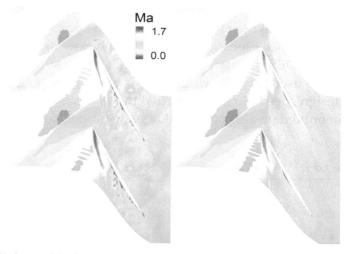

Figure 8. Relative Mach contour at 99.3% span at NS (**left**: DES, **right**: RANS).

Moreover, the correct switch of LES and RANS is critical in DES calculations and needs to be checked in the present study to reduce the impact of grid-induced separation(GIS) and grey area problems of the model itself. A criterion to distinguish LES region from RANS region is as follows:

$$\xi_{DES} = \frac{\tilde{d} - d}{C_{DES}\Delta - d} \in [0, 1] \tag{11}$$

where $\zeta_{DES} = 1$ indicates LES method is switched on and $\zeta_{DES} = 0$ for RANS. Figure 9 shows LES and RANS regions in DES calculations at 90% span, with LES switched on in the mainstream and RANS used inside the boundary layer as expected. Other locations such as 10%, 50% spans and axial cuts were also checked at different working conditions. The switch is appropriate, and no considerable separation was found in the flow field, which contributes to the credibility of present calculations.

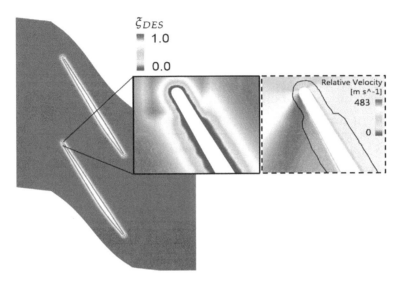

Figure 9. LES and RANS regions in DES results at 90% span.

3.2. Unsteadiness at NS Point

The tip leakage flow in high-speed compressors has self-induced unsteadiness features [4,43]. In present calculation, by monitoring nine static pressure points, it is found that the fluctuations at NS point is much stronger than PE point and NC point. Therefore, the following part will mainly focus on the unsteadiness at NS condition.

Figure 10 shows the static pressure convergence history of different monitoring points. Point 1, 2 and 5 experienced weaker pressure fluctuation than the rest. It is worth noting that point 1 and 2 are exactly in the initial trajectory of the TLV and in front of the shock near suction side, while point 5 is at the leading edge and after the bow shock. For transonic rotors, it is typical that the TLV starts from the leading edge of tip blade, traveling towards the adjacent pressure side. It passes through and interacts with the bow shock, then imprints on the adjacent pressure side, and finally develops downstream along the blade surface. In other words, these three points mentioned above are far away from the interaction region of TLV and the shock, indicating that neither the TLV nor the shock alone is the root cause of the unsteadiness. In addition, points 6 to 9, which experienced strong fluctuations in static pressure, are also located along the trajectory of TLV. However, what makes a difference is that these points are all located after the shock wave. From this aspect, it is exactly the interaction between TLV and the shock that leads to the unsteadiness of the tip region. The detailed reasons will be explained in next section.

Figure 11 shows the spectrum of some representative monitor points at near stall point, which is the result of the fast Fourier transform of the time series data. The frequency characteristics of each point are not the same. Wide though the frequency bands are, they do share some similarities, that is, peak values appeared near two specific frequencies 0.64 BPF and 1.80 BPF. It is worth noting that the former is close to the characteristic frequency of tip leakage flow in typical transonic compressors, for instance, 0.6 BPF [44] for NASA Rotor 37 and 0.57 BPF [4,20] for Darmstadt Rotor 1. Other monitoring points with large fluctuation, such as point 8 and point 9, have similar results as well.

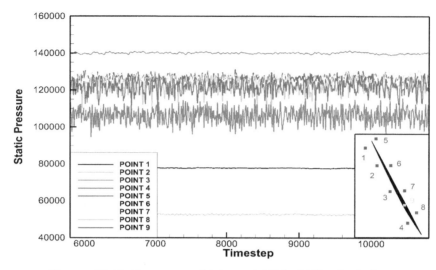

Figure 10. Static pressure history at NS in DES calculation.

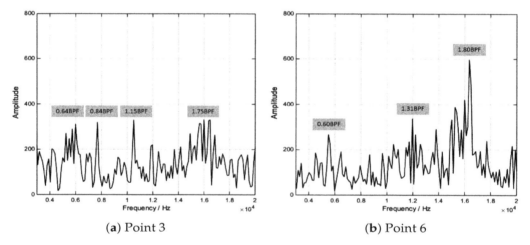

(a) Point 3   (b) Point 6

Figure 11. Amplitude spectrum of certain monitor points at NS.

The unsteadiness near the tip region inevitably affected the performance of the compressor and is the critical reason for the fluctuation in overall parameters at near stall point, as shown in Figure 12. T1 to T6 are six different moments with the same time interval 5×10^{-5} s, whose corresponding frequency is about 2.2 BPF. Obvious unsteady feature exists above 80% span with a fluctuation range of about 10 degrees for the outlet flow angle and 0.1 (about 6.5% of the time-averaged total pressure ratio) for the total pressure ratio, while the unsteadiness is weak in the middle and root span, indicating that tip leakage flow is an essential factor causing instability at NS point. Nevertheless, there is no obvious fluctuation of the static pressure rise coefficient among the whole blade height range, which means the unsteadiness of the top flow field is mainly caused by the fluctuation of kinetic energy and can be a proof for the wake-like nature of the TLV when passing through the shock [45].

Figure 13 is the top view of R1 shroud with blade profile imprinted. We chose a certain axial location (red line) and obtained its circumferential distribution of static pressure in one period, as shown in Figure 14, taking circumferential angle as the abscissa. There are two low pressure regions, one is at about -1 degrees (indicated by the black arrow) and the other is from -4.5 to -7 degrees (indicated by the red arrow). The former region is at the tip clearance region, very little influenced by the unsteadiness, while the latter experienced a strong oscillation at different time steps, with the valley traveling from the left side to the right side then returning to the left side. According to Figure 13, the latter region is near the pressure side of the tip blade, exactly located in the TLV trajectory after the shock, on its way to hit the adjacent pressure side and to develop downstream. The oscillation of pressure valley indicated that the TLV was no longer stable at NS point and was oscillating in the blade passage.

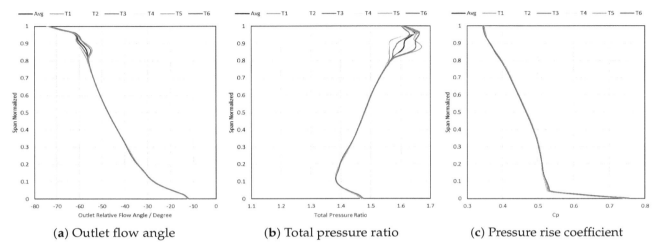

| (a) Outlet flow angle | (b) Total pressure ratio | (c) Pressure rise coefficient |

Figure 12. Parameter distributions at NS.

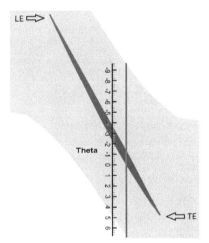

Figure 13. Circumferential line location on the shroud surface.

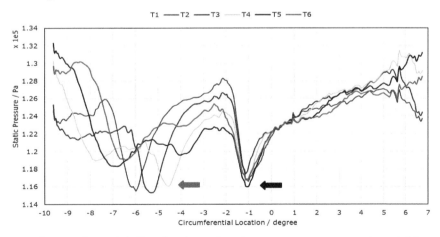

Figure 14. Circumferential distribution of casing pressure at a certain axial location.

Figure 15 can illustrate the above phenomena more clearly, with black dash-dotted line indicated the same axial location in Figure 13. Interacted with the shock and the parallel small vortex, TLV started to swing tangentially at a distance after the shock. This resulted in a small oscillation, developing downstream with an increasing amplitude. As is known, TLV has a lower static pressure value and a lower axial velocity than the mainstream. Every time the trajectory of TLV swept over a certain point, that location would experience a drop in static pressure as well as axial velocity, which is the cause of the oscillation in Figure 14. Eventually, it caused the oscillation of overall parameters such as outlet flow angle, pressure ratio, or efficiency.

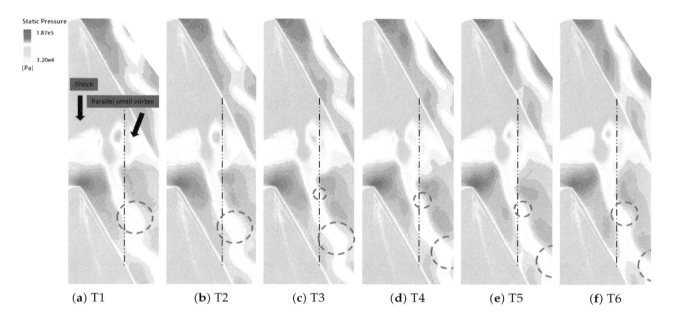

| (a) T1 | (b) T2 | (c) T3 | (d) T4 | (e) T5 | (f) T6 |

Figure 15. Casing Pressure Contour in One Period at NS.

In general, the unsteadiness of the rotor at NS point is mainly manifested at the tip region, closely related to the TLV. The unsteady characteristics in middle span or root span can be totally ignored in comparison. Neither the TLV nor the shock alone is the origin of the unsteadiness. It was found that the TLV became unstable after the interaction with the shock and experienced a stronger oscillation when developing downstream. The most unstable region in this transonic rotor is near the pressure side of the tip blade, which is similar to the conclusions [4] for Darmstadt Rotor 1.

3.3. Detailed Structure of Tip Leakage Flow

DES calculation can obtain detailed information of the top flow field than RANS and may provide some clues for the unsteady characteristic mentioned above. The general structure of tip leakage flow can be clearly observed in Figure 16, with shock position indicated by dashed line. The vortex structures are in three different forms as the operating condition changes.

At the NC point, there are two shocks, one is the bow shock near the leading edge and the other is the passage shock near the trailing edge. Under this condition, the pressure gradient in the tip clearance is relatively small and the TLV started from the leading edge is weak as well. As a result, the vortex core was slenderer in Figure 16a and vanished directly after the bow shock. Meanwhile, the TLV was too weak to draw downstream leakage fluid into, which led to another vortex system after the middle chord of the blade. At PE point, however, the blade load increased and the leading-edge TLV became stronger, so that most leakage fluid can be drawn into the main leakage vortex. The characteristics of the TLV were kept so well in the developing process that the vortex core remained continuous when passing through the shock with a slight expansion in volume, which indicates a stable flow state at PE point. At NS point, the entire top flow field had a very rich flow pattern, and there were many small vortex structures after the shock. Some of the small vortices were developed from the main leakage vortex. The other parts were from the secondary leakage, that is, after imprinting the pressure surface of the adjacent blade, the TLV leaked to the next blade passage again through the tip clearance of the adjacent blade.

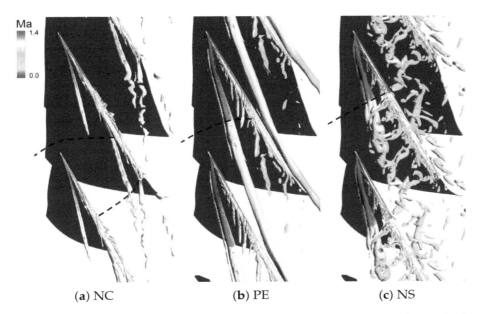

(a) NC   (b) PE   (c) NS

Figure 16. λ_2 Isosurface at tip region of R1 (colored by relative Mach number).

The complex flow field at top region is consistent with the strong unsteadiness mentioned above at NS point. The static pressure oscillation in Figure 15 was caused by the "shedding" of small vortices after the shock. The small vortices did not exist at NC point or PE point and mainly came from the main leakage vortex, which is probably caused by the interaction with the shock. When passing though the bow shock, the TLV at NS point changed from a single solid core to many separate vortices.

Figure 17 is the helicity distribution of different crossflow planes at near stall point, which is used to characterize the intertwining degree of the fluid around TLV core, with the blade surface colored by static pressure, the shock position indicated by the black dotted line, and the beginning of the TLV indicated by the black arrow. The vortex core before the shock was concentrated and slowly increasing in volume. After the shock, the helicity distribution was dispersed and not concentrated anymore, which indicated that the fluid no longer moved around the vortex core tightly after the shock. The distance between the four crossflow planes can be approximated as isometric, but the helicity distribution before and after the shock surface is quite different, which means that the vortex core experienced an abrupt change in its internal structure. Combined with the vortex structure in Figure 16c, we can conclude that a vortex breakdown took place after the interaction between TLV and the shock.

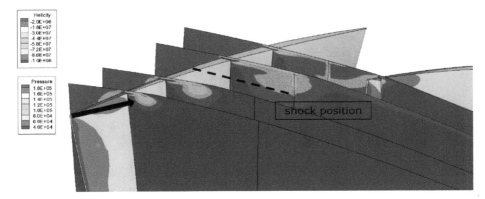

Figure 17. Helicity distribution on crossflow planes at NS (with blade surface colored by static pressure).

The breakdown process can be clearly illustrated in Figure 18, which is a transient λ_2 isosurface colored by relative Mach number. It is confirmed that the TLV changed to a three-dimensional structure and was not axisymmetric anymore when passing through the shock. The vortex core changed its direction to a perpendicular path and started to rotate around the original one when developing downstream, which is the typical structure of a spiral-type breakdown. The structure after the onset of the breakdown process is not stable, with rotating phase related to the flow time, thus causing strong unsteadiness. This contributes a lot to our understanding of the oscillation after the shock in Figure 15 that the previous view on the unsteady flow behavior of the leakage vortex may be incomplete. The unsteadiness is not a two-dimensional phenomenon inside the S1 plane, but a three-dimensional structure in nature. The underlying reason behind is the spiral-type vortex breakdown after the interaction with the shock.

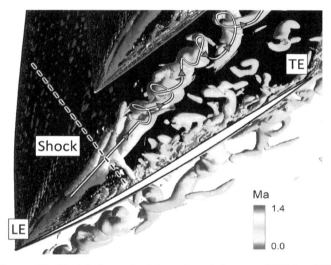

Figure 18. Sketch for spiral-type breakdown of TLV at NS.

To the author's knowledge, this is probably the first time that a detailed structure of the breakdown process for the TLV in a transonic compressor is obtained in numerical investigations. In addition, PIV (particle image velocimetry) measurements of another transonic compressor near stability limit were conducted [46], whose results are in great agreement with the present calculation results, as shown in Figure 19.

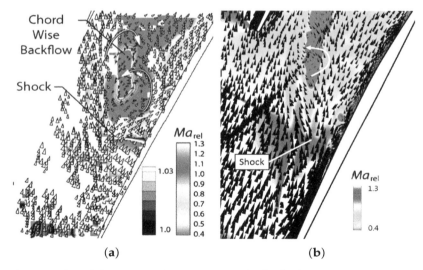

Figure 19. Spiral vortex breakdown at NS.. (**a**) PIV results [46]; (**b**) DES results in the present study.

4. Discussion

Vortex breakdown is a very complex flow phenomenon and is an independent branch in fluid mechanics. The interaction of a streamwise vortex and a normal or oblique shock under supersonic conditions could lead to a vortex breakdown phenomenon. The interacting process and mechanisms were widely studied [47–50]. Recently, lots of efforts [51–55] have been made extending the state of knowledge regarding onset, internal structure and mode selection of vortex breakdown. In the present study, a shock-induced spiral-type vortex breakdown was found at NS. We will deal with the related structure of the breakdown process in this section.

In transonic compressors, the shock wave in the rotor passage can provide a large adverse pressure gradient, with a great influence on the streamwise velocity in the vortex core. As shown in Figure 20, subfigure (a) is the λ_2 isosurface, which illustrates the breakdown process. Subfigure (b) is streamwise velocity contour at the same viewpoint, which indicates the component of relative velocity in the vortex trajectory direction, and subfigure (c) demonstrates the surface streamline. The corresponding cut plane is represented by the red point line in the top view and the location of the bow shock near the suction surface is characterized by the black dashed line. The vortex breakdown is clearly observed in the upper figure with its initial location indicated by the red circle. The breakdown location was not at the shock surface exactly, but at a certain distance downstream, which is in qualitative agreement with experimental results in Figure 19a. Leibovich [56] pointed out that this axial interval is several vortex-core diameters in length. It is worth noting that the initial location of vortex breakdown in (a) coincided with the location S1 in (b) where the streamwise velocity of the vortex core was zero, which indicates that the breakdown of the leakage vortex is closely related to the stagnation point in the center of vortex core. Please note that there were not only one stagnation point in (b). Along with subfigure (c), we could observe a recirculation zone between the two stagnation points S1 and S2. This is consistent with the direct numerical simulation of vortex breakdown in swirling jets and wakes by Ruith et al. [52].

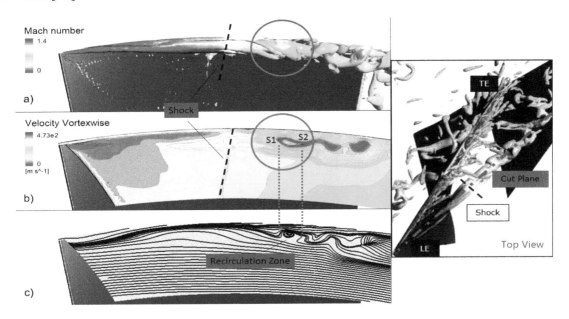

Figure 20. Structures Related to the Breakdown of TLV at NS.

Figure 21 is the swirl velocity vector of the TLV, demonstrating the rotating direction of the TLV before and after the breakdown. For clarity, the adjacent blade was hidden, and the larger images of the dashed zones might have a different view point. Whether before or after the breakdown point, the vortices were temporally rotating in the same direction as the initial TLV, except for the induced vortex in (b). However, the helices after breakdown were coiling in the opposite direction (more clear in Figure 18). This is similar to the findings of Pasche et al. [57] who conducted experiments on obstacle-induced spiral vortex breakdown. The breakdown helices originating at a locally wake-like profile have negative winding sense [52]. Of course, the TLV had a wake-like profile in nature and the breakdown structure in the present study contributed to this view.

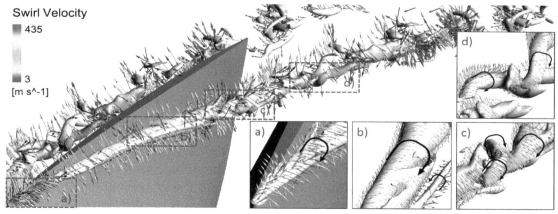

Figure 21. Swirl velocity vector on the λ_2 isosurface.

The causes of vortex breakdown are very complicated. The generation of the recirculation bubble of the vortex breakdown remains unclear but the spiraling motion of the flow behind the recirculation bubble comes from a global unstable mode of the flow [53]. The same mechanism is also observed without recirculation bubble [57], the spontaneous spiraling motion is due to a self-sustained instability. At present, it is consistent to conclude that the reverse pressure gradient and the strength of the vortex itself are crucial factors in the vortex breakdown process [48]. The adverse pressure gradient characterizes the deceleration of the vortex core along the streamwise direction. Criteria for vortex breakdown based on the interaction between the Rankine vortex and the normal shock waves were proposed by Mahesh [58], Smart and Kalkhoran [49] and other scholars. As shown in Figure 22, the abscissa is the freestream Mach number, reflecting adverse pressure gradient in the streamwise direction. The ordinate is the swirl ratio before the shock, which reflects the intensity of the vortex. The swirl ratio is defined as follows:

$$\tau = \frac{\Lambda_{max}}{V_a} \tag{12}$$

where Λ_{max} is the maximum swirl velocity component and V_a is the streamwise velocity component in the vortex core.

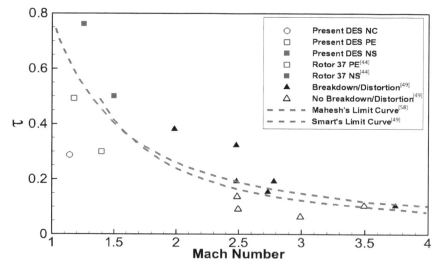

Figure 22. Limit curve for normal shock wave/vortex interaction.

The interaction between the TLV and the bow shock in the transonic compressor could be simplified to the interaction of normal shock and vortex. Both the compressor in the present paper (red marks) and Rotor 37 (blue marks, [44]) exceeded the breakdown limit at NS conditions, and the breakdown of leakage vortex did occur in the detailed flow field. In addition, at PE point and NC point, the vortex breakdown phenomenon was not observed in the present calculation, which is still in good agreement with the breakdown criterion. As the operating point moves to the NS point, the swirl ratio will increase sharply with the increase of the leakage vortex intensity, while the change of Mach number in the axial direction of the vortex core is not obvious. Therefore, transonic rotors with heavy blade loading are likely to experience leakage vortex breakdown at NS point.

We could use the mechanism for the shock-induced breakdown to illustrate the interacting process between the TLV and the bow shock or the breakdown process of the TLV in transonic compressors. When the vortex passes through the shock, the streamwise velocity decreases rapidly due to the strong adverse pressure gradient produced by the shock. The swirl component, however, is orthogonal to the shock surface and is not much affected. In other words, the attenuation of the streamwise component is far greater than that of the swirl component [48]. When the shock is strong enough, the streamwise velocity will decay to zero after the interaction. Under this circumstance, the breakdown phenomenon of the TLV is likely to occur.

The small vortices produced by the TLV vortex breakdown are distributed in the tip region, causing a large blockage effect in the rotor passage. The low energy fluid might subsequently result in leading-edge spillage or induce a spike-type stall inception. Therefore, how to postpone or eliminate the breakdown of TLV may become one of the important ways to enhance the stability at NS point and enlarge the surge margin of transonic compressors.

5. Conclusions

In this paper, a numerical investigation of a transonic compressor rotor using DES method is conducted at three working conditions, focused on the structure of tip leakage flow, the interaction with shock wave and the unsteady characteristics.

Strong unsteadiness was found at NS point with the characteristic frequencies of 0.64 BPF and 1.80 BPF. The most unstable region for this transonic rotor is in the rotor passage near the pressure side of the tip blade, which is the result of the interaction between the TLV and the shock. The affected area is mainly located at the tip region, causing a considerable fluctuation in total pressure distribution and flow angle distribution above 80% span. The unsteadiness at NC point and PE point is not obvious.

Detailed structures of the tip leakage flow were captured, closely related to the working condition. At the NS point, a vortex breakdown can be observed downstream the bow shock. While, at NC point or PE point, the TLV breakdown did not take place. The breakdown process at NS point was confirmed in a spiral-type form. The vortex core changed its direction to a perpendicular path and started to rotate around the original one when developing downstream. Whether before or after the breakdown point, the vortices were temporally rotating in the same direction as the initial leakage vortex. Whereas, the helices after breakdown were coiling in the opposite direction. The breakdown of the TLV generated unstable small vortices and contributed greatly to the unsteadiness at NS point.

Author Contributions: X.S. acquired the numerical data and wrote the paper; and X.R., X.L. and C.G. revised the paper and offered useful suggestions to write the paper.

Acknowledgments: The author thanks Senior Engineer JianBai Li for his critical comments and helpful advice. This work is supported by the "Explorer 100" HPC Platform, State Laboratory for Information Science and Technology, Tsinghua University.

Nomenclature

C_{DES} the coefficient in DES97 model
C_p specific heat capacity at the constant pressure
C_s the coefficient in Smagorinsky-Lilly model
$\tilde{d}$ DES length scale

e inner energy
E total energy, $E = e + u_i u_i / 2$
H total enthalpy, $H = E + p/\rho$
k turbulent kinetic energy
M molecular weight
Ma Mach number
p pressure
R_m gas constant, $R_m = 8.314\,\text{J}/(\text{mol}\cdot\text{K})$
S deformation rate, $\bar{S} = \sqrt{2\bar{S}_{ij}\bar{S}_{ij}}$
S_{ij} deformation rate tensor
T temperature
u velocity
x Cartesian coordinates
y^+ nondimensional wall distance
Δ local grid scale
Δx^+ nondimensional grid scale in x direction
Δy^+ nondimensional grid scale in y direction
Δz^+ nondimensional grid scale in z direction
λ thermal conductivity
λ' turbulent thermal conductivity
μ dynamic viscosity
μ_t turbulent eddy viscosity
ν kinematic viscosity
$\tilde{\nu}$ the working viscosity in S-A model
ν_{SGS} sub-grid-scale kinematic viscosity
ν_t turbulent kinematic viscosity
ξ_{DES} a criterion to distinguish LES region from RANS region
ρ density
τ_{ij} viscous stress tensor
τ'_{ij} Reynolds stress tensor or sub-grid-scale stress tensor

Abbreviations

BPF   Blade passing frequency, $1\,BPF = n\cdot NB/60$ (Hz), where n is the rotation speed of the axis with the unit rpm
DES   Detached-eddy simulation
LES   Large eddy simulation
NB   Number of rotor blades
NC   Near choke point
NS   Near stall point
PE   Peak efficiency point
PIV   Particle image velocimetry
RANS   Reynolds-averaged Navier–Stokes
SGS   sub-grid-scale stress
TLV   Tip leakage vortex
UDF   User-defined function

References

1.   Storer, J.A.; Cumpsty, N.A. Tip leakage flow in axial compressors. *J. Turbomach.* **1991**, *113*, 252. [CrossRef]
2.   Gerolymos, G.A.; Vallet, I. Tip-clearance and secondary flows in a transonic compressor rotor. *J. Turbomach.* **1999**, *121*, 751. [CrossRef]
3.   Tan, C.; Day, I.; Morris, S.; Wadia, A. Spike-type compressor stall inception, detection, and control. *Ann. Rev. Fluid Mech.* **2010**, *42*, 275–300.
4.   Du, J.; Lin, F.; Chen, J.; Nie, C.; Biela, C. Flow structures in the tip region for a transonic compressor rotor. *J. Turbomach.* **2013** *135*, 031012. [CrossRef]
5.   Liu, Y.; Tan, L.; Wang, B. A review of tip clearance in propeller, pump and turbine. *Energies* **2018**, *11*, 2202. [CrossRef]

6.   Tab, L.; Xie, Z.; Liu, Y.; Hao, Y.; Xu, Y. Influence of T-shape tip clearance on performance of a mixed-flow pump. *Proc. Inst. Mech. Eng. Part A J. Power Energy* **2018**, *232*, 386–396. [CrossRef]

7.   Liu, M.; Tan, L.; Cao, S. Cavitation-vortex–turbulence interaction and one-dimensional model prediction of pressure for hydrofoil ALE15 by large eddy simulation. *J. Fluids Eng.* **2018**, *141*, 021103. [CrossRef]

8.   Hao, Y.; Tan, L. Symmetrical and unsymmetrical tip clearances on cavitation performance and radial force of a mixed flow pump as turbine at pump mode. *Renew. Energy* **2018**, *127*, 368–376. [CrossRef]

9.   Denton, J.D. Loss mechanisms in turbomachines. *J. Turbomach.* **1993**, *115*, 621–656. [CrossRef]

10. Vo, H.D. Role of Tip Clearance Flow on Axial Compressor Stability. Ph.D. Thesis, Massachusetts Institute of Technology, Cambridge, MA, USA, 2001.

11. Suder, K.L.; Hathaway, M.D.; Thorp, S.A.; Strazisar, A.J.; Bright, M.B. Compressor stability enhancement using discrete tip injection. *J. Turbomach.* **2001**, *123*, 14. [CrossRef]

12. Eveker, K.M.; Gysling, D.L.; Nett, C.N.; Sharma, O.P. Integrated control of rotating stall and surge in high-speed multistage compression systems. *J. Turbomach.* **1998**, *120*, 440–445. [CrossRef]

13. Müller, M.W.; Schiffer, H.P.; Hah, C. Effect of circumferential grooves on the aerodynamic performance of an axial single-stage transonic compressor. In *Proceedings of the ASME Turbo Expo 2007: Power for Land, Sea, and Air, Montreal, QC, Canada, 14–17 May 2007*; ASME: New York, NY, USA; Volume 6, pp. 115–124. [CrossRef]

14. Vo, H.D. Rotating stall suppression in axial Compressors with casing plasma actuation. *J. Propuls. Power* **2010**, *26*, 808–818. [CrossRef]

15. Hoying, D.A.; Tan, C.S.; Vo, H.D.; Greitzer, E.M. Role of blade passage flow structurs in axial compressor rotating stall inception. *J. Turbomach.* **1999**, *121*, 735. [CrossRef]

16. Vo, H.D.; Tan, C.S.; Greitzer, E.M. Criteria for spike initiated rotating stall. In *Proceedings of the ASME Turbo Expo 2005: Power for Land, Sea, and Air, Peno, NV, USA, 6–9 June 2005*; ASME: New York, NY, USA, 2005; Volume 6, pp. 155–165. [CrossRef]

17. Deppe, A.; Saathoff, H.; Stark, U. Discussion: "Criteria for spike initiated rotating stall" (Vo, H. D., Tan, C. S., Greitzer, E. M., 2008, ASME J. Turbomach., 130, p. 011023). *J. Turbomach.* **2008**, *130*, 015501. [CrossRef]

18. Yamada, K.; Kikuta, H.; Iwakiri, K.i.; Furukawa, M.; Gunjishima, S. An explanation for flow features of spike-type stall inception in an axial compressor rotor. *J. Turbomach.* **2012**, *135*, 021023. [CrossRef]

19. Hah, C. Effects of double-leakage tip clearance flow on the performance of a compressor stage with a large rotor tip gap. *J. Turbomach.* **2017**, *139*, 061006. [CrossRef]

20. Biela, C.; Müller, M.W.; Schiffer, H.P.; Zscherp, C. Unsteady pressure measurement in a single stage axial transonic compressor near the stability limit. In *ASME Turbo Expo 2008: Power for Land, Sea, and Air*; American Society of Mechanical Engineers: New York City, NY, USA, 2008; pp. 157–165.

21. Zhang, Y.; Lu, X.; Chu, W.; Zhu, J. Numerical investigation of the unsteady tip leakage flow and rotating stall inception in a transonic compressor. *J. Therm. Sci.* **2010**, *19*, 310–317. [CrossRef]

22. Hah, C.; Rabe, D.C.; Wadia, A.R. Role of tip-leakage vortices and passage shock in stall inception in a swept transonic compressor rotor. In *Proceedings of the ASME Turbo Expo 2004: Power for Land, Sea, and Air, Vienna, Austria, 14–17 June 2004*; ASME: New York, NY, USA, 2004; Volume 5, pp. 545–555. [CrossRef]

23. Hoeger, M.; Fritsch, G.; Bauer, D. Numerical simulation of the shock-tip leakage vortex interaction in a HPC front stage. *J. Turbomach.* **1999**, *121*, 456–468. [CrossRef]

24. Bergner, J.; Kinzel, M.; Schiffer, H.P.; Hah, C. Short length-scale rotating stall inception in a transonic axial compressor: Experimental investigation. In *Proceedings of the ASME Turbo Expo 2006: Power for Land, Sea, and Air, Barcelona, Spain, 8–11 May 2006*; ASME: New York, NY, USA, 2006; Volume 6, pp. 131–140. [CrossRef]

25. Iim, H.; Chen, X.Y.; Zha, G. Detached-eddy simulation of rotating stall inception for a full-annulus transonic rotor. *J. Propuls. Power* **2012**, *28*, 782–798. [CrossRef]

26. Spalart, P.R. Detached-eddy simulation. *Ann. Rev. Fluid Mech.* **2009**, *41*, 181–202. [CrossRef]

27. Riéra, W.; Castillon, L.; Marty, J.; Leboeuf, F. Inlet condition effects on the tip clearance flow with zonal detached eddy simulation. *J. Turbomach.* **2013**, *136*, 041018. [CrossRef]

28. Riéra, W.; Marty, J.; Castillon, L.; Deck, S. Zonal detached-eddy simulation applied to the tip-clearance flow in an axial compressor. *AIAA J.* **2016**, *54*, 2377–2391. [CrossRef]

29. Yamada, K.; Furukawa, M.; Tamura, Y.; Saito, S.; Matsuoka, A.; Nakayama, K. Large-scale detached-eddy

simulation analysis of stall inception process in a multistage axial flow compressor. *J. Turbomach.* **2017**, *139*, 071002. [CrossRef]

30. Lin, D.; Su, X.; Yuan, X. DDES analysis of the wake vortex related unsteadiness and losses in the environment of a high-pressure turbine stage. *J. Turbomach.* **2018**, *140*, 041001. [CrossRef]

31. Nie, C.; Xu, G.; Cheng, X.; Chen, J. Micro air injection and its unsteady response in a low-speed axial compressor. In *Proceedings of the ASME Turbo Expo 2002: Power for Land, Sea, and Air, Amsterdam, The Netherlands, 3–6 June 2002*; ASME: New York, NY, USA, 2002; Volume 5, pp. 343–352. [CrossRef]

32. Shah, P.N.; Tan, C.S. Effect of blade passage surface heat extraction on axial compressor performance. In *Proceedings of the ASME Turbo Expo 2005: Power for Land, Sea, and Air, Reno, NV, USA, 6–9 June 2005*; ASME: New York, NY, USA, 2005; Volume 6, pp. 327–341. [CrossRef]

33. Zhang, H.; Deng, X.; Lin, F.; Chen, J.; Huang, W. A study on the mechanism of tip leakage flow unsteadiness in an isolated compressor rotor. In *Proceedings of the ASME Turbo Expo 2005: Power for Land, Sea, and Air, Barcelona, Spain, 8–11 May 2006*; ASME: New York, NY, USA, 2005; Volume 6, pp. 435–445. [CrossRef]

34. Liu, Y.; Yu, X.; Liu, B. Turbulence models assessment for large-scale tip vortices in an axial compressor rotor. *J. Propuls. Power* **2008**, *24*, 15–25. [CrossRef]

35. Sun, L.; Zheng, Q.; Li, Y.; Bhargava, R. Understanding effects of wet compression on separated flow behavior in an axial compressor stage using CFD analysis. *J. Turbomach.* **2011**, *133*, 031026. [CrossRef]

36. Suman, A.; Kurz, R.; Aldi, N.; Morini, M.; Brun, K.; Pinelli, M.; Ruggero Spina, P. Quantitative computational fluid dynamics analyses of particle deposition on a transonic axial compressor blade—Part I: Particle zones impact. *J. Turbomach.* **2014**, *137*, 021009. [CrossRef]

37. Cameron, J.D.; Bennington, M.A.; Ross, M.H.; Morris, S.C.; Du, J.; Lin, F.; Chen, J. The influence of tip clearance momentum flux on stall inception in a high-speed axial compressor. *J. Turbomach.* **2013**, *135*, 051005. [CrossRef]

38. Sutherland, W. The viscosity of gases and molecular force. *Philos. Mag. Ser. 5* **1893**, *36*, 507–531. [CrossRef]

39. Spalart, P.; Jou, W.; Strelets, M.; Allmaras, S. Comments of feasibility of LES for wings, and on a hybrid RANS/LES approach. In *Advances in DNS/LES*; Greyden Press: Columbus, OH, USA, 1997; p. 1.

40. Shur, M.; Spalart, P.; Strelets, M.; Travin, A. Detached-eddy simulation of an airfoil at high angle of attack. In *Engineering Turbulence Modelling and Experiments 4*; Elsevier: Amsterdam, The Netherlands, 1999; pp. 669–678. [CrossRef]

41. Spalart, P.; Allmaras, S. A one-equation turbulence model for aerodynamic flows. In Proceedings of the 30th Aerospace Sciences Meeting and Exhibit, Reno, NV, USA, 6–9 January 1992; pp. 5–21. [CrossRef]

42. Lilly, D.K. A proposed modification of the Germano subgrid-scale closure method. *Phys. Fluids Fluid Dyn.* **1992**, *4*, 633–635. [CrossRef]

43. Du, J.; Lin, F.; Zhang, H.; Chen, J. Numerical investigation on the self-induced unsteadiness in tip leakage flow for a transonic fan rotor. *J. Turbomach.* **2010**, *132*, 021017. [CrossRef]

44. Yamada, K.; Furukawa, M.; Nakano, T.; Inoue, M.; Funazaki, K. Unsteady three-dimensional flow phenomena due to breakdown of tip leakage vortex in a transonic axial compressor rotor. In *Proceedings of the ASME Turbo Expo 2004: Power for Land, Sea, and Air, Vienna, Austria, 14–17 June 2004*; ASME: New York, NY, USA, 2004; Volume 5, pp. 515–526. [CrossRef]

45. Puterbaugh, S.L.; Brendel, M. Tipclearance flow-shock interaction in a transonic compressor rotor. *J. Propuls. Power* **1997**, *13*, 24–30. [CrossRef]

46. Brandstetter, C.; Schiffer, H.P. PIV measurements of the transient flow structure in the tip region of a transonic compressor near stability limit. *J. Global Power Propuls. Soc.* **2018**, *2*, 303–316. [CrossRef]

47. Cattafesta, L.; Settles, G. Experiments on shock/vortex interactions. In Proceedings of the 30th Aerospace Sciences Meeting and Exhibit, Reno, NV, USA, 6–9 January 1992. [CrossRef]

48. Delery, J.M. Aspects of vortex breakdown. *Prog. Aerosp. Sci.* **1994**, *30*, 1–59. [CrossRef]

49. Smart, M.K.; Kalkhoran, I.M. Flow model for predicting normal shock wave induced vortex breakdown. *AIAA J.* **1997**, *35*, 1589–1596. [CrossRef]

50. Kalkhoran, I.M.; Smart, M.K. Aspects of shock wave-induced vortex breakdown. *Progress Aerosp. Sci.* **2000**, *36*, 63–95. [CrossRef]

51. Delbende, I.; Chomaz, J.M.; Huerre, P. Absolute/convective instabilities in the Batchelor vortex: A numerical study of the linear impulse response. *J. Fluid Mech.* **1998**, *355*, 229–254. [CrossRef]

52. Ruith, M.R.; Chen, P.; Meiburg, E.; Maxworthy, T. Three-dimensional vortex breakdown in swirling jets and wakes: Direct numerical simulation. *J. Fluid Mech.* **2003**, *486*, 331–378. [CrossRef]

53. Gallaire, F.; Ruith, M.; Meiburg, E.; Chomaz, J.M.; Huerre, P. Spiral vortex breakdown as a global mode. *J. Fluid Mech.* **2006**, *549*, 71–80. [CrossRef]

54. Ortega-Casanova, J.; Fernandez-Feria, R. Three-dimensional transitions in a swirling jet impinging against a solid wall at moderate Reynolds numbers. *Phys. Fluids* **2009**, *21*. [CrossRef]

55. Meliga, P.; Gallaire, F.; Chomaz, J.M. A weakly nonlinear mechanism for mode selection in swirling jets. *J. Fluid Mech.* **2012**, *699*, 216–262. [CrossRef]

56. Leibovich, S. The structure of vortex breakdown. *Ann. Rev. Fluid Mech.* **1978**, *10*, 221–246. [CrossRef]

57. Pasche, S.; Gallaire, F.; Dreyer, M.; Farhat, M. Obstacle-induced spiral vortex breakdown. *Exp. Fluids* **2014**, *55*. [CrossRef]

58. Mahesh, K. A model for the onset of breakdown in an axisymmetric compressible vortex. *Phys. Fluids* **1996**, *8*, 3338–3345. [CrossRef]

7

Gas–Liquid Two-Phase Upward Flow through a Vertical Pipe: Influence of Pressure Drop on the Measurement of Fluid Flow Rate

Tarek A. Ganat [1,*] and **Meftah Hrairi** [2]

[1] Department of Petroleum Engineering, Universiti Teknologi PETRONAS, Seri Iskandar, Perak 32610, Malaysia

[2] Department of Mechanical Engineering, International Islamic University Malaysia, P.O. Box 10, Kuala Lumpur 50728, Malaysia; meftah@iium.edu.my

* Correspondence: tarekarbi.ganat@utp.edu.my

Abstract: The accurate estimation of pressure drop during multiphase fluid flow in vertical pipes has been widely recognized as a critical problem in oil wells completion design. The flow of fluids through the vertical tubing strings causes great losses of energy through friction, where the value of this loss depends on fluid flow viscosity and the size of the conduit. A number of friction factor correlations, which have acceptably accurate results in large diameter pipes, are significantly in error when applied to smaller diameter pipes. Normally, the pressure loss occurs due to friction between the fluid flow and the pipe walls. The estimation of the pressure gradients during the multiphase flow of fluids is very complex due to the variation of many fluid parameters along the vertical pipe. Other complications relate to the numerous flow regimes and the variabilities of the fluid interfaces involved. Accordingly, knowledge about pressure drops and friction factors is required to determine the fluid flow rate of the oil wells. This paper describes the influences of the pressure drop on the measurement of the fluid flow by estimating the friction factor using different empirical friction correlations. Field experimental work was performed at the well site to predict the fluid flow rate of 48 electrical submersible pump (ESP) oil wells, using the newly developed mathematical model. Using Darcy and Colebrook friction factor correlations, the results show high average relative errors, exceeding ±18.0%, in predicted liquid flow rate (oil and water). In gas rate, more than 77% of the data exceeded ±10.0% relative error to the predicted gas rate. For the Blasius correlation, the results showed the predicted liquid flow rate was in agreement with measured values, where the average relative error was less than ±18.0%, and for the gas rate, 68% of the data showed more than ±10% relative error.

Keywords: pressure loss; pressure drop; friction factor; multiphase flow; flow rate; flow regime

1. Introduction

In the oil and gas industry, multiphase flow in vertical pipes often occurs. The flow of fluids through the vertical pipe string causes a loss of energy through friction losses, where the value of this loss depends on the fluid flow viscosity and the size of the conduit. Often, the friction loss is an important part of the oil well completion design [1]. The pressure drop occurs as a result of the changes in potential and kinetic energy of the fluid due to the friction on the pipe walls [2]. Generally, the total pressure drop in the vertical conduit is basically related to four main components: frictional, hydrostatic, acceleration, and pressure drop. Among these four components, calculation of the pressure drop is the most complex component and has received extensive attention by researchers [3,4]. Many researchers have attempted to determine the two-phase frictional pressure drop over the whole range of flow patterns through a vertical pipe. A substantial number of experiments have been carried out to determine fluid flow friction losses in both Newtonian [1,2,5–11] and non-Newtonian systems [5,12]. A large number of experimental works was made in short tubes. Consequently, a lot of engineering problems come up when efforts are made to extend these experimental results to real oil field conditions where a longer pipe is used. In those experiments, the data shows only a limited number of variables, and as a result, imprecisions are introduced when the friction correlations are applied outside the limitations of the experimental data. As a consequence of the limited amount of data available for these experiments, the effects of some significant variables were ignored in the early studies [13–17]. The accuracy of the pressure drop prediction in flowing wells has a significant influence on the fluid flow measurement. There are many particular solutions, but they are valid only for some specific conditions. This is due to the complexity of two-phase flow analysis. In some conditions, the gas travels at a much higher velocity than the liquid. Accordingly, the flowing density of the gas–liquid mixture is higher than the corresponding density. Moreover, the liquid's velocity inside the pipe wall can be different over a short distance and can cause a variable friction loss. The difference in velocity and flow regime of the two phases strongly affect pressure drop computations [13], meaning that slippage is a consequence of the difference between the combined velocities of the two phases, which is caused by the physical properties of the fluids involved. For single-phase flow, the frictional pressure losses do not normally increase with a decrease in the tubing size or an increase in well production flow rate. This refers to the existence of a gas phase, which tends to slip by the liquid phase without essentially contributing to its lift. Many researchers have tried to show a relationship between the slippage losses and the friction losses [15–18]. A method for the estimation of gas–liquid flow rates in the vertical pipe has been proposed [19]. The method was used to calibrate a differential pressure sensor to predict the flow rates of both phases in air–water flow. The estimations were in good agreement with real flow rate measurements. A study by Daev and Kairakbaev [20] proposed a new model of the liquid flow through pipes that incorporated flow straighteners. The prediction of the flow rate of liquid was studied and the parameters affecting the process of measuring the flow rate of liquid were considered. An experimental study of the two-phase flow regime and frictional pressure drop inside the pipe was done by Cai et al. [21]. The flow patterns were defined and recorded by a high-speed camera. A new empirical correlation was proposed based on the experimental results to predict the liquid multiplier factor of the test channel. A two-phase flow measurement applying a resistive void fraction meter combined to a venturi, or orifice plate, was suggested by Oliveira et al. [22]. This method was applied to determine the fluid mass flow rates using an air–water experimental apparatus. The results showed that the flow path has no important effect on the meters in relation to the frictional pressure drop in the experimental process range. The outcomes of the experimental work displayed a mean slip ratio of less than 1.1, when slug and bubbly flow patterns were lower than 70%.

This research work aims to evaluate the influence of a pressure drop on the measurement of the fluid flow rate in ESP oil wells. A new mathematical model was developed to determine the fluid flow rate of the oil wells through the prediction of multiphase flow parameter variations inside a vertical pipe based on local temperature and pressure changes with depth and applying multiphase flow physics equations and empirical correlations. The objective of this study was to obtain data from well tests conducted in a long vertical pipe and utilize this data to evaluate the effects of slippage and friction factor, in different flow regimes, on the calculation accuracies of the fluid flow rate of the oil wells. The approach measured the liquid hold-up along the conduit and used different friction correlations such as Blasius, Darcy, and Colebrook friction factor correlations to compare the predicted fluid flow rate with the measured fluid flow rate for each oil well. Generally, the results show that any errors in pressure drop calculation will generate inaccuracies in the prediction of fluid flow rate.

2. Experimental Arrangements and Measurement Procedure

The experiments conducted in the present study were carried out for two-phase flow through a vertical pipe of 48 oil wells using ESP pumps. A schematic of the experimental system is shown in Figure 1. The flow measurement starts at the surface wellhead, and then down to the bubble point pressure location depth in the well. Wellhead flowing pressure was measured at normal production conditions before and after the wing valve shut-in, leaving the ESP pump running, to measure the build-up of pressure at the wellhead. The total shut-in time period of the wellhead valve was then recorded. The first free gas bubbles started liberating from the bubble point location depth inside the tubing string. This occurred in the production flowing well before and after the wellhead wing valve shut-in, and the changes of flow patterns inside the pipe were reallocated once again, due to variations of temperature and pressure along the conduit. As a consequence, the liberated gas was dissolved in the oil phase, and the location depth of the bubble point pressure relocated to another position after the wellhead wing valve shut-in. The column of liquid that replaced the liberated gas column space, during the shut-in time period, was the difference between the first and second bubble point location depths. Figure 2 shows the bubble point location depths before and after the wellhead wing valve shut-in. A conceptual basis of physics for prediction of fluid flow rate in the conduit was employed along with multiphase empirical correlations to compute the variations of fluid flow parameters inside the tubing.

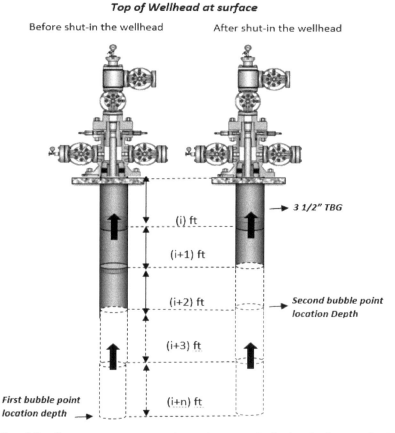

Figure 1. Schematic of the flow measurement stages in a vertical pipe before and after the well head wing valve shut-in.

Several assumptions were made to conduct the calculations such as: assumed one-dimensional flow in the conduit, assumed uniform cross-sectional area of the pipe, the phase's properties varied with depth, the frictional factor varied along the conduit, and the effect of the liquid compressibility was neglected.

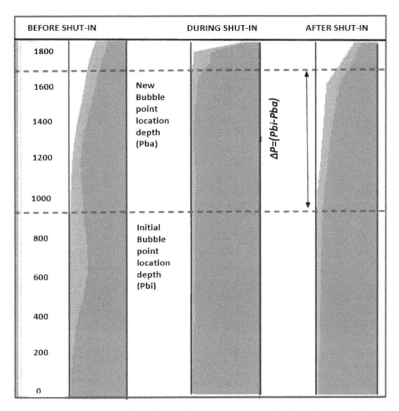

Figure 2. Bubble point location depths before and after closing the well head wing valve.

2.1. Required Input Data

The input data required were the well parameters data and physical properties data of the fluid, as seen in Table 1. To carry out this study, 48 ESP oil wells were selected where the wells were producing from four different reservoirs using same production pipe diameter. Also, these reservoirs had almost the same reservoir fluid properties: the bubble point pressure ranged from 924 psi to 1124 psi, the American Petroleum Institute (API) oil gravity ranged from 36 to 37 @ 60 °F, the oil viscosity ranged from 0.784 cP to 1.0119 cP, and the reservoir temperature ranged from 157 °F to 186 °F. Furthermore, Figure 3 classifies the input data required.

Table 1. Well and physical properties of the fluid.

Well Name	WHPb	WHPa	WHT	GOR	WC	Total Shut-in Time
	(PSIA)	(PSIA)	(F)	(SCE/STB)	(%)	(min)
A33	140	200	98	360.92	93	1.55
A125	100	200	95	360.92	91.62	0.83
A64	180	250	107	360.92	81.52	1.06
A29	250	270	107	360.92	84.88	0.80
A23	210	260	127.7	360.92	82.11	0.24
A135	210	250	100	360.92	59.88	0.56
A126	250	300	98	360.92	66.91	0.20
A12	175	270	107	360.92	82.9	0.84
A108	260	300	97	360.92	81.31	0.28
5J5	150	300	95	360.92	4	3.36
5J2	100	170	101	360.92	52	1.39
5J4	250	300	101.6	360.92	58.95	0.27
5J7	250	300	98	360.92	30	0.47
E89	150	190	140	300	79	0.27
E210	80	120	110	300	83	0.33

Table 1. *Cont.*

Well Name	WHPb (PSIA)	WHPa (PSIA)	WHT (F)	GOR (SCE/STB)	WC (%)	Total Shut-in Time (min)
E211	80	120	129	300	74	0.96
E286	70	100	146	300	90	0.25
E192	80	110	146	300	83	0.24
E327	80	110	115.5	300	77	0.35
E325	70	110	124.6	300	83	0.44
E197	90	110	146.1	300	82	0.11
E208	95	110	146.8	300	81	0.07
E226	80	110	138.4	300	91	0.25
E284	80	120	124.5	300	76	0.33
E258	65	90	142	300	86	0.23
E326	60	100	113	300	82	0.48
E227	100	150	142	300	84	0.36
4E_3	130	300	146	300	87	1.18
B56	120	170	120	384	42	2.3
B70	160	230	120	384	29.9	2.9
B121	100	160	120	364	67.29	0.5
B119	100	160	120	364	76.18	1.1
B50	180	250	110	364	68.65	0.55
B88	100	160	110	364	63.59	2.1
B14	250	270	110	364	76.28	0.15
B151	180	230	110	364	55.78	0.66
B164	100	170	120	364	26.3	2.1
B51	240	310	120	364	59.05	0.44
Q89	100	150	120	364	0	3.7
Q21	80	150	120	364	79.22	2.1
Q53	80	150	120	364	71.41	2.3
Q14	75	130	120	364	74.83	1.4
Q100	80	130	110	364	78.27	0.55
Q12	80	150	110	364	80.33	0.58
Q85	100	150	110	364	18.18	2.5
Q82	100	150	110	364	75.3	0.5
Q78	80	150	120	364	37.27	2.5
Q76	80	150	120	364	80.5	1.3

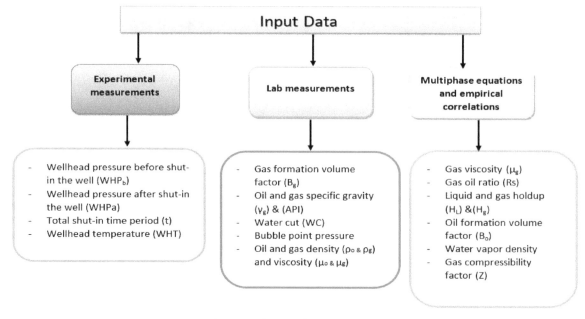

Figure 3. Input data required.

2.2. Computational Algorithm

Figure 4 shows the algorithm steps to evaluate the mathematical model. The algorithm classified all the main stages and sub-steps in the model. In this process, the calculations were performed to obtain the bubble point pressure location depth before and after the wellhead wing valve shut-in. The fluid flowing pressure gradient could be calculated anywhere inside the pipe. All the variables needed to be identified to correctly evaluate the physics interactions between all the fluid parameters using the suitable multi-physics equations and empirical correlations.

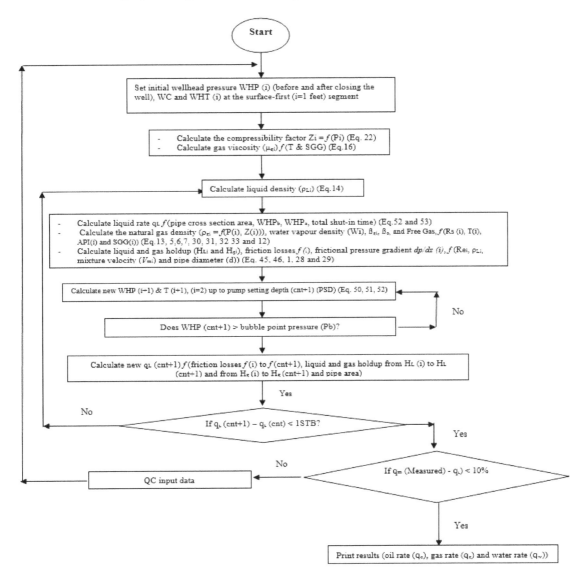

Figure 4. Flowchart of the new mathematical model algorithm.

The calculation starts at the surface wellhead and then down to the location depth of the bubble point pressure as a function of temperature and pressure variations with depth. To consider the fact that flow regimes vary depending on the in situ flow rates of gas/liquid, the model calculates, at each foot along the vertical pipe, the variations of supercritical velocities, viscosities, and densities for both phases (liquid and gas). The in situ flow rate can also be calculated by the mathematical model at any flow regime at any depth. As shown in Figures 5 and 6, the calculation iteration can stop at any depth $(i, ..., i + n)$ using all the equations (from Equation (1) to Equation (56)), where there is a different flow regime along the vertical pipe.

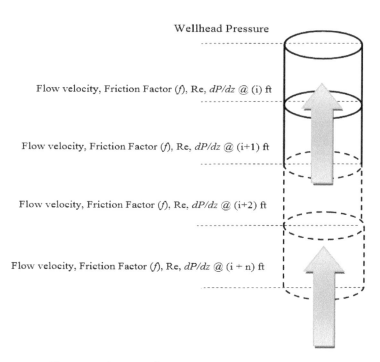

Figure 5. Stages of computational methodology.

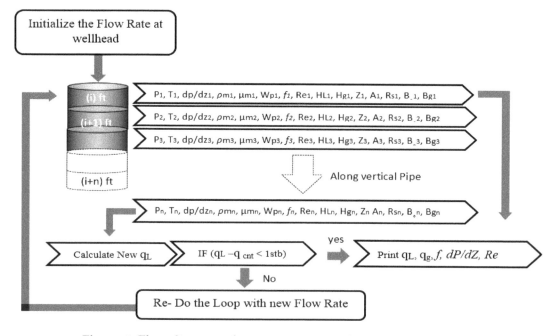

Figure 6. Flow diagram of new computational method procedure.

The following are the physics equations and the correlations applied to determine each independent variable at every single foot.

Total pressure losses expressed as

$$\Delta P_{Total} = \Delta P_{HH} + \Delta P_{Frictional} \qquad (1)$$

Hydrostatic head is expressed as

$$\Delta P_{HH} = \frac{\rho_m g \Delta Z}{144 g_c} \qquad (2)$$

Darcy–Weisbach equation [23] was used to calculate the frictional pressure loss

$$\Delta P = f \frac{L}{D} \frac{\rho V^2}{2} \tag{3}$$

Reynolds number is given by

$$Re = \frac{2.2 \times 10^{-2} m_t}{D \mu_L{}^{H_L} \mu_g{}^{(1-H_L)}} \tag{4}$$

Three different friction factor correlations were applied to evaluate the impact of the friction on the computation of the fluid flow rate. The first correlation is the Blasius empirical correlation for turbulent flow [24].

$$f = 0.316 \, (Re)^{-0.25} \tag{5}$$

The second friction correlation applied is Darcy correlation [23]

$$f = \frac{64}{Re} \tag{6}$$

The third friction factor correlation applied is from Colebrook [25]

$$\frac{1}{\sqrt{f}} = 2 log_{10} \left(\frac{\varepsilon / D_h}{3.7} + \frac{2.51}{Re\sqrt{f}} \right) \tag{7}$$

for $Re <\approx 2300$ and $Re >\approx 4000$.

The gas density is expressed as

$$\rho_g = \frac{m_g}{V_R} = \frac{M_g P}{ZRT} \tag{8}$$

The density at wellbore condition, is given by

$$\rho_g = \frac{\rho_{gs}}{B_g} \tag{9}$$

The oil density is expressed as

$$\rho_o = \frac{62.428 \gamma_o + 0.014 \gamma_g R_s}{B_o} \tag{10}$$

where

$$\gamma_o = \rho_o / \rho_w \tag{11}$$

$$\gamma_g = \frac{\rho_g}{\rho_{air}} = \frac{\rho_g}{0.077} \tag{12}$$

$$\rho_g = 0.077 \gamma_g \tag{13}$$

Liquid density is given by

$$\rho_L = \rho_W WC + \rho_o (1 - WC) \tag{14}$$

Mixture density is expressed as

$$\rho_m = \rho_L H_L + \rho_g (1 - H_L) \tag{15}$$

The gas viscosity is determined by the following equation [26]:

$$\mu_g = K_1 exp \left(X \rho^Y \right) \tag{16}$$

where

$$\rho = \frac{pM_g}{zRT} = 0.0015\frac{pM_g}{zT} \tag{17}$$

$$K_1 = \frac{(0.001 + 2 \times 10^{-6}M_g)T^{1.5}}{(209 + 19M_g + T)} \tag{18}$$

$$X = 3.5 + \frac{986}{T} + 0.01M_g \tag{19}$$

$$Y = 2.4 - 0.2X \tag{20}$$

Mixture viscosity is given by

$$\mu_m = \mu_L^{H_L} + \mu_g^{(1-H_L)} \tag{21}$$

Beggs and Brill equation [27] was applied to estimate the gas compressibility factor (Z)

$$Z = A + \frac{(1-A)}{e^B} + CPr^D \tag{22}$$

Using Standing and Katz equations [28] to obtain the pseudo critical temperature and pressure of the gas mixture

$$Pr = 688.634 - 21.983\gamma_g - 13.886\gamma_g^2 \tag{23}$$

$$Tr = 158.01 + 342.12\gamma_g - 16.04\gamma_g^2 \tag{24}$$

and

$$A = 1.39(Tr - 0.92)^{0.5} - 0.36Tr - 0.101 \tag{25}$$

$$B = (0.62 - 0.23Tr)Pr + \left(\frac{0.066}{Tr - 0.86} - 0.037\right)Pr^2 + \frac{0.32}{10^{9(Tr-1)}}Pr^2 \tag{26}$$

$$C = 0.132 - 0.32log(Tr) \tag{27}$$

$$D = 10^{(0.302-0.49Tr+0.182Tr^2)} \tag{28}$$

Superficial gas velocity is expressed as

$$V_{sg} = \frac{4q_g B_g}{\pi D^2} \tag{29}$$

Superficial liquid velocity is expressed as

$$V_{sL} = \frac{4q_L}{\pi D^2} \tag{30}$$

The water vapor density using the Sloan correlation [29] is expressed as

$$W = exp\left(c_1 + \frac{c_2}{T} + c_3ln(p) + \frac{c_4}{T^2} + \frac{c_5ln(P)}{T} + c_6(ln(P))^2\right) \tag{31}$$

where the values of constants c_1 to c_6 are shown in Table 2.

Table 2. Constants c_1 to c_6.

Constants	Value
c_1	28.911
c_2	−9668.146
c_3	−1.663
c_4	−130,823.5
c_5	205.323
c_6	0.0385

The gas formation volume factor is expressed as

$$B_g = \frac{P_{sc}ZT}{T_{sc}P} = 0.028\frac{ZT}{P} \tag{32}$$

Using the Vasquez and Beggs equation [30] to obtain the oil formation volume factor

$$B_{ob} = 1 + C_1 R_{sb} + C_2(T - 60)\left(\frac{\gamma_{API}}{\gamma_g}\right) + C_3 R_{sb}(T - 60)\left(\frac{\gamma_{API}}{\gamma_g}\right) \tag{33}$$

and oil gas ratio

$$R_{sb} = \frac{\gamma_g Pb^{C_2}}{C_1}10^{\left(\frac{C_3\gamma_{API}}{T + 459.67}\right)} \tag{34}$$

where the coefficients C_1, C_2 and C_3 are given by

Coefficient	$^\circ$API $\leq$ 30	$^\circ$API $\geq$ 30
C_1	27.64	56.060
C_2	1.0937	1.187
C_3	11.172	10.393

To make sure that the obtained liquid and gas hold-up is accurate, some popular correlations, used by the industry and are included in almost every commercial software package, were considered to predict the liquid and gas hold-up inside each well. The correlations considered in this study are the ones developed by Hagedorn and Brown [31], Duns and Ros [32], Orkiszewski [33], and Aziz et al. [34]. The statistical results for the various prediction methods when applied to all 25 well tests are shown in Figure 7 and Table 3. These results indicate that the Hagedorn and Brown correlation seems to predict liquid and gas hold-up better than the other correlations selected in this study. However, the overall results show minor differences between the different correlations. This is because each correlation was developed based on certain assumption and for a particular range of data.

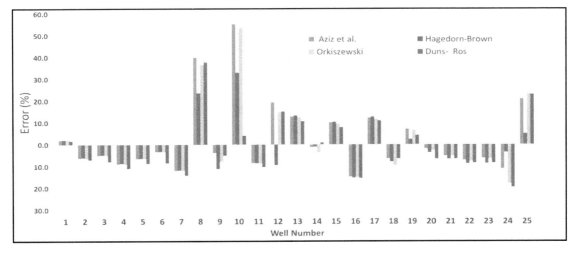

Figure 7. Hold-up prediction accuracy using some popular correlations.

Table 3. Statistical results for the various prediction correlations.

Prediction Method	Average Error (%)	Standard Deviation (%)
Duns and Ros	−1.06	13.06
Hagedorn and Brown	−0.86	11.57
Orkiszewski	1.8	16.52
Aziz et al.	2.9	16.66

Using the Hagedorn-Brown empirical correlation [31] to obtain liquid and gas hold-up (H_L and H_g)

$$N_{Lv} = 1.938 v_{sL} \frac{\sqrt[4]{\rho}}{\sigma} \tag{35}$$

$$N_{gv} = 1.938 v_{sg} \sqrt[4]{\frac{\rho_L}{\sigma}} \tag{36}$$

$$N_d = 120.872 D \frac{\sqrt{\rho_L}}{\sigma} \tag{37}$$

$$N_L = 0.157 \mu_L \sqrt[4]{\frac{1}{\rho_L \sigma^3}} \tag{38}$$

$$Y = -2.699 + 0.158 X_1 - 0.551 X_1{}^2 + 0.548 X_1{}^3 - 0.122 X_1{}^4 \tag{39}$$

where

$$X_1 = log(N_L + 3) \tag{40}$$

$$CN_L = 10^Y \tag{41}$$

$$\frac{H_L}{\psi} = -0.103 + 0.618(log X_2 + 6) - 0.633(log X_2 + 6)^2 + 0.296(log X_2 + 6)^3 - 0.04(log X_2 + 6)^4 \tag{42}$$

where

$$X_2 = \frac{N_{VL} P^{0.1} CN_L}{N_{Vg}{}^{0.575} Pa^{0.1} N_D} \tag{43}$$

$$\psi = 0.912 - 4.822 X_3 + 1232.25 X_3^2 - 22253.6 X_3^3 + 116174.3 X_3^4 \tag{44}$$

where

$$X_3 = \frac{N_{Vg} N_L{}^{0.38}}{N_D{}^{2.14}} \tag{45}$$

The liquid hold-up is

$$H_L = \psi \left(\frac{H_L}{\psi} \right) \tag{46}$$

and

$$H_g = (1 - H_L) \tag{47}$$

The liquid flow rate is expressed as

$$q_L = \frac{\Delta H \cdot A}{t} \tag{48}$$

the cross section of the conduit is given by

$$A = \frac{\pi r^2}{4} \tag{49}$$

$$\Delta H = H_2 - H_1 \tag{50}$$

as

$$\Delta P = \Delta H \cdot \rho_L \qquad (51)$$

then

$$\Delta H = \frac{\Delta P}{\rho_L} \qquad (52)$$

and

$$\Delta P = WHP_a - WHP_b \qquad (53)$$

then

$$q_L = \Delta P \frac{A}{\rho_L} t \qquad (54)$$

The flow rates for gas, oil, and water are expressed as

$$q_o = q_L(1 - WC) \qquad (55)$$

$$q_g = q_o\, Rs \qquad (56)$$

$$q_w = q_o\, WC \qquad (57)$$

3. Results and Discussion

The experiments were run on 48 ESP oil wells from four different reservoirs. For each friction factor correlation, the measured oil flow rate values for each oil well were compared against the predicted flow rate values. It should be noted that as the points near the dotted straight line drawn at 45° (i.e., $y = x$) in the graph, the more accurate the prediction was. The results show that the pressure drop value was the significant parameter that had the main influence on the fluid flow rate computation. Indeed, any errors in pressure drop values would lead to high uncertainty errors of fluid flow rate prediction. For this reason, the properties of independent variables needed to be considered. Likewise, the interactions between each phase needed to be taken into account along with mixture properties and in situ volume fractions of oil and gas inside the conduit. Each multiphase flow correlation found the friction factor differently. Typically, each friction correlation made its own assumptions and modifications to make them useable to multiphase conditions. The prediction of frictional pressure drop in two-phase flow was usually complicated due to pressure and temperature variations along the flow path. When estimating the friction factor, there were a number of methods for calculating the Reynolds number depending on how much of the two-phase flow mixture was defined. Therefore, the oil and water were considered as a single liquid phase while the gas was considered as a separate phase.

By using the Blasius friction factor correlation, the differences between the predicted flow rate and the measured flow rate were very small. R-squared (R^2) explained exactly how the data points were fitted close to the regression line ($y = x$). Figures 8–10 displayed the regression model for oil, water, and gas flow rate measurements. It can be seen that the plots show that most data points lie on or close to the unit slope line (e.g., best fit line), indicating that the predicted and actual values were in excellent agreement and illustrated an accurate flow rate prediction for oil, water, and gas with good correlating coefficients of 0.994, 0.993, and 0.966, respectively. This means that 99.4%, 99.3%, and 96.6% of the variance in the oil, water, and gas data, respectively, was explained by the line and 0.6%, 0.7%, and 3.4% of the variance was due to unexplained effects. The figures show that the predicted wells flow rates fell within the accepted uncertainty when compared with the measured flow rates.

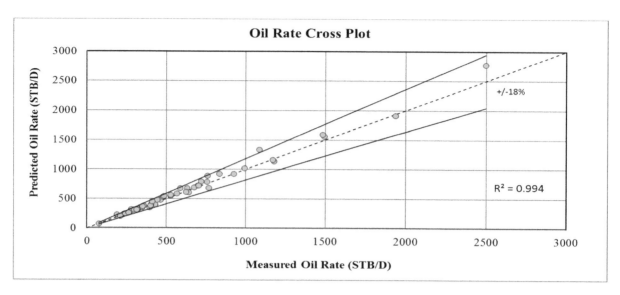

Figure 8. Predicted vs measured oil rate using the Blasius correlation.

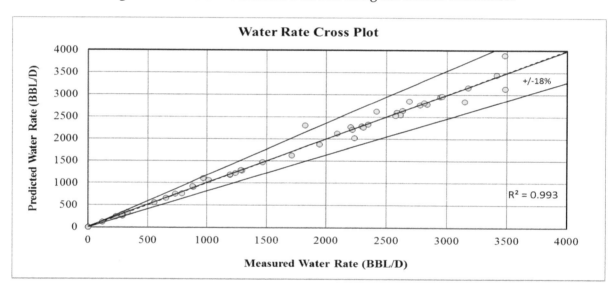

Figure 9. Predicted vs measured water rate using the Blasius correlation.

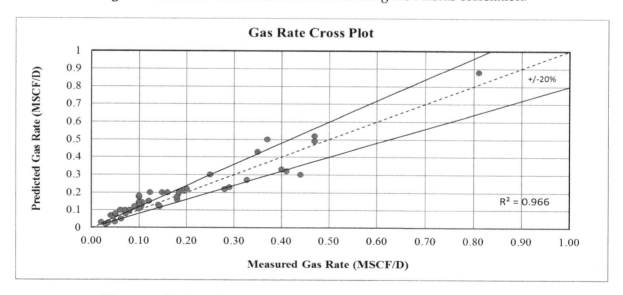

Figure 10. Predicted vs measured gas rate using the Blasius correlation.

By using the Darcy friction factor correlation, the differences between the predicted fluid flow rates with the measured flow rates were larger than those of the Blasius correlation. Figures 11–13 displayed the regression model for oil, water, and gas flow rate measurements. From these figures, one can easily recognize that the data plotted is under-estimated for oil and water flow rates and scattered around the best fit line for gas flow rates. This discrepancy was more evident for high flow rates where the correlation coefficients for oil, water, and gas flow rates accounted for 90.6%, 86.6%, and 78.7% of the variance, respectively. The figures show that the predicted well flow rates did not fall within the accepted uncertainty when compared with the measured flow rates.

By using the Colebrook friction factor correlation, the differences between the predicted fluid flow rates with the measured flow rates were slightly better than the Darcy correlation performance, but still less than the Blasius correlation performance. Figures 14–16 displays the data fitting for oil, water, and gas flow rate measurements. Similar to the performance of the Darcy correlation, one can easily recognize that the data plotted is under-estimated for oil and water flow rates and scattered around the best fit line for gas flow rates. This discrepancy was more evident for high flow rates where the correlation coefficients for oil, water, and gas flow rates accounted for 93.0%, 87.1%, and 80.8% of the variance, respectively. The figures showed that the predicted wells flow rates did not fall within the accepted uncertainty when compared with the measured flow rates.

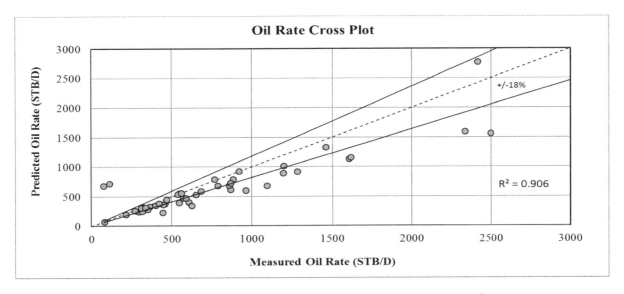

Figure 11. Predicted vs measured oil rate using the Darcy correlation.

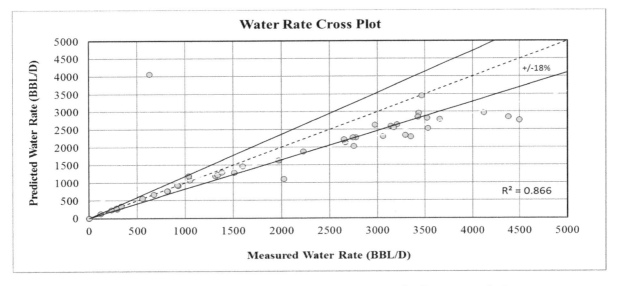

Figure 12. Predicted vs measured water rate using the Darcy correlation.

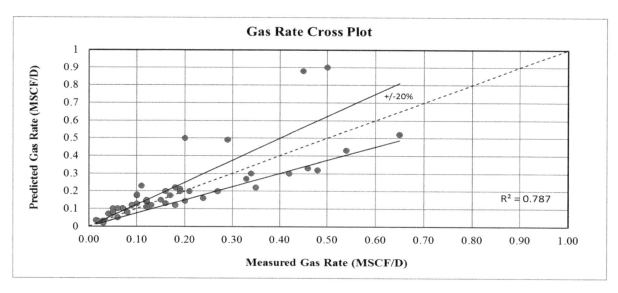

Figure 13. Predicted vs measured gas rate using the Darcy correlation.

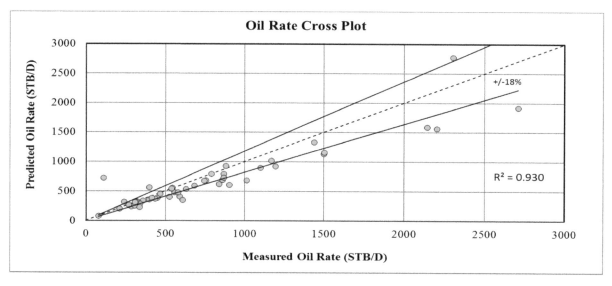

Figure 14. Predicted vs measured oil rate using the Colebrook correlation.

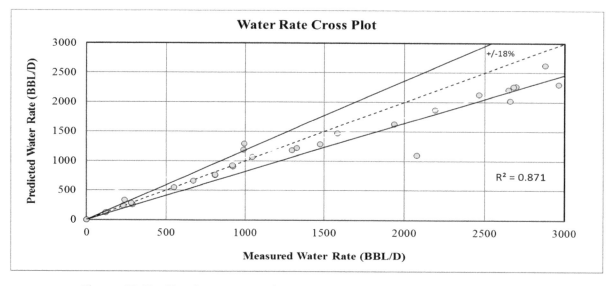

Figure 15. Predicted vs measured water rate using the Colebrook correlation.

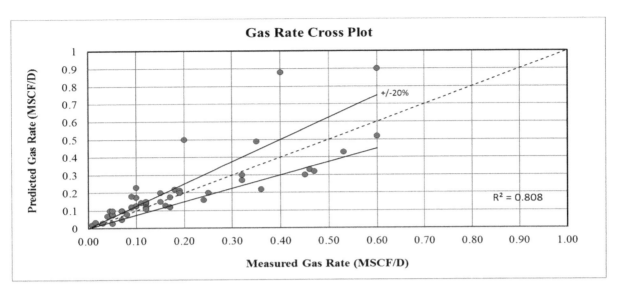

Figure 16. Predicted vs measured gas rate using the Colebrook correlation.

In general, the validation results of the predicted fluid flow rates were satisfactory when using the Blasius correlation rather than the Darcy or Colebrook correlations, where 96% and 98% of the predicted fluid flow rates were in good agreement with the real measured oil and water flow rate, respectively. Furthermore, the relative errors were less than ±18%, which were still within the reasonable uncertainty, as shown in Figures 17 and 18. For the predicted and measured gas rates, 68% of the wells showed about ±10% relative errors, as shown in Figure 19. By using the Darcy correlation, 63% and 70% of the wells were not in good agreement with the predicted and measured oil and water rate, respectively, with more than ±18% for relative errors, as shown in Figures 20 and 21. For predicted and measured gas rates, 79% of the wells showed more than ±10% relative errors, as shown in Figure 22. By using the Colebrook correlation, 67% and 75% of the wells were not in good agreement with the predicted and measured oil and water rate, respectively, with more than ±18% relative errors, as shown in Figures 23 and 24. For predicted and measured gas rates, 77% of the wells showed more than ±10% relative errors, as shown in Figure 25.

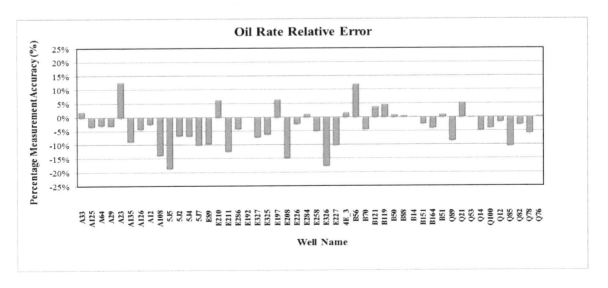

Figure 17. Oil rate measurement accuracy using the Blasius correlation.

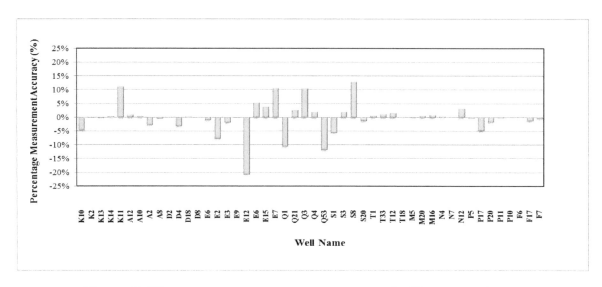

Figure 18. Water rate measurement accuracy using the Blasius correlation.

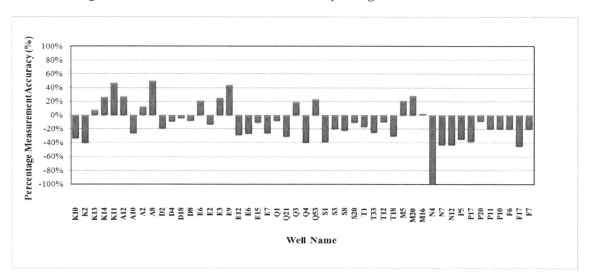

Figure 19. Gas rate measurement accuracy using the Blasius correlation.

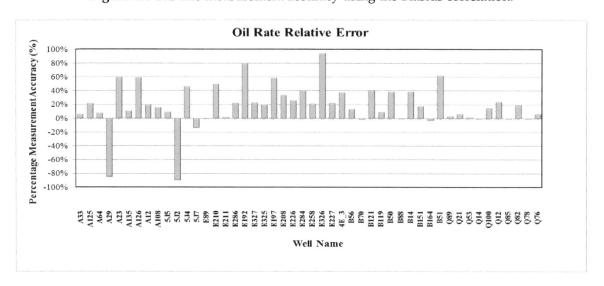

Figure 20. Oil rate measurement accuracy using the Darcy correlation.

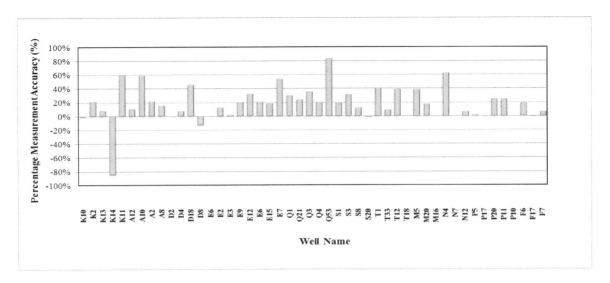

Figure 21. Water rate measurement accuracy using the Darcy correlation.

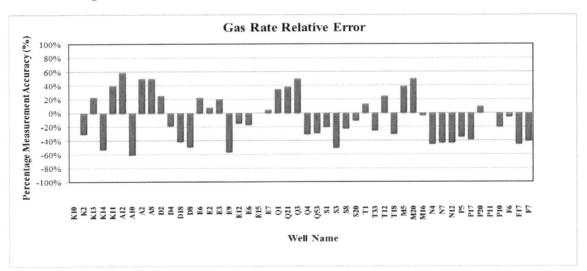

Figure 22. Gas rate measurement accuracy using the Darcy correlation.

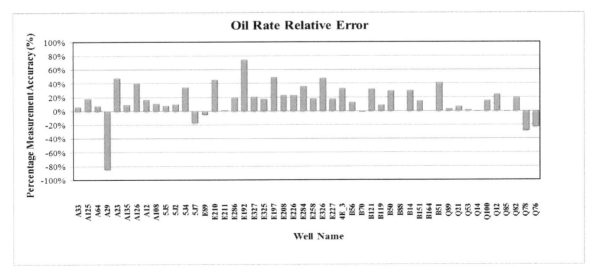

Figure 23. Oil rate measurement accuracy using the Colebrook correlation.

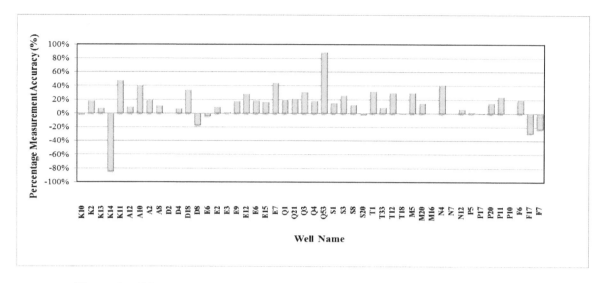

Figure 24. Water rate measurement accuracy using the Colebrook correlation.

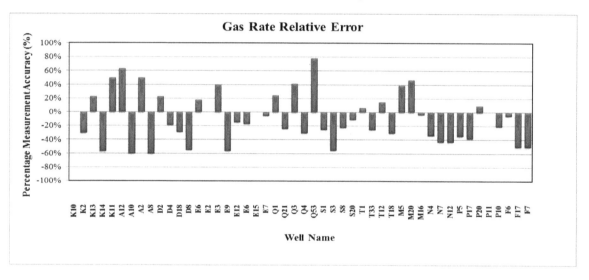

Figure 25. Gas rate measurement accuracy using the Colebrook correlation.

The results showed high relative errors in gas rate prediction which can happen due to the oil separator meters being insufficiently accurate. Also, these errors may occur due to fixed orifice plate meters used to measure the gas flow rate despite the fact that orifice plates are not appropriate to measure low gas rates. Besides, wear and corrosion can increase the orifice size and cause excessive loss.

4. Summary and Conclusions

The prediction of the fluid flow rate of oil wells using the new mathematical model has been made and validated with experimentally measured fluid flow rate data. To evaluate the influence of the frictional pressure drop value on the measurement of fluid flow rate of oil wells, Blasius, Darcy, and Colebrook friction correlations were applied. Using the Blasius correlation, the analysis showed that the predicted fluid flow rate values were in accord with the measured values, while by using the Darcy and Colebrook friction correlations, the results were not in good agreement with the measured values. This discrepancy was due to the fact that each friction correlation found the friction factor differently. To determine the friction factor, many expressions were used to compute the Reynolds number. Essentially, each empirical correlation states its own assumptions and modifications to defend the variable components in order to be applicable to multiphase conditions. The two-phase flow significantly complicated the pressure drop calculations, where any errors in determining the frictional pressure drop values would generate some inaccuracies in predicting the fluid flow rate of the oil wells. Consequently, mixture properties and the interactions between the existing phase's properties must be considered. Therefore, the gas and liquid volume fractions throughout the conduit needed to be determined. Overall, the performance of the new mathematical model indicated that the selection of the appropriate friction factor correlation would lead to predicting the gas and liquid flow rate within the acceptable accuracy. However, the friction loss dominated only with very high flow rates. For relatively small flow rates, the hydrostatic pressure played the key role in the overall pressure drop in the vertical tubing. Thus, different multiphase flow models, either empirical or mechanistic model, used in the computation would output different predictions. That being said, the Blasius equation may be superior to other models coupled with the Hagedorn-Brown empirical correlation, as it has been shown in this work. Indeed, a very reasonable average relative error of 4.6% was observed between the predicted and measured flow rates. However, it may not be as good as it is when coupled with other mechanistic models that may further reduce this error. Further research is needed to further validate the developed model by accounting for other sophisticated multiphase models.

Author Contributions: T.A.G. derived the mathematical model, carried out the experimental validation and drafted the manuscript. M.H. contributed to data presentation, analysis of results and revision preparation.

Acknowledgments: Special thanks are due to production engineering staff of Waha Oil Company for their support. The authors also gratefully acknowledge assistance received from Lynn Mason for editing this manuscript.

Nomenclature

A	cross-sectional area, (sq ft)
API	American Petroleum Institute
B_o	oil formation volume factor, (bbl/stb)
B_{ob}	oil formation volume (at bubble point pressure), (bbl/STB)
B_g	gas formation volume factor, (cf/scf)
Cnt	count
dp/dz	pressure gradient, (psi/ft)
d	inside diameter, (ft)
ESP	electrical submersible pump
f	friction factor, (unitless)
g	Gravity, (ft/s^2)
H_L	liquid hold-up
H_G	gas hold-up
$H1$	bubble point pressure (at location depth before shut-in the well head valve), (ft)
$H2$	bubble point pressure (at location depth after shut-in the well head valve), (ft)
mt	mass flow rate, (lb/day)
N_{gv}	gas velocity number, (unitless)
N_{Lv}	liquid velocity number, (unitless)
N_d	pipe diameter number, (unitless)
N_{CL}	coefficient number of viscosity correction, (unitless)
N_L	liquid viscosity number, (unitless)
q_o	oil rate, (stb/day)
q_g	gas rate, (stb/day)
q_w	water rate, (stb/day)
q_L	liquid rate, (stb/day)
q_m	measured flow rate, (stb/day)
Q_C	quality check

P	pressure, (psia)
P_r	pseudo-critical pressure (for gas mixture), (psia)
P_b	bubble point pressure, (psia)
P_{sc}	pressure at standard conditions (P = 14.7 atm, T = 60 °F), (psia)
PSD	pump setting depth
SGG	gas specific gravity
STB	stock tank barrel (for liquid)
r_w	wellbore radius, (ft)
R_s	gas-oil ratio, (scf/stb)
R_{sb}	gas oil ratio at bubble point pressure, (cf/scf)
R_e	Reynolds number, (unitless)
T	temperature, (°F)
t	total shut-in time, (min)
T_r	pseudo-critical temperature (for gas mixture), (psia)
T_{sc}	temperature at standard condition, (°R)
T_r	reservoir fluid temperature, (°F)
VR	gas volume at reservoir conditions, (ft^3)
V_{sc}	gas volume at standard condition, (ft^3)
V_{SL}	superficial velocity for liquid, (ft/sec)
V_{Sg}	superficial velocity for gas, (ft/sec)
V_m	total mixture velocity, (ft/sec)
WHP_a	well head pressure (after shut-in the well), (psia)
WHP_b	well head pressure (before shut-in the well), (psia)
WC	water cut (unitless)
WHT	well head temperature, (°F)
W	water vapor density, (unitless)
Z	gas compressibility factor (unitless)

Greek Symbols

ΔP	pressure drop, (psia)
H_L/ψ	hold-up correlation factor
γ_o	oil gravity
γ_w	water gravity
γ_g	gas gravity
σ	surface tension, (dyne/m)
ΔH	differences between bubble point pressure location depths (before and after shut-in the well head valve), (ft)
ρ_o	oil density, (lbm/ft^3)
ρ_g	gas density, (lbm/ft^3)
ρ_w	water density, (lbm/ft^3)
ρ_L	liquid density, (lb/ft^3)
ρ_m	mixture density, (lbm/ft^3)
μ_o	oil viscosity, cP
μ_g	gas viscosity, cP
μ_L	liquid viscosity, cP

Subscripts

g_{sc}	gas (at standard condition)
h	hydrostatic
L	liquid
m	mixture (liquid and gas)
o	oil
sc	standard condition
w	water

References

1. Saeb, M.B.; Philip, D.M.; David, C.C.; Ali, J. Modeling friction factor in pipeline flow using a GMDH-type neural network. *Cogent Eng.* **2015**, *2*. [CrossRef]

2. Shannak, B.A. Frictional pressure drop of gas liquid two-phase flow in pipes. *Nuclear Eng. Des.* **2008**, *238*, 3277–3284. [CrossRef]

3. Jiang, J.Z.; Zhang, W.M.; Wang, Z.M. Research progress in pressure-drop theories of gas-liquid two-phase pipe flow. In Proceedings of the China International Oil & Gas Pi (CIPC 2011), Langfang, China, 5 September 2011.

4. Xu, Y.; Fang, X.D.; Su, X.H.; Zhou, Z.R.; Chen, W.W. Evaluation of frictional pressure drop correlations for two-phase flow in pipes. *Nuclear Eng. Des.* **2012**, *253*, 86–97. [CrossRef]

5. Mittal, G.S.; Zhang, J. Friction factor prediction for Newtonian and non-Newtonian fluids in pipe flows using neural networks. *Int. J. Food Eng.* **2007**, *3*, 1–18. [CrossRef]

6. Fadare, D.A.; Ofidhe, U.I. Artificial neural network model for prediction of friction factor in pipe flow. *J. Appl. Sci.* **2009**, *5*, 662–670.

7. Bilgil, A.; Altun, H. Investigation of flow resistance in smooth open channels using artificial neural networks. *Flow Meas. Instrum.* **2008**, *19*, 404–408. [CrossRef]

8. Özger, M.; Yildirim, G. Determining turbulent flow friction coefficient using adaptive neuro-fuzzy computing techniques. *Adv. Eng. Softw.* **2009**, *40*, 281–287. [CrossRef]

9. Shayya, W.H.; Sablani, S.S.; Campo, A. Explicit calculation of the friction factor for non-Newtonian fluids using artificial neural networks. *Dev. Chem. Eng. Miner. Process.* **2005**, *13*, 5–20. [CrossRef]

10. Yuhong, Z.; Wenxin, H. Application of artificial neural network to predict the friction factor of open channel flow. *Commun. Nonlinear Sci. Numer. Simul.* **2009**, *14*, 2373–2378. [CrossRef]

11. Yazdi, M.; Bardi, A. Estimation of friction factor in pipe flow using artificial neural networks. *Can. J. Autom. Control Intell. Syst.* **2011**, *2*, 52–56.

12. Sablani, S.S.; Shayya, W.H. Neural network based non-iterative calculation of the friction factor for power law fluids. *J. Food Eng.* **2003**, *57*, 327–335. [CrossRef]

13. Griffith, P. Two-Phase Flow in Pipes. In *Special Summer Program*; Massachusetts Institute of Technology: Cambridge, MA, USA, 1962.

14. Raxendell, P.B. The Calculation of Pressure Gradients in High-Rate Flowing Wells. *J. Pet. Technol.* **1961**, *13*, 1023. [CrossRef]

15. Fancher, G.H.; Brown, K.E. Prediction of Pressure Gradients for Multiphase Flow in Tubing. *Soc. Pet. Eng. J.* **1963**, *3*, 59. [CrossRef]

16. Hagedorn, A.R.; Brown, K.E. The Effect of Liquid Viscosity in Vertical Two-Phase Flow. *J. Pet. Technol.* **1964**, *16*, 203. [CrossRef]

17. Poettmann, F.H.; Carpenter, P.G. The Multiphase Flow of Gas, Oil and Water through Vertical Flow Strings with Application to the Design of Gas-Lift Installations. In Proceedings of the Drilling and Production Practice, New York, NY, USA, 1 January 1952; Volume 52, p. 257.

18. Tek, M.R. Multiphase Flow of Water, Oil and Natural Gas Through Vertical Flow Strings. *J. Pet. Technol.* **1961**, *13*, 1029. [CrossRef]

19. Shaban, H.; Tavoularis, S. Identification of flow regime in vertical upward air–water pipe flow using differential pressure signals and elastic maps. *Int. J. Multiph. Flow* **2014**, *61*, 62–72. [CrossRef]

20. Daev, Z.A.; Kairakbaev, A.K. Measurement of the Flow Rate of Liquids and Gases by Means of Variable Pressure Drop Flow Meters with Flow Straighteners. *Meas. Tech.* **2017**, *59*, 1170–1174. [CrossRef]

21. Cai, B.; Guo, D.X.; Jing, F.W. Study on gas-liquid two-phase flow patterns and pressure drop in a helical channel with complex section. In Proceedings of the 23rd International Compressor Engineering Conference, West Lafayette, IN, USA, 11–14 July 2016.

22. Oliveira, J.L.G.; Passos, J.C.; Verschaeren, R.; Van der Geld, C. Mass flow rate measurements in gas-liquid flows by means of a venturi or orifice plate coupled to a void fraction sensor. *Exp. Therm. Fluid Sci.* **2009**, *33*, 253–260. [CrossRef]

23. Brown, G.O. The history of the Darcy-Weisbach equation for pipe flow resistance. *Environment.* Available online: https://ascelibrary.org/doi/abs/10.1061/40650 (accessed on 17 September 2018).

24. Blasius, H. *Das Ähnlichkeitsgesetz bei Reibungsvorgängen in Flüssigkeiten, Mitteilung 131 über Forschungsarbeiten auf dem Gebiete des Ingenieurwesens*; Springer: Berlin, Germany, 1913.

25. Colebrook, C.F. Turbulent flow in pipes, with particular reference to the transition region between the smooth and rough pipe laws. *J. Inst. Civ. Eng.* **1939**, *11*, 133–156. [CrossRef]

26. Lee, A.L.; Gonzalez, M.H.; Eakin, B.E. The Viscosity of Natural Gases. *J. Pet. Technol.* **1966**, *18*, 997–1000. [CrossRef]

27. Beggs, D.H.; Brill, J.P. A study of two-phase flow in inclined pipes. *J. Pet. Technol.* **1973**, *25*, 607–617. [CrossRef]

28. Standing, M.B.; Katz, D.L. Density of natural gases. *Trans. AIME* **1942**, *146*, 140–149. [CrossRef]

29. Sloan, E.; Khoury, F.; Kobayashi, R. Measurement and interpretation of the water content of a methane-propane mixture in the gaseous state in equilibrium with hydrate. *Ind. Eng. Chem. Fundam.* **1982**, *21*, 391–395.

30. Vazquez, M.; Beggs, H.D. Correlations for Fluid Physical Property Prediction. *J. Pet. Technol.* **1980**, *32*, 968–970. [CrossRef]

31. Hagedorn, A.R.; Brown, K.E. Experimental Study of Pressure Gradients Occurring during Continuous Two-Phase Flow in Small Diameter Vertical Conduit. *J. Pet. Technol.* **1965**, *17*, 475–484. [CrossRef]

32. Duns, H., Jr.; Ros, N. Vertical Flow of Gas and Liquid Mixtures in Wells. In Proceedings of the 6th World Petroleum Congress, Frankfurt am Main, Germany, 19–26 June 1963; p. 451.

33. Orkiszewski, J. Predicting Two-Phase Pressure Drops in Vertical Pipe. *J. Pet. Technol.* **1967**, *19*, 829–838. [CrossRef]

34. Aziz, K.; Govier, G.; Fogarasi, M. Pressure Drop in Wells Producing Oil and Gas. *J. Can. Pet. Technol.* **1972**, *11*, 38. [CrossRef]

8

Experimental Study of Particle Deposition on Surface at Different Mainstream Velocity and Temperature

Fei Zhang⬚, Zhenxia Liu *, Zhengang Liu and Yanan Liu

School of Power and Energy, Northwestern Polytechnical University, Xi'an 710129, China;
zhangfei089@mail.nwpu.edu.cn (F.Z.); zgliu@nwpu.edu.cn (Z.L.); liuyanan1993@mail.nwpu.edu.cn (Y.L.)
* Correspondence: zxliu@mail.nwpu.edu.cn

Abstract: The effect of mainstream velocity and mainstream temperature on the behavior of deposition on a flat plate surface has been investigated experimentally. Molten wax particles were injected to generate particle deposition in a two-phase flow wind tunnel. Tests indicated that deposition occurs mainly at the leading edge and the middle and backward portions of the windward side. The mass of deposition at the leading edge was far more than that on the windward and lee sides. For the windward and lee sides, deposition mass increased as the mainstream velocity was increased for a given particle concentration. Capture efficiency was found to increase initially until the mainstream velocity reaches a certain value, where it begins to drop with mainstream velocity increasing. For the leading edge, capture efficiency followed a similar trend due to deposition spallation and detachment induced by aerodynamic shear at high velocity. Deposition formation was also strongly affected by the mainstream temperature due to its control of particle phase (solid or liquid). Capture efficiency initially increased with increasing mainstream temperature until a certain threshold temperature (near the wax melting point). Subsequently, it began to decrease, for wax detaches from the model surface when subjected to the aerodynamic force at the surface temperature above the wax melting point.

Keywords: gas turbine engine; particle deposition; capture efficiency; multiphase flow

1. Introduction

Aero-engines would encounter particle laden air flows when operating in environments with a high concentration of airborne particles during extended service. Sand, fly ash, debris and other external particles may flow into the engine combustion, once introduced into the inlet airflow. A majority of particles flow through the combustion along the main channel flow, and subsequently attack the hot component surface through deposition. The temperature is in excess of 2400 K in the primary combustion zone, far above the particle melting temperature (T_{melt}). Particles may flow through the engine without any effect, or impact on the surface by means of rebounding, spreading, spattering or adhering [1–3], possibly experiencing phase transition and deposition afterwards. Deposition is a complex physical and chemical process. Deposition on the turbine blade would dramatically increase the surface roughness and block film cooling holes in certain conditions, which could lead to a loss in aerodynamic and cooling efficiency [4,5]. For land-based gas turbines, trace amounts of foreign matter in fuels and carbon microparticles generated from the burning of raw energy sources can be injected into the turbine along with the main flow and subsequently deposited on the blade surface, which will affect its heat transfer characteristics and aerodynamic performance. In addition, a growing number of impurities are found in the urban environment, indicating more particles flow into the combustion along the turbine cascade passage once ingested into the engine. This could shorten part life and increase the risk of operation failure [6,7].

To date, the deposition of sand and coal ash on gas turbine engines has been studied by multiple researchers. Kim et al. [8] conducted an experimental study of volcanic ash particle deposition. They found that blockage of film cooling holes could cause damage to vanes at the normal inlet temperature. Koenig et al. [9] explored the negative effect of gas-particle two phase flow on gas turbine blade surfaces by experimental investigation. They found that if particle-laden gas flowed over the blade surface at high temperatures, molten particles would be deposited in the proximity of the blade cooling holes, causing a degradation in turbine blade cooling performance.

However, deposition study on a turbine is neither cost nor time efficient. To improve efficiency and enable cost savings, Jensen et al. [10] designed the Turbine Accelerated Deposition Facility (TADF) to generate deposition in a 4 h test that could simulate 10,000 h of turbine operation. They conducted validation tests to simulate the ingestion of foreign particles typically found in an urban environment by seeding a combustion with large concentrations of airborne particles. Crosby et al. [11] studied the effects of particle size, inlet temperature, and impingement cooling on deposition. They noted that deposition growth increased as particle mass mean diameter, inlet temperature, and target coupon temperature increased, based on a series of 4-h deposition tests. Furthermore, deposition thickness became more uniform at a lower test model surface temperature for a given inlet temperature. Wammack et al. [12] studied the behaviors of deposition on a flat plate with three kinds of surface treatments. They observed extensive spallation of the thermal barrier coated coupons caused by successive deposition. Ai et al. [13] utilized the TADF to generate particle deposition on a test model with film cooling holes. Hot gas flow impinged on the film-cooled target surface at 45 deg. The coupon surface temperature was measured by a red, green, bule (RGB) camera, allowing measurement of spatial temperature distribution to evaluate the influence of deposition thickness on surface heat transfer and film cooling performance. Bonilla et al. [14] studied how particle size affected ash deposition on nozzle guide vanes (NGVs) from a CFM56-5B aero engine. Lundgreen et al. [15] constructed a new turbine cascade to study the deposition on an actual blade surface at turbine mainstream temperatures of 1350 °C, 1265 °C, and 1090 °C. They found that deposition mass on the pressure surface increased as the mainstream temperature increased. The largest amount of deposition occurred on the leading edge. Whitaker et al. [16] noted that the mass of deposition on the blade surface was dependent on particle size distribution, indicating that microparticles with larger sizes were easier to deposit on the surface. They concluded that a reduction of inflow turbulence intensity could slow down the deposition buildup and extend the service life of the turbine. Laycock et al. [17] investigated the independent effects of mainstream temperature and surface temperature on fly ash deposition at elevated operating temperatures. They found capture efficiency increased as the mainstream temperature increased. It showed an increase until a certain threshold temperature and then a decrease, with increasing initial deposition surface temperature.

Recently, a considerable effort has been devoted to developing computational models for the prediction of particle deposition behavior. Numerical models of deposition mainly include two types of models, the critical velocity model and the critical viscosity model. As proposed by Brach and Dunn [18] in the former model, a particle will adhere to a surface when the incoming velocity of the particle (critical velocity) is larger than the particle normal impact velocity. Based on the former model, Zhou and Zhang [19] studied the characteristics of particle deposition on the film-cooled wall numerically and the effect of particle size and blowing ratio on deposition efficiency of shaped holes. Bons et al. [20] subsequently improved the critical velocity model and simulated the particle deposition on a turbine vane surface [21]. They found that capture efficiency increased initially with Stokes number (S_t), and then decreased rapidly when $S_t > 1.5$. In the latter model, the particle viscosity at the softening temperature is considered as the reference viscosity and the sticking probability value is dependent on the relationship between particle viscosity and the reference value. When the particle viscosity is less than or equal to the critical value, the sticking probability value is assumed to be 100%, whereas for higher particle viscosity, capture efficiency decreased with increasing particle viscosity. Based on the latter model, Sreedharan and Tafti [22] investigated the deposition of ash particles impinging on a flat of 45° wedge-shape geometry numerically. They found that the majority of the ash deposition occurred near the stagnation region and capture efficiency increased with increase in jet temperature. Yang and Zhu [23] investigated the particle deposition in the vane passage numerically according to the model. They found that the mainstream temperature and capture efficiency were positively related, but deposition distribution was less sensitive to temperature. Considering the coupled effect that deposition has on the flow geometry, Forsyth et al. [24,25] took advantage of the dynamic mesh morphing technique to make an accurate prediction of particle deposition and found that varied surface topology caused by deposition could be the reason for marked change in fluid and particle velocity. With a similar method, Liu et al. [26] modelled the deposition of particles of different sizes numerically. They investigated the independent effect of particle size on the deposition mass and found that the deposition mass increased with the increasing particle size.

Model-size scaling and test condition scaling are required, due to the stringent test conditions and limitations of necessary measurements at the engine-representative temperature. Lawson et al. [27] constructed a low-speed open loop wind tunnel facility to simulate wax deposition on an endwall with film cooling holes. They modeled the sand and coal ash with atomized molten wax droplets. Particle Stokes number and thermal scaling parameter (TSP) were matched based on the similarity laws. Albert et al. [28,29] adopted a similar method to investigate the wax microparticle deposition on a blade leading edge and pressure side in a closed loop wind tunnel at low operating temperatures. The effect of particle deposition on surface heat transfer characteristics has been thoroughly investigated. However, few studies have been performed to shed light on deposition distribution under different inlet operating conditions.

A deeper understanding of the particle deposition mechanism and characteristics would provide a more accurate quantification of deposition effect on the efficiency and performance of engines. The objectives of the study were to investigate wax particle deposition behavior experimentally in an open loop wind tunnel facility and explore the individual effects of mainstream velocity and temperature on spatial distribution of particle deposition on a flat plate.

2. Experimental Facilities and Procedures

2.1. Experimental Facilities and Model

Tests for this study were carried out in an open loop wind tunnel. The wind tunnel inlet mainstream is driven by two axial fans, after which the flow passes through an air heater, turbulence grids, and a wax sprayer before entering the test section (Figure 1). Mainstream flow is warmed in the air heater to simulate gas from engine combustors. Turbulence grids are situated behind the air heater and comprised of a series of vertically-oriented bars. They help to generate turbulence, thus simulating the actual operating conditions for engines. A wax sprayer is positioned at the center of the wind tunnel behind the turbulence grids. Wax particles could be seeded across the span of test section from the sprayer head. Atomized molten wax particles are delivered to the test section by the wind tunnel mainstream flow. They then impact the model surface to simulate the deposition phenomenon that occurs for sand particles in an actual gas turbine flow-path. The surface temperature is measured by an IR camera and calibrated in situ by a thermocouple (error $\leq 1\,°C$). The mass of deposition on the test model is measured by an analytical balance with a high precision degree of 0.005 mg (METTLER TOLEDO XPE206DR).

The test section is 0.3 m in width, 0.3 m in height, and 2.5 m in length (Figure 2). The model could be viewed through glass windows in the test section walls. The mainstream approach velocity could be maintained in a range of 5–40 m/s. The mainstream flow temperature is measured by a thermal resistor (uncertainty: $\pm0.15\,°C$). The thermal resistor sends temperature feedback signals to a temperature controller, which adjusts power output of the air heater based on the input current. With this method, mainstream flow could attain a stabilized temperature during testing. The mainstream temperature could be increased by 70 °C at most. Both operating parameter range and sensor accuracy are summarized and listed in Table 1.

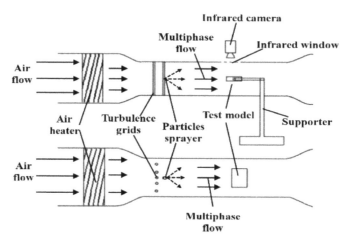

Figure 1. Experimental system schematic.

Figure 2. Photograph of the test section.

Table 1. Operating parameter range and sensor accuracy.

Mainstream Velocity (m/s)	Mainstream Temperature (°C)	Mainstream Velocity Measurement Error (%)	Mainstream Temperature Measurement Precision (°C)	Deposition Mass Measurement Precision (mg)	Surface Temperature Measurement Error (°C)
5–40	25–95	2	±0.15	0.005	±0.5

The wax spray system schematic is shown in Figure 3. The system serves to heat the wax particles above their melting point temperature and spray liquid wax particles into the test section mainstream flow. Wax mass flow rate was in the range of 5–80 g/min with a relative uncertainty of 1.6%. Calibration experiments were carried out based on the relationship between the wax mass flow rate and gear pump speed to acquire calibrated data. A laser particle analyzer is used to measure particle size with a precision degree of 3% (OMEC Instrument CO. DP-02). Wax particle size and flow rate can be controlled by adjusting the air supply pressure and gear pump speed, respectively. The adjustable range for mean particle size is 8–100 μm.

A photograph of the spray system in the wind tunnel is shown in Figure 4. The spray nozzle is located in a hole at the middle of a hollow column (Spray Co. 1/8JJAUCO). The wax heating system is composed of a molten wax reservoir, water-bath vessel, resistive heater and control system. The molten wax reservoir is made up of stainless steel. It was heated uniformly and stably in the water-bath vessel. The control system is used to send temperature feedback signals and minimize fluctuations in the water temperature. The precision uncertainty for the wax temperature is 1 °C. The wax can be heated to 80 °C by the heating system. The purposes of compressed air are to atomize the molten wax and provide the start-stop control of the wax spray system. The wax has a nominal solidification temperature around 41 °C, above the ambient temperature. Consequently, air heating is required to avoid the solidification of wax after its mixing with the atomized air in the spray nozzle. The air temperature can reach 100 °C by using the air heater. All lines are wrapped with electrical heating belts and foaming polyurethane materials as an outer layer to keep the liquid wax above the molten point and achieve thermal insulation and energy saving. A thermocouple is applied to measure the pipeline surface temperature for temperature control purposes. The working conditions of the wax spray system are listed in Table 2.

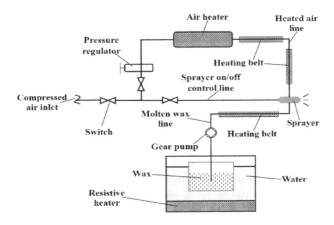

Figure 3. Schematic of the wax spray system.

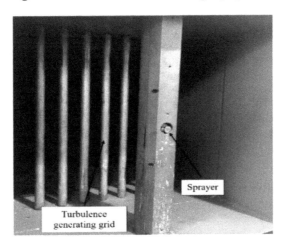

Figure 4. Picture of the wax sprayer and turbulence generating.

Table 2. Wax spray system working conditions.

Particle Size (μm)	Wax Mass Flow Rate (g/min)	Wax Heating Temperature (°C)	Atomized Air Temperature (°C)	Atomized Air Pressure (MPa)
8–100	5–80	30–80	25–100	0.1–0.8

The test model and its fixing holder are shown in Figure 5. The test model is a flat plate, which is 150 mm in length, 150 mm in width, and 13 mm in thickness. The test model is fixed by a holding device. Flow fields around the test model might be affected, in that the holding device is thicker than the test model. To mitigate the possible effect, the holding device adopt streamlined outer surface and three air guide holes. Both the pitch angle and rotating angle of the test model can be adjusted. Schematic of the particle-laden flow and angle of attack (AOA) is shown in Figure 6. The adjustable ranges for AOA, pitch angle and rotating angle are $-15°\sim+15°$, $-20°\sim+20°$, $-90°\sim+90°$, respectively. The height of the test model can be adjusted by a mechanical screw lift.

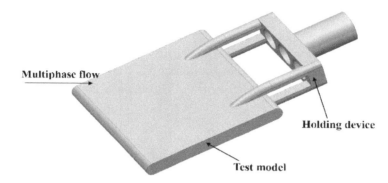

Figure 5. Three-dimensional maps of the test model and its fixing holder.

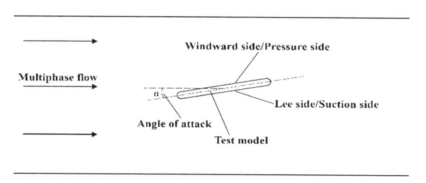

Figure 6. Schematic of multiphase flow and angle of attack.

2.2. Experimental Parameters

The experimental parameters in this paper are shown in Table 3. The mainstream velocity range was 10–30 m/s and the temperature range was 32–48 °C. Wax with a molten point of 41 °C was selected as the particle for injection to observe its deposition behavior. This enables an experimental simulation of particle deposition in a wind tunnel facility. The wax particle size distribution was measured by a laser particle analyzer, as shown in Figure 7. Its main range was 10–20 μm. According to the study conducted by Dring et al. [30], the trajectory of the particles is dominated by the Stokes number. The Stokes number of particles in air flow can be defined as:

$$St_k = \frac{\rho_p d_p^2 U_p}{18 \mu_\infty L} \tag{1}$$

where ρ_p is the wax particle density, d_p is the wax particle diameter, U_p is the wax particle velocity, L is the characteristic length and μ_∞ is the gas dynamic viscosity. The Stokes number characterizes the behavior of particles suspended in a fluid flow, which provides an indicator of how well the particles follow the mean fluid streamlines. A particle with a $St_k < 1$ follows fluid streamlines closely, indicating a perfect convection-diffusion. On the contrary, a particle with a $St_k > 1$ is determined by the inertia force and it is likely to continue along its initial trajectory.

The density of molten wax used in this paper was approximately 800 kg/m³. The corresponding St_k, in the range of 0.04–0.12, could be estimated based on the length of the test model. The distance between the sprayer and the test model was chosen as about 10 times length of the test model and the corresponding St_k is 0.004–0.012, ensuring that the atomized wax particles could diffuse in mainstream flow as uniformly as possible.

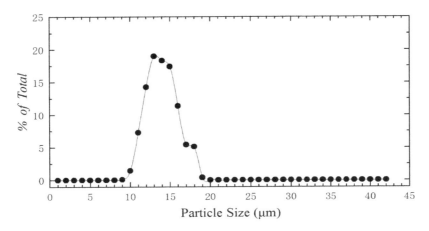

Figure 7. Wax particle size distribution.

Table 3. Test parameters for deposition testing.

Case No.	Mainstream Velocity U_∞ (m/s)	Mainstream Temperature T_∞ (°C)	Test Time t (min)	Wax Volume Concentration c	AOA α (deg)
1	10	40			
2	15	40			
3	20	40			
4	25	40			
5	30	40	5	4.21×10^{-7}	-5
6	10	32			
7	10	37			
8	10	44			
9	10	48			

2.3. Experimental Procedures

The test procedure was as follows: (1) For a typical experiment, the test model should be fixed by the holder at first. The status of all test devices was checked before testing, including the wax spray system, air heater, and AOA of the flat plate. Full test preparations are required. (2) The wax heating system was held at the desired temperature to keep wax in a molten state. The gear pump for starting the wax spray system was closed at the time. (3) The axial fan was opened at the front of the wind tunnel. The air speed was brought to the required value and then held stable. (4) Then the air heater for the wind tunnel was opened and adjusted to a certain temperature so that the test model surface and air can reach a thermal steady state. (5) The gear pump for starting the wax spray system was opened and adjusted according to the required wax mass flow rate in order to achieve different test conditions. The surface temperature of the test model was detected precisely with an infrared radiation thermometer. The data were recorded every second. (6) At the end of an experiment, the wax spray system, air heater for the wind tunnel and axial fan need to be turned off successively. (7) The test model was moved away from the wind tunnel to be photographed. The wax depositions from the entire leading edge, windward side, and lee side were collected to measure the wax deposition mass, respectively. Finally, the remaining wax on the test model was removed in preparation for the next experiment.

2.4. Uncertainty Analysis

An uncertainty analysis serves to validate the measurement technique effectiveness and repeatability of test results. The test Case 1 in Table 3 was performed five times for uncertainty analysis.

148

Handbook of Heat Transfer and Fluid Flow

Deposition mass standard deviation σ and uncertainty $\Delta\overline{m}_{dep}$ were calculated by Equations (2) and (3), respectively.

$$\sigma = \sqrt{\frac{\sum\limits_{i=1}^{n}\left(m_i - \overline{m}_{dep}\right)^2}{n-1}}, \tag{2}$$

$$\Delta\overline{m}_{dep} = \frac{t_\alpha \sigma}{\sqrt{n}}, \tag{3}$$

where m_i is an observed value for deposition mass, $\overline{m}_{dep}$ is the mean deposition mass, n is the sample size, and t_α is the upper quantile of t-distribution. When the confidence interval is 0.95, t_α is 2.766.

The standard deviation σ of the deposition mass was calculated as 0.046 with a sample size of five (Table 4). Uncertainty in the deposition mass was calculated as 0.056 g (relative uncertainty 3.3%). The resulting deposition mass showed sufficient repeatability between experiments to support the discussions and conclusions.

Table 4. Test results of the deposition mass.

i	m_{dep} (g)	$\overline{m}_{dep}$ (g)	σ (g)	$\Delta\overline{m}_{dep}$ (g)	$\frac{\Delta\overline{m}_{dep}}{\overline{m}_{dep}}$ (%)
1	1.71				
2	1.67				
3	1.66	1.70	0.046	0.056	3.3
4	1.77				
5	1.67				

3. Results and Discussion

3.1. The Effect of Mainstream Velocity

Mainstream velocity through the wind tunnel was set to investigate its effect on the particle deposition mass and efficiency. Mainstream temperature and particle concentration were held constant in tests.

Deposition distributions created on the model leading edge, windward side and lee side are presented in Figures 8–10, respectively. The particle deposition reached the maximum at the leading edge and they became sparser on the windward and lee sides. The forward portions of the windward and lee sides were not visible, with very thin and sparse depositions, while the middle and backward portions of both the windward and lee sides were visible with increasing amounts of deposition towards the trailing edge. Significant depositions at the leading edge region were attributed to particle inertia force. When the particle-laden flow approached the stagnation line, change to the motion direction occurred and it was more evident for the air phase. However, particles cannot follow fluid well or flow around the semi-circular surface due to their inertia. They impacted and then deposited on the leading edge surface. Small particles with a low Stokes numbers tended to follow the flow readily along the model surface and they were less prone to deposition under the drag force. This resulted in a much larger amount of deposition at the leading edge than that on the windward and lee sides. Sreedharan and Tafti [22] conducted numerical simulations of ash deposition on a flat plate oriented at 45 deg to an impinging ash laden jet. The obtained results were validated by experiments from Crosby et al. [11]. Sreedharan and Tafti presented that most particle deposition occurred near the stagnant region. Herein, the stagnant region is located at the leading edge of the test model where maximum deposition mass occurred. Liu et al. [26] conducted numerical simulations of particle deposition on a flat plate under operating conditions similar to those of this study. They noted that deposition mass at the leading edge was far larger than that on the windward side. The result is consistent with that obtained in this study. Deposition buildup at the leading edge affected its geometry structure, serving as a shield from the mainstream, which resulted in less deposition on the forward portions of the windward and lee sides. On the other hand, a deposition layer was generated first at the leading edge with a faster deposition rate. The model surface became rough gradually, which increased the deposition at the leading edge. Sparsely distributed spotted particle deposition was observed on the windward and lee sides. Compared with the leading edge, surface roughness changed less for the windward and lee sides, which caused a lower deposition growth rate. Wax deposition was almost symmetric about the stream-wise direction for both sides. It was less at each end than that in the middle portion along the span-wise direction, due to the flow field edge effect.

Deposition behavior at the leading edge changed evidently with the increasing mainstream velocity (Figure 8). At low velocity, particle deposition was less at the stagnant line. Deposition thickness increased as an evident central depression belt extended to the ends of the windward and lee surfaces. Deposition flaking occurred at the boundary lines between the leading edge and windward side or lee side. The deposition had an irregular shape with a slightly serrated and rough border. With velocity increasing, the mass of deposition adjacent to the stagnant line increased fast and the depressed belt was filled up gradually. Also, deposition on the boundary lines became smooth. Deposition flaking in long strips was observed at the same position in Figure 8c,d. The shedding region expanded with the increasing velocity. Inspection of Figure 8e revealed the spallation of deposition that occurred at the maximum velocity, which caused an irregular physical appearance of the deposition and changed the streamlined shape of the leading edge. At low velocity, deposition was sparsely distributed in spots at the windward and lee sides. The deposition pattern became uniform and the deposition range was almost constant with increasing velocity.

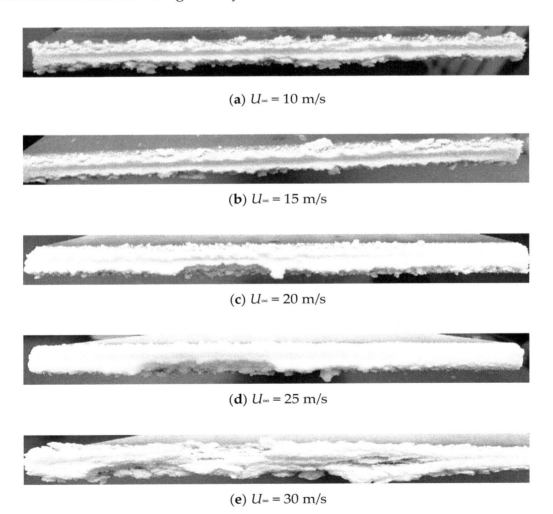

(a) $U_\infty = 10$ m/s

(b) $U_\infty = 15$ m/s

(c) $U_\infty = 20$ m/s

(d) $U_\infty = 25$ m/s

(e) $U_\infty = 30$ m/s

Figure 8. Photographs of wax deposition on leading edge with different mainstream velocities.

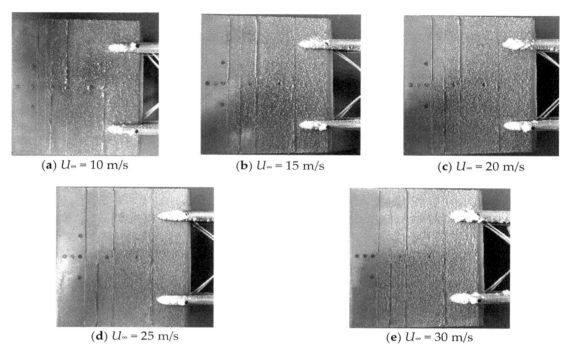

Figure 9. Photographs of wax deposition on windward surface with different mainstream velocities.

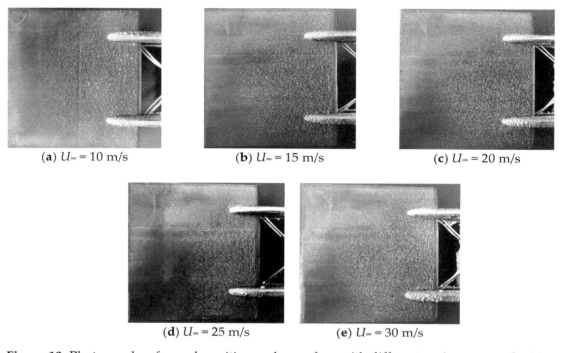

Figure 10. Photographs of wax deposition on lee surface with different mainstream velocities.

The deposition mass at different mainstream velocities is presented in Figure 11. Deposition mass increased gradually with increasing velocity. This was explained by the increased mainstream velocity, which indicated an increased number of impinging particles in unit time for a constant particle concentration. Deposition began to decrease just above a threshold velocity of 25 m/s. This was caused by the deposition spallation occurring at the maximum velocity (Figure 8e). A large number of particles detached from the deposition at the leading edge under the shear force, which resulted in a loss of deposition mass. The windward and lee sides of the test model experienced a decreasing deposition mass difference as the velocity increased.

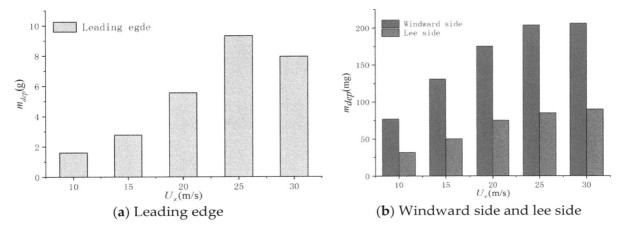

Figure 11. Deposition mass with different mainstream velocities.

Capture efficiencies were measured and calculated in numerous literatures related to effects on deposition to characterize the deposition behavior. The capture efficiencies can be defined as:

$$\eta_{cap} = \frac{m_{dep}}{m_{in}} = \frac{m_{dep}}{q_v \times c \times t}, \tag{4}$$

where m_{dep} is the mass of particle deposition, m_{in} is the total mass of particle injected, q_v is the mainstream volume flow rate, c is the particle volume concentration and t is the test time. Figure 12 shows the effects of the mainstream velocity on deposition capture efficiency. For the leading edge, capture efficiency increased over the first four tests and then decreased as the velocity increased. The dominant force for wax particles to impinge on the test model is inertial force. The increasing capture efficiency could be attributed to increased particle Stokes number, indicating a relative increase in the inertial force. Deposition mass increased with increasing velocity. The increase in deposition mass was related to deposition thickness with a given exposed coupon surface area for 5 min of exposure time. As the deposition thickness increased, the resulting vane contour was no longer in the streamline shape, which enhanced the capture efficiency. Capture efficiency decreased at a velocity of 30 m/s. The main reason for this is that the inlet air mass flow rate increased with increasing mainstream velocity, resulting in increased deposition gravity. The leading edge could have been 'saturated' with wax particles. Meanwhile, the increase in velocity causes the aerodynamic shear force to increase. When the sum of the gravity and aerodynamic force is greater than the bonding force of the deposition, splitting and detachment may occur. For the windward and lee sides, the capture efficiency increases first and then decreases with increasing mainstream velocity. The capture efficiency is at maximum at a mainstream velocity of 20 m/s. Deposition was easier to generate with higher mainstream Reynolds number and particle Stokes number, which resulted in the enhanced capture efficiency. The effect of particle impingement on the wall surface could not be neglected [20]. Due to increased mainstream Reynolds number, particles impacted the test model at the AOA of 5 deg. with more normal impact energy and momentum transfer, and subsequently rebounded more easily. This could lead to the loss in capture efficiency. The change trend of the capture efficiency on the windward and lee sides could be the result of both particle tracing fidelity to the mainstream and rebounding against the wall surface.

The total capture efficiency for a test model was consistent with the capture efficiency for the leading edge (Figure 12c). This could be attributed to the fact that the mass of the deposition at the leading edge occupied a large proportion of the total deposition mass. Variations in mainstream velocity would account for change in the Stokes number. Apart from the Stokes number, particle collision and rebound characteristics combined with the deposition splitting and detachment due to aerodynamic force could account for the change of the capture efficiency. For a flat plate, total capture efficiency was dominated by the mass of deposition at the leading edge. Figure 13 shows the capture efficiency as a function of mainstream velocity and its comparisons with that reported by Taltavull [31] and Bowen [32]. The horizontal coordinate is the particle kinetic energy (E_{kin}), a function of velocity. The expression for E_{kin} is shown in Equation (5):

$$E_{kin} = \frac{1}{2}m_p U_p^2,\tag{5}$$

where m_p is the particle mass, U_p is the particle velocity, which is equal to U_∞. The variation trend of capture efficiencies with E_{kin} is consistent to that reported by Taltavull and Bowen, but the value of capture efficiency is not consistent, mainly due to the different geometric structure. Taltavull et al. and Bowen et al. studied the behavior of deposition on flat plates oriented at 30 deg, 90 deg to an impinging gas flow, respectively. The area of windward sides is bigger than that herein, accounting for the difference in the capture efficiency.

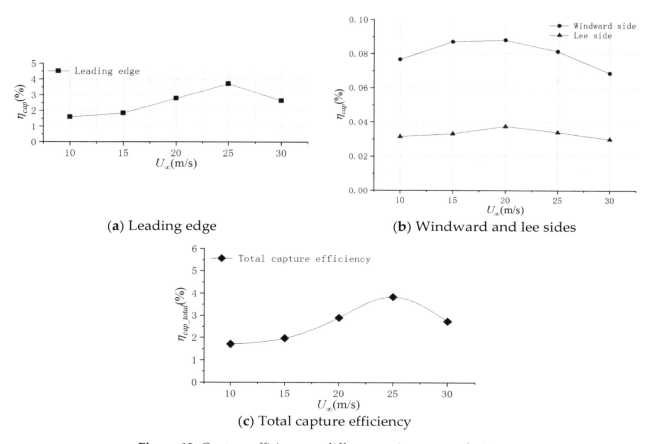

(a) Leading edge   **(b)** Windward and lee sides

(c) Total capture efficiency

Figure 12. Capture efficiency at different mainstream velocities.

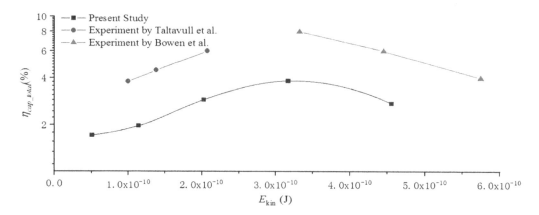

Figure 13. Total capture efficiency in this study compared with data from Taltavull [31] and Bowen [32].

3.2. The Effect of Mainstream Temperature

Five series of tests were set up to investigate the individual effects of mainstream temperature (T_∞) on particle deposition for a given mainstream Reynold number and particle concentration. The tests were run with a mainstream velocity of 10 m/s and particle concentration of 4.21×10^{-7}. The temperature was increased from 32 °C up to a maximum of 48 °C.

Wax deposition distribution at the leading edge, windward and lee sides are presented in Figures 14–16, respectively. The maximum deposition amount was observed at the leading edge, while the amount of deposition was less in the forward portions of the windward and lee sides than that in the middle and backward portions. At the T_∞ of 32 °C, far below the wax melting temperature, deposition near the stagnation line reached maximum and it became gradually less towards the two ends, in a V shaped pattern. The deposition patterns at the two ends were uniform and no flaking was observed. As T_∞ increased, a nonuniform deposition structure occurred, indicating softened particle behavior and the effect of the mainstream flow along the surface. When the test was conducted at the T_∞ of 37 °C, particle deposition was less than that at 32 °C in the vicinity of the stagnation line. An evident depressed belt at the leading edge could be observed from Figure 14b. Particle deposition increased gradually from the middle to the two ends. It was also observed during tests that larger bulks of wax detached from the deposition at the bottom side of the leading edge. At the T_∞ of 40 °C, close to the wax melting temperature, the depressed belt appeared to be more evident. The deposition pattern at both ends was nonuniform, for particles there were blown off from the deposition. This could mean that the deposition stickiness increased with more trapping particles at the leading edge. Deposition decreased significantly as T_∞ increased from 40 °C to 43 °C, above the wax melting temperature. A transparent deposition layer was observed at the leading edge at 43 °C, while no deposition was observed on the surface of the leading edge at 48 °C. When T_∞ was below the wax melting point, sparse individual deposition was observed at 32 °C on the windward and lee sides, whereas more deposition could be discerned at the higher T_∞. However, deposition decreased significantly when the temperature increased to 43 °C. Sparse deposition was observed at the backward portions of the windward and lee sides at the T_∞ of 43 °C, slightly above the melting point. At the T_∞ of 48 °C, a transparent wax deposition layer was observed only at the trailing edge.

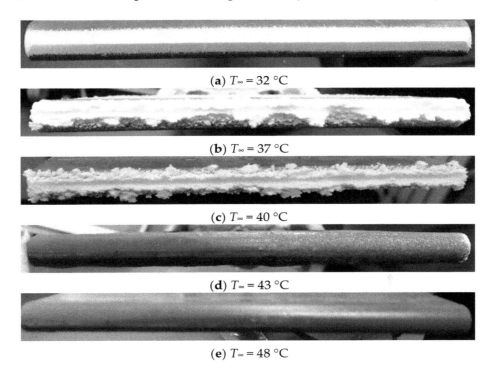

(a) T_∞ = 32 °C

(b) T_∞ = 37 °C

(c) T_∞ = 40 °C

(d) T_∞ = 43 °C

(e) T_∞ = 48 °C

Figure 14. Photographs of wax deposition on leading edge with different mainstream temperature.

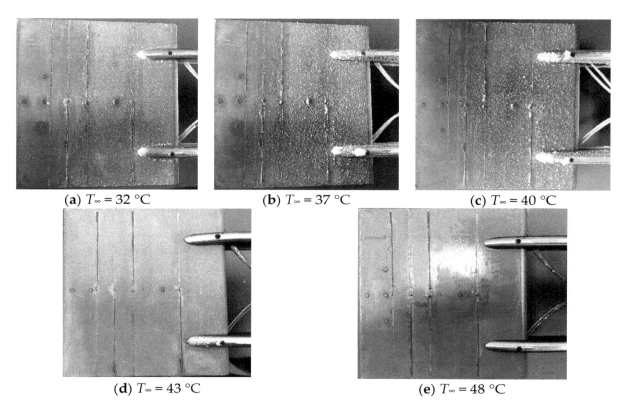

Figure 15. Photographs of wax deposition on windward surface with different mainstream temperatures.

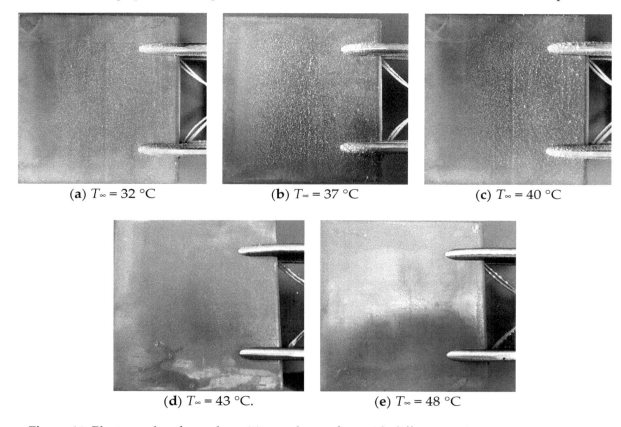

Figure 16. Photographs of wax deposition on lee surface with different mainstream temperatures.

The deposition mass variation with T_∞ was shown in Figure 17. Deposition mass initially increased and then began to decrease at a mainstream temperature of 40 °C for the leading edge, windward sides and lee sides. The threshold temperature was found to be approximately equal to the wax melting point. Once a molten wax particle was injected from the nozzle head at the T_∞ of 32 °C, it was very likely to quickly cool down and solidify just before it impacted the surface. Lower T_∞ increased the probability of a particle rebounding upon impact and not sticking to the surface. As T_∞ approached the wax melting point, particles attached to the test model in a molten, or partially molten state, making them less susceptible to rebounding or detachment. Instead, deposition appeared to spread and solidify. This increased deposition mass was attributed to the larger number of molten particles in the larger temperature mainstream. Liquid wax spatter could occur due to the model surface temperature above the wax melting point. The uncoagulated particles flew along the surface in the form of a liquid film. The aerodynamic stress caused the liquid particles to detach from the surface of the test model, resulting in less deposition.

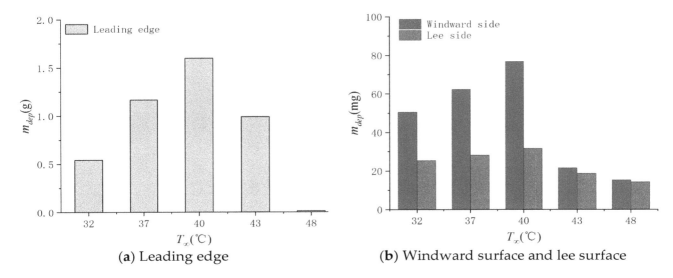

(a) Leading edge   (b) Windward surface and lee surface

Figure 17. Deposition mass with different mainstream temperature.

Capture efficiency at different T_∞ was shown in Figure 18. Capture efficiency increased to a certain threshold temperature and then decreased with increasing T_∞. It reached the maximum when T_∞ was close to the melting point. When T_∞ was below the melting temperature, the trend of capture efficiency in this paper was consistent with that reported by Lundgreen and Crosby [11,15]. They noted that capture efficiency increased with the mainstream temperature increasing. However, for T_∞ above the wax melting temperature, the capture efficiency decreased with T_∞ increasing. This behavior was also observed by Laycock et al. [17] They concluded that capture efficiency increased until a certain temperature, where capture efficiency began to decrease as surface temperature increased. This was attributed to the increased number of particles in molten form as they encountered the flat plate and increased probability of liquid wax spatter at increased T_∞. Since the test model was non-cooled, its surface temperature was highly dependent on T_∞. Increasing T_∞ could elevate the model surface temperature. When the surface temperature of the test model exceeded the wax melting point, liquid wax film would detach from the structure surface under the aerodynamic force and thus it would not solidify upon reaching it. Figure 19 shows total capture efficiency reported by this paper, Crosby and Laycock. The variation trend of the capture efficiency reported in the paper is basically consistent with the data from Crosby and Laycock. The result is more closely matched to the data from Crosby, mainly for Stokes number of wax particles adopted herein is close to that of sand particles adopted by Crosby. For the particles of larger sizes used by Laycock, capture efficiency is higher due to a larger Stokes number.

Temperature increases were found to decrease particle capture efficiency at inlet temperatures above the particle melting point. However, excessively high inlet temperatures may cause damage to the component surface [12,33]. Therefore, it is not reasonable to achieve low capture efficiency and normal engine performance only by increasing gas temperature. Further investigation is still required to decrease the negative effects of deposition on plate surface heat transfer characteristics and aerodynamic performance.

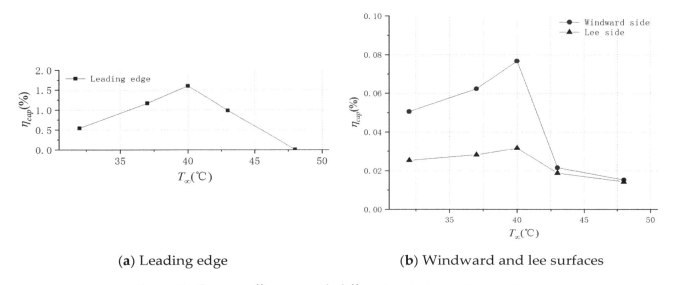

(a) Leading edge   (b) Windward and lee surfaces

Figure 18. Capture efficiency with different mainstream temperature.

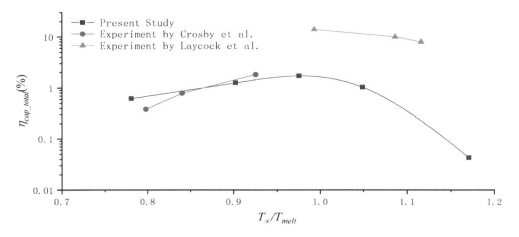

Figure 19. Total capture efficiency herein compared to data from Crosby [11] and Laycock [17].

4. Conclusions

Nine series of tests were conducted in a two-phase flow wind tunnel to investigate the independent effects of mainstream Reynolds number and mainstream temperature on particle deposition mass and distribution. Based on the results presented in this paper, the following conclusions can be drawn:

(1) Maximum wax deposition occurred at the leading edge, which increased levels of surface roughness and affected the vane contour. For the windward and lee sides, deposition was distributed in sparse spots, mainly in the middle and backward portions. No position was observed in the forward portion.

(2) Increased mainstream velocity denoted more particles reaching the surface in unit time for a given particle concentration. For the windward and lee sides, deposition mass increased with increasing mainstream velocity. Capture efficiency increased initially to a certain value and then decreased as mainstream velocity increased, for particles were easier to rebound with more kinetic energy. For the leading edge, capture efficiency increased and then began to decrease at a mainstream velocity of 25 m/s.

(3) T_∞ was adjusted to change the phase of particles impacting the model surface. When T_∞ was below the particle melting point, deposition mass and capture efficiency increased with increase in T_∞

for both the leading edge and windward and lee sides. The mainstream temperature, higher than the particle melting point, would result in lower deposition mass and capture efficiency.

The study focused on particle deposition on a flat plate. Further work should consider the effect of geometric structure on the particle deposition and the effect of particle deposition on surface heat transfer characteristics.

Author Contributions: Investigation, Writing-Original Draft Preparation, F.Z.; Resources, Z.L. (Zhenxia Liu); Supervision, Z.L. (Zhengang Liu); Data Curation, Y.L.

References

1. Dean, J.; Taltavull, C.; Clyne, T. Influence of the composition and viscosity of volcanic ashes on their adhesion within gas turbine aeroengines. *Acta Mater.* **2016**, *109*, 8–16. [CrossRef]
2. Shin, D.; Hamed, A. *Advanced High Temperature Erosion Tunnel for Testing TBC and New Turbine Blade Materials*; ASME Paper 2016, No. 2016-57922; ASME: New York, NY, USA, 2016. [CrossRef]
3. Tabakoff, W. Measurements of particles rebound characteristics on materials used in gas turbines. *J. Propuls. Power* **2015**, *7*, 805–813. [CrossRef]
4. Chambers, J.C. The 1982 Encounter of British Airways 747 with the Mt. Galuggung Eruption Cloud. In Proceedings of the 23rd Aerospace Science Meeting, Reno, NV, USA, 14–17 January 1985.
5. Tabakoff, W. Review—Turbomachinery performance deterioration exposed to solid particulates environment. *J. Fluids Eng.* **1984**, *106*, 125–134. [CrossRef]
6. Dunn, M.G.; Moller, J.C.; Moller, J.E.; Adams, R.M. Performance deterioration of a turbofan and a turbojet engine upon exposure to a dust environment. *J. Eng. Gas Turbines Power* **1987**, *109*, 336–343. [CrossRef]
7. Boulanger, A.; Patel, H.; Hutchinson, J.; DeShong, W.; Xu, W.; Ng, W.; Ekkad, S. Preliminary experimental investigation of initial onset of sand deposition in the turbine section of gas turbines. In Proceedings of the ASME Turbo Expo 2016, Seoul, Korea, 13–17 June 2016. [CrossRef]
8. Kim, J.; Dunn, M.G.; Baran, A.J. Deposition of volcanic materials in the hot sections of two gas turbine engines. *J. Eng. Gas Turbines Power* **1993**, *115*, 641–651. [CrossRef]
9. Koenig, P.; Miller, T.; Rossmann, A. Damage of High Temperature Components by Dust-Laden Air. In Proceedings of the Agard Conference, Rotterdam, The Netherlands, 25–28 April 1994; p. 25.
10. Jensen, J.W.; Squire, S.W.; Bons, J.P.; Fletcher, T.H. Simulated Land-based Turbine Deposits Generated in an Accelerated Deposition Facility. *J. Turbomach.* **2005**, *127*, 462–470. [CrossRef]
11. Crosby, J.M.; Lewis, S.; Bons, J.P.; Ai, W.; Fletcher, T.H. Effects of temperature and particle size on deposition in land based turbines. *J. Eng. Gas Turbines Power* **2008**, *130*, 819–825. [CrossRef]
12. Wammack James, E.; Crosby, J.; Fletcher, D.; Bons, J.P.; Fletcher, T.H. Evolution of surface deposits on a high-pressure turbine blade-Part I: Physical Characteristics. *J. Turbomach.* **2008**, *130*, 021020. [CrossRef]
13. Ai, W.; Murray, N.; Fletcher, T.H.; Harding, S.; Lewis, S.; Bons, J.P. Deposition near film cooling holes on a high pressure turbine vane. *J. Turbomach.* **2012**, *134*, 041013. [CrossRef]
14. Bonilla, C.; Webb, J.; Clum, C.; Casaday, B.; Brewer, E.; Bons, J.P. The effect of particle size and film cooling on nozzle guide vane deposition. *J. Eng. Gas Turbines Power* **2012**, *134*, 101901. [CrossRef]
15. Lundgreen, R.; Sacco, C.; Prenter, R.; Bons, J.P. Temperature Effects on Nozzle Guide Vane Deposition in a New Turbine Cascade Rig. In Proceedings of the ASME Turbo Expo 2016, Seoul, Korea, 13–17 June 2016. [CrossRef]
16. Whitaker, S.M.; Prenter, R.; Bons, J.P. The effect of freestream turbulence on deposition for nozzle guide vanes. *J. Turbomach.* **2015**, *137*, 121001. [CrossRef]
17. Laycock, R.; Fletcher, T.H. Independent effects of surface and gas temperature on coal fly ash deposition in gas turbines at temperatures up to 1400 °C. *J. Eng. Gas Turbines Power* **2016**, *138*, 021402. [CrossRef]
18. Brach, R.; Dunn, P. A mathematical model of the impact and adhesion of microspheres. *Aerosol Sci. Technol.* **1992**, *16*, 1–14. [CrossRef]
19. Zhou, J.H.; Zhang, J.Z. Numerical investigation of particle deposition on converging slot-hole film-cooled wall. *J. Cent. South Univ.* **2017**, *24*, 2819–2828. [CrossRef]
20. Bons, J.P.; Prenter, R.; Whitaker, S. A Simple Physics-Based Model for Particle Rebound and Deposition in Turbomachinery. *J. Turbomach.* **2017**, *139*, 081009. [CrossRef]

21. Prenter, R.; Ameri, A.; Bons, J.P. Computational simulation of deposition in a cooled high-pressure turbine stage with hot streaks. *J. Turbomach.* **2017**, *139*, 091005. [CrossRef]

22. Sreedharan, S.S.; Tafti, D.K. Composition dependent model for the prediction of syngas ash deposition in turbine gas hotpath. *Int. J. Heat Fluid Flow* **2011**, *32*, 201–211. [CrossRef]

23. Yang, X.J.; Zhu, J.X. Numerical simulation of particle deposition process inside turbine cascade. *Acta Aeronaut. Astronaut. Sin.* **2017**, *38*, 120530. [CrossRef]

24. Forsyth, P. High Temperature Particle Deposition with Gas Turbine Applications. Ph.D. Dissertation, University of Oxford, Oxford, UK, 2017.

25. Connolly, J.; Forsyth, P.; McGilvray, M.; Gillespie, D. The Use of Fluid-Solid Cell Transformation to Model Volcanic Ash Deposition within a Gas Turbine Hot Component. In Proceedings of the ASME Turbo Expo 2018, Oslo, Norway, 11–15 June 2018. [CrossRef]

26. Liu, Z.; Zhang, F.; Liu, Z. A Numerical Model for Simulating Liquid Particles Deposition on Surface. In Proceedings of the ASME Turbo Expo 2018, Oslo, Norway, 11–15 June 2018. [CrossRef]

27. Lawson, S.A.; Thole, K.A. Simulations of Multiphase Particle Deposition on Endwall Film-Cooling. *J. Turbomach.* **2012**, *134*, 011003. [CrossRef]

28. Albert, J.E.; Keefe, K.J.; Bogard, D.G. Experimental Simulation of Contaminant Deposition on a Film Cooled Turbine Airfoil Leading Edge. In Proceedings of the ASME Turbo Expo 2009, Lake Buena Vista, FL, USA, 13–19 November 2009. [CrossRef]

29. Albert, J.E.; Bogard, D.G. Measurements of Adiabatic Film and Overall Cooling Effectiveness on a Turbine Vane Pressure Side with a Trench. In Proceedings of the ASME Turbo Expo 2011, Vancouver, BC, Canada, 6–10 June 2011. [CrossRef]

30. Dring, R.P.; Caspar, J.R.; Suo, M. Particle trajectories in turbine cascades. *J. Energy* **1979**, *3*, 161–166. [CrossRef]

31. Taltavull, C.; Dean, J.; Clyne, T.W. Adhesion of volcanic ash particles under controlled conditions and implications for their deposition in gas turbines. *Adv. Eng. Mater.* **2016**, *18*, 803–813. [CrossRef]

32. Bowen, C.P.; Libertowski, N.D.; Mortazavi, M.; Bons, J.P. Modeling Deposition in Turbine Cooling Passages with Temperature-Dependent Adhesion and Mesh Morphing. *J. Eng. Gas Turbines Power* **2019**, *141*, 071010. [CrossRef]

33. Davidson, F.T.; Kistenmacher, D.A.; Bogard, D.G. A Study of Deposition on a Turbine Vane with a Thermal Barrier Coating and Various Film Cooling Geometries. *J. Turbomach.* **2014**, *136*, 1769–1780. [CrossRef]

9

Macro and Meso Characteristics of In-Situ Oil Shale Pyrolysis using Superheated Steam

Lei Wang [1,2], **Dong Yang** [2], **Xiang Li** [2], **Jing Zhao** [1,2], **Guoying Wang** [2] and **Yangsheng Zhao** [1,2,*]

[1]   College of Mining Engineering, Taiyuan University of Technology, Taiyuan 030024, China;
      leiwang0327@163.com (L.W.); zhaojing19860207@163.com (J.Z.)

[2]   Key Laboratory of In-situ Property Improving Mining of Ministry of Education, Taiyuan University of
      Technology, Taiyuan 030024, China; ydscience@hotmail.com (D.Y.); ivanobstinate@163.com (X.L.);
      wangguoyingscience@gamil.com (G.W.)

*   Correspondence: y-s-zhao@263.net

Abstract: The efficiency of oil shale pyrolysis is directly related to the feasibility of in-situ mining technology. Taiyuan University of Technology (China) proposed the technology of in-situ convective heating of oil shale, which uses superheated steam as the heat carrier to heat the oil shale's ore-body and transport the pyrolysis products. Based on the simulated experiments of in-situ oil shale pyrolysis using superheated steam, the changes in fracture characteristics, pyrolysis characteristics and mesoscopic characteristics of the oil shale during the pyrolysis have been systematically studied in this work. The Xinjiang oil shale's pyrolysis temperature ranged within 400–510 °C. When the temperature is 447 °C, the rate of pyrolysis of kerogen is the fastest. During the pyrolysis process, the pressure of superheated steam changes within the range of 0.1–11.1 MPa. With the continuous thermal decomposition, the horizontal stress difference shows a tendency to first increase and then, decrease. The rate of weight loss of oil shale residue at various locations after the pyrolysis is found to be within the range of 0.17–2.31%, which is much lower than the original value of 10.8%, indicating that the pyrolysis is more adequate. Finally, the number of microcracks (<50 μm) in the oil shale after pyrolysis is found to be lie within the range of 25–56 and the average length lies within the range of 53.9636–62.3816 μm. The connectivity of the internal pore groups is satisfactory, while the seepage channel is found to be smooth. These results fully reflect the high efficiency and feasibility of in-situ oil shale pyrolysis using superheated steam.

Keywords: superheated steam; triaxial stress; thermogravimetry; X-ray microtomography; thermal cracking

1. Introduction

As an unconventional oil and gas resource, oil shale is a fine-grained sedimentary rock, which is rich in solid organic matter (kerogen) and has fine bedding [1,2]. Oil shale can generate shale oil through retorting. After shale oil's hydrocracking, refined oil, such as gasoline, kerosene and diesel oil can be obtained, which is of great significance to alleviate the current oil shortage. The reserves of oil shale resources in China are huge and can be converted into 47.6 billion tons of shale oil [3,4].

At present, most countries around the world use in-situ retorting to exploit oil shale [5,6]. In-situ mining of oil shale only needs to pass the heat-injection well to the ore-body and directly heat the ore-body. After the oil shale ore-body is fully pyrolyzed, the organic matter is also pyrolyzed to generate oil and gas, whereas the hydrocarbon is discharged to ground through production well [7]. According to different forms of heating, the in-situ mining of oil shale can be divided into three classifications, namely heat conduction, convection heating and radiation [8,9]. Furthermore, in-situ conversion process [10] uses high temperature of electrode to heat the ore-body, whereas the heater temperature can reach up to 1000 °C. However, the thermal conductivity of the oil shale is extremely poor, and the heating efficiency is low. Therefore, it takes a long time for the oil shale to reach the effective pyrolysis temperature. Han et al. [11] found that it takes about 10 years or more to reach the initial pyrolysis temperature of oil shale in the area of 400 m^2. Kyung et al. [12] simulated the effect of electric heating on the behavior of pyrolysis products and concluded that, after pyrolysis, almost 60% of the shale oil was trapped in mineral matrix because of poor fluidity. Raytheon's radio frequency/critical flow (RF/CF) technology [13] uses (RF) transmitters to heat the oil shale's ore-body, and then, extract oil and gas, which is produced from pyrolysis, using supercritical carbon dioxide. Yang et al. [14] proposed an in-situ oil shale recovery method, which combines microwave heating and hydraulic fracturing, and then, simulated the thermal decomposition of oil shale under microwave irradiation. Compared with the conventional heating method, microwave heating requires shorter time, and has lower energy consumption, higher oil production and quality. Radiation produces strong heat penetration and faster heating, though the technology is not yet mature enough [15,16]. The in-situ fracturing and heating technology using nitrogen injection [17,18] uses high-temperature and high-purity nitrogen to pyrolyze the oil shale's ore-body. Therefore, high-temperature nitrogen can play an important role in the recovery of oil and gas. Allawzi et al. [19,20] found that the solubility of organic matter increased due to the interaction of supercritical carbon dioxide and cosolvent. Zhang et al. [21] studied the changes in shale oil composition and yield after bioleaching the oil shale, and found that, after bioleaching, the yield of shale oil increased by 15.38%, whereas the contents of high molecular weight and low molecular weight hydrocarbons in shale oil also increased.

It can be said that, regardless of the mining technology used, the most important thing is to find a way to maximize the efficiency of pyrolysis, whereas the pyrolysis temperature and the development of pores and fractures directly determine the efficiency of pyrolysis. At present, studies have focused on describing the evolution of pore construction in the pyrolysis of oil shale [22–26]. Geng et al. [27] have systematically analyzed the evolution of pores and the structure of fractures in oil shale under high temperature and high pressure using a combination of X-ray microtomography (μCT) and mercury intrusion porosimetry. It is considered that 300–500 °C is the stage, where the porosity and the number and aperture of fractures increase significantly. Bai et al. [28–30] studied the evolution characteristics of pore structure during the pyrolysis of Huadian oil shale at the temperature of 100–800 °C and found that the permeability of oil shale significantly increases within the temperature range of 350–450 °C. Kang et al. [9,31] calculated the percolation probability of true three dimensional (3D) digital CT cores of oil shale specimens under different temperatures. The results showed that, when the porosity is higher than 12%, the connectivity of pore-connected clusters is very good and the connection of seepage channels is smooth, which is favorable for oil and gas production and high temperature fluid injection. Saif et al. [32–34] studied the evolution of pores and fractures during pyrolysis of Green River oil shale and found that the critical temperature for a sharp increase in the porosity of oil shale lies within the range of 390–400 °C. After the critical temperature, the porosity of oil shale rapidly increased to 22–25%. Liu et al. [35] analyzed the evolution of pore structure of Fushun oil shale under pressure and temperature conditions and found that the lithostatic pressure would significantly inhibit the development of pores. Pan et al. [36,37] reported that the mineral matters have an insignificant effect on the pyrolysis reactions of kerogen in Jimsar oil shale. Barshefsky et al. [38] reported that the isolated kerogen of Russian oil shale was completely pyrolyzed at 420 °C, while the raw oil shale decomposition was only 65% complete.

In short, many experts and scholars have done a lot of research on the relationship between temperature and pyrolysis characteristics; however, there is little research on the study of pyrolysis characteristics of oil shale under stress constraints. In 2005, the in-situ convection heating of oil shale was put forward by Zhao Yangsheng's team at the Taiyuan University of Technology, China [39]. The technology used superheated steam as the heat carrier to heat the oil shale ore-body, while the produced oil and gas were transported using steam. During the in-situ mining of oil shale using superheated steam, the internal pores and fractures of oil shale are not only a channel for the migration of steam and kerogen pyrolysis products but also a location for heat exchange and transfer in the rock mass, which is directly related to the efficiency of pyrolysis. In this work, based on the simulated experiments of in-situ oil shale pyrolysis using superheated steam, the cracking and pyrolysis characteristics of oil shale samples are thoroughly studied during the pyrolysis and after the pyrolysis, the evolutionary characteristics of pores and fractures inside the oil shale are carefully discussed. The study provides a necessary prerequisite for the application and commercialization of in-situ oil shale mining technology using superheated steam.

2. Experimental

2.1. Thermogravimetric Experiments

The experimental sample was taken from Jimsar County, Xinjiang, China. The oil shale was crushed and sieved to a particle size of ≤180 μm for pyrolysis experiments. Table 1 summarizes the results of oil shale industrial analysis and low-temperature carbonization.

Table 1. Proximate and Fischer assay analyses of the Xinjiang oil shale.

Analysis	Composition
Proximate analysis (wt %, ad)	
Moisture	0.56
Ash	77.89
Volatile matter	17.78
Fixed carbon	3.77
Fischer assay analysis (wt %, ad)	
Oil yield	9.08
Water yield	1.50
residue	86.48
Gas + loss	2.94

The pyrolysis weight loss experiment of oil shale was conducted using DTU-2B thermogravimetric analyzer. The device has a temperature measuring accuracy of 0.1 °C and a sensitivity of less than 1 μg. The ground oil shale samples were evenly spread in the crucible. The cooling water was turned on and high-purity nitrogen was slowly passed into the crucible. The temperature was increased from 70 °C to 900 °C at the rate of 3.5 °C/min. The thermogravimetric (TG) and differential thermogravimetric (DTG) curves of the oil shale were obtained using the thermogravimetric experiments. The TG curve reflects the change in sample's mass with temperature, while the DTG curve reflects the relationship between the rate of change of sample's mass and the temperature.

2.2. Simulated Experiments for In-Situ Oil Shale Pyrolysis Using Superheated Steam

For the simulated experiments of in-situ oil shale pyrolysis using superheated steam, the process of preparing the samples is shown in Figure 1. Large oil shale samples were cast through concrete, making the fabricated specimen a cube with the dimensions of 300 mm × 300 mm × 300 mm. After the sample was fully dried, a well-shaped diversion trough was ground on the surface of the specimen using the polisher, which facilitates the outflow of oil shale pyrolysis products. Meanwhile, core drilling was carried out in the middle of the specimen. The diameter of the drill hole was 32 mm, whereas the depth was 200 mm. The drill hole was used as the location for the insertion of heat injection tube. The heat injection tube was mainly composed of a lower-end flower tube and an upper-end sleeve tube. The flower tube was used as the channel of oil shale pyrolysis using the superheated steam and the sleeve tube played the role of sealing and insulation.

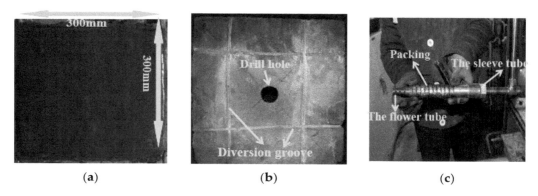

(a)   (b)   (c)

Figure 1. The process of specimen preparation: (**a**) Casting large the oil shale sample with concrete; (**b**) Drill hole and diversion groove of the processed sample; (**c**) Structure of the heat injection tube.

The vertical stress of 3 MPa and the horizontal stress of 4 MPa (Figure 2) were applied to the specimen using a large-size true triaxial press (Figure 3) to simulate the geo-stress environment where the oil shale was located. The press was mainly composed of test loading frame, axial and lateral hydraulic cylinder loading system, numerically controlled hydraulic instrument and other auxiliary devices. The superheated steam, generated by the steam generator, pyrolyzed the specimen under the condition of stress constraint using the heat injection tube. The numerically controlled hydraulic instrument can monitor the stress characteristics of the specimen in real time during the pyrolysis process.

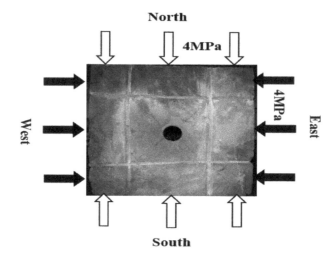

Figure 2. Schematic of the applied horizontal stress.

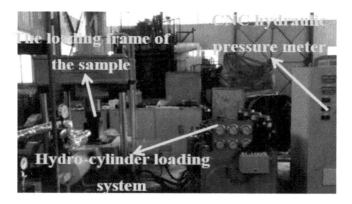

Figure 3. Large-size true three-axes press.

2.3. Micro-CT Scan and the Analysis of Pyrolysis Effect

According to the difference of vertical distance between the outlet of heat injection tube and the pyrolysis oil shale, the oil shale residue was divided into three locations. The thermogravimetric analyses of the oil shale residues at different locations were carried out to analyze the pyrolysis of oil shale. The oil shale residues at different locations were processed into cylindrical samples having the length and diameter of 10 mm and 5 mm, respectively. The internal structure of the oil shale residue was scanned using a µCT225kVFCB high-precision CT analysis system (Figure 4) and the mesoscopic characteristics of the oil shale after pyrolysis were obtained. In this experiment, the scanning voltage was 90 kV and the electric current was 70 µA. The scans were obtained in with 400 frames, the superimposed frame rate was 2 fps and a plane image was generated after the reconstruction. There were 1500 scanned layers and the size of the scanning cell was 2.66 µm × 2.66 µm.

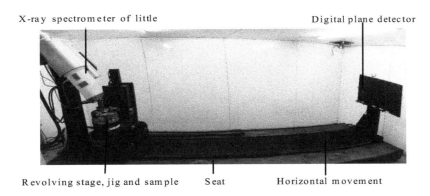

Figure 4. X-ray microtomography experimental system.

3. Results and Discussion

3.1. Thermogravimetric Analysis of Xinjiang Oil Shale

Figure 5 shows the TG and DTG curves of Xinjiang oil shale obtained using thermogravimetric analysis. As can be seen from Figure 5, the major stage of weight loss of Xinjiang oil shale occurs in a relatively small temperature range of 400–510 °C, while the rate of weight loss reaches 10.8%. During this stage, the pyrolysis of oil shale kerogen is closer to completion and the rate of pyrolysis is faster. The pyrolysis of oil shale is considered to be a process in which oil shale is pyrolyzed into oil, gas and semi-coking products in two steps. Firstly, the oil shale was pyrolyzed into tar, which then, was further pyrolyzed to obtain the final products. When the temperature is 447 °C, the rate of mass loss of oil shale is the highest and the pyrolysis rate of kerogen is the fastest.

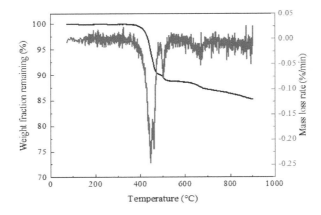

Figure 5. Thermogravimetric (TG) and differential thermogravimetric (DTG) curves of Xinjiang oil shale.

3.2. Stress Characteristics of the Specimen during Pyrolysis

In the process of in-situ oil shale pyrolysis using superheated steam, the weak cementing surface inside the oil shale will break under the action of high-temperature and high-pressure steam, due to which, the heat exchange area inside the rock mass increases. The superheated steam will heat the rock mass along the fracture surface. After the decomposition of organic matter, more pores and fractures are formed inside the oil shale and the hydrocarbon generated by the pyrolysis will further widen the pores and fractures in the migration process, which forms a huge seepage channel. At the same time, due to the continuous development of internal fractures inside oil shale and the continuous injection of superheated steam, the cohesive force of the molecular bonds of oil shale is reduced, which reduces the tensile strength of oil shale and makes the oil shale more prone to tensile fracture. The variation in steam pressure with pyrolysis time during the pyrolysis process is shown in Figure 6.

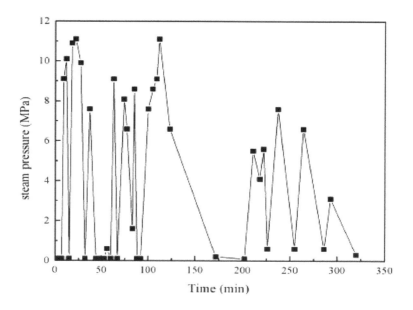

Figure 6. Variation in steam pressure with pyrolysis time.

It can be seen from Figure 6 that the superheated steam pressure varies between 0.1–11.1 MPa as the oil shale pyrolysis proceeds. This is due to the reason that the thermal cracking of oil shale is a process of gradual expansion. With the continuous injection of superheated steam, the stress at the tip of the fracture gradually increases. When the stress value reaches the threshold point of crack initiation, the fracture expands. The expansion process is also the process of energy release, which shows the decrease in stress. Once the stress is lower than the threshold point of the crack, the fracture stops expanding. Meanwhile, the stress concentration is produced again in the tip of the fracture and the fracture continues to expand. Therefore, the fracture expansion process inside the oil shale is characterized by the continuous cycle of stress concentration—fracture expansion—stress reduction and stress re-centralization.

The critical state of tensile fracture of oil shale is:

$$p - \sigma_v \geq T_0 \tag{1}$$

where T_0 is the tensile strength of oil shale (*MPa*), p is the superheated steam pressure (MPa) and σ_v is the vertical stress (MPa).

When the tensile fracture of the specimen occurs, the fracture direction is perpendicular to the vertical principal stress direction, thereby forming multiple horizontal fractures around the heat injection tube, as shown in Figure 7.

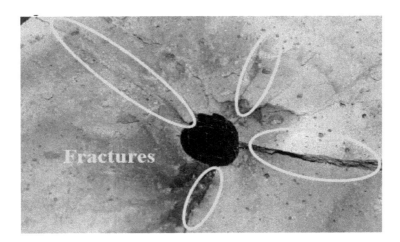

Figure 7. Characteristics of the oil shale fracture after pyrolysis.

The anisotropy of the oil shale is obvious. In the process of pyrolysis, the thermal expansion coefficient of particles in different positions inside the oil shale is different, which leads to the change in stress state in the horizontal direction.

Figure 8 shows the variation in horizontal stress difference with time during the pyrolysis process.

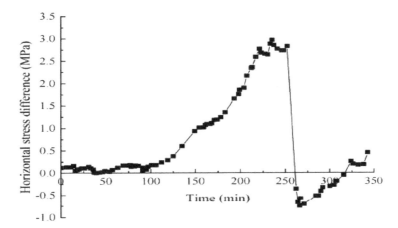

Figure 8. Variation in horizontal stress difference with time in pyrolysis process.

In Figure 8, the horizontal stress difference is the difference between the horizontal stress in the north-south direction and the horizontal stress in the east-west direction. When the pyrolysis time is within the range of 0–252 min, the horizontal stress difference increases with the increase in pyrolysis time and the maximum horizontal stress difference is found to be 2.95 MPa. During this period of time, the north-south direction is the direction of the maximum principal stress and the fracture will expand perpendicular to the north-south direction. Additionally, the degree of cracking will become more obvious during this time period. This is due to the reason that, in the experiment, the direction of oil shale bedding is perpendicular to the north-south direction, whereas the bedding plane undergoes tensile brittle fractures, which appear to be open perpendicular to the north-south direction, resulting in an increase in the stress in the north-south direction. When the pyrolysis time changes from 256 min to 316 min, the maximum horizontal stress difference is 0.75 MPa. During this time period, the rock mass between the oil shale bedding surface in in-situ condition undergoes shear failure under the action of high-temperature and high-pressure steam, which results in larger horizontal stress in the east-west direction than that in the north-south direction. Under these conditions, the east-west direction is the direction of the maximum principal stress. Overall, the extent of shear failure of rock mass between the bedding planes is lower than that of the brittle fractures of the bedding plane. From the macroscopic point of view, a large number of cracks formed by thermal fracturing and pyrolysis of oil shale will cause the injected heat-carrying fluid to seep into the ore-body from the injection well, which continues to pyrolyze the oil shale and continuously extract oil and gas. Therefore, the results of this study provide a scientific basis and technical support for the implementation of in-situ retorting of oil shale.

3.3. Microscopic Characteristics of the Specimen after Pyrolysis

The pyrolyzed sample is shown in Figure 9a. The oil shale around the heat injection tube has been broken into many small pieces, indicating that the oil shale has been fully pyrolyzed and its color has changed from yellow brown to black. Figure 9b shows the sampling position of the oil shale after pyrolysis. The distance between Location A and the outlet of the heat injection tube is the smallest, followed by Location B, whereas the distance from Location C to the outlet of the heat injection tube is the largest.

(a)   (b)

Figure 9. Pyrolized sample: (**a**) Morphology of the oil shale after pyrolysis; (**b**) Sampling locations.

The TG curves of oil shale at different locations after pyrolysis are shown in Figure 10. At the temperature of 510 °C, the rates of weight loss of oil shale at Locations A, B and C are 0.17%, 0.72% and 2.31%, respectively, while that of the original oil shale is 10.8%, indicating that the oil shale at each location has been fully pyrolyzed. At the same time, the pyrolysis effect decreases with the increase in distance from the outlet of the heat injection tube. This is because the process of oil shale pyrolysis using superheated steam is an energy consuming process. Farther from the outlet of the heat injection tube, lower is the temperature and worse is the pyrolysis effect.

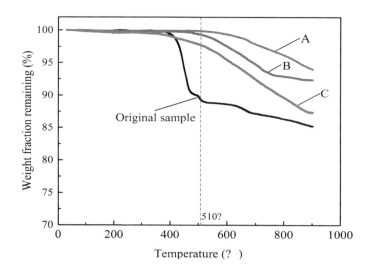

Figure 10. TG curves of oil shale after pyrolysis.

Xue et al. [40,41] used thermogravimetry to analyze the oil shale residue, pyrolyzed using low-temperature dry distillation technology and found that the rate of weight loss of oil shale residue was 6.52%.

The thermogravimetric results of oil shale at different locations after pyrolysis show that the rates of weight loss of oil shale at Locations A, B and C are 0.63%, 2.49% and 4.61%, respectively, which are lower than that of the oil shale residue after low-temperature dry distillation. Therefore, it can be said that the method of oil shale pyrolysis using superheated steam can achieve a high degree of pyrolysis of organic matter in oil shale. In order to further study the pyrolysis properties of oil shale and its residue, the Coats-Redfern method [42] was used to analyze the pyrolysis kinetics of oil shale and its residue. The expression of the Coats-Redfern method is given by Equation (2).

$$In\left[-\frac{In(1-\alpha)}{T^2}\right] = In\frac{AR}{\beta E} - \frac{E}{R} \cdot \frac{1}{T} \tag{2}$$

The equation represents a straight line with as $In\frac{AR}{\beta E}$. the intercept and $-\frac{E}{R}$ as the slope. The activation energy E can be obtained by fitting the least square method. The activation energy of oil shale and its residue in the main weight loss stage (400–510 °C) is presented in Table 2.

Table 2. Analysis of the activation energy of oil shale and its residue.

	Activation Energy (kJ/mol)	
Temperature Range	**400 °C–450 °C**	**450 °C–510 °C**
Original sample	24.804	25.396
Sample A	10.582	6.294
Sample B	9.532	4.151
Sample C	5.935	7.177

CT scanning technology uses the principle that X-rays have different penetration capabilities for different density materials and therefore, the density is reflected in voxel of different gray levels. In the grayscale image of CT, greater the brightness, higher is the density of the material. As the density of the pores and fractures is the lowest, it appears black in the CT image [43,44]. Figure 11 shows a micro-CT reconstructed image of the internal structure of the 500th and 1000th layers of the cross-sections of oil shale at Locations A, B and C.

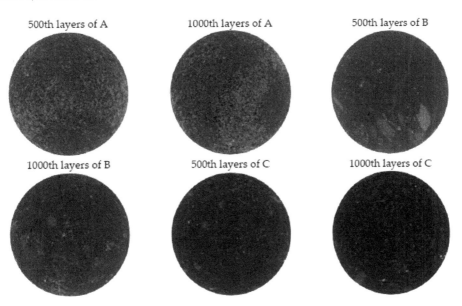

Figure 11. CT-scan grayscale imaging of oil shale after pyrolysis.

In Figure 11, the scattered white areas are the undecomposed minerals during pyrolysis. There are less pores and fractures inside the oil shale. This is because the pores and fractures, formed by the pyrolysis of oil shale under triaxial stress constraint, are constrained in the process of expansion and the fracture surface may be closed. Due to these reasons, the thermal cracking of oil shale under in-situ condition is the result of the combined effect of thermal stress of superheated steam and triaxial stress. In order to visually obtain the distribution of pores and fractures in oil shale after pyrolysis, the CT grayscale images of Figure 11 need to be subjected to "image segmentation" (binarization processing). To conduct segmentation (separate the image into pore and solid phases), 288 the maximum entropy method proposed by Kapur et al [45] was adopted. The Kapur et al [46] 289 method quantitatively considers the gray values of all pixels of an image, and assigns a unique 290 threshold to each image (Figure 12). Because there are a lot of noises in the binary image, the cracks less than 3 voxel are cleared by MATLAB software. Figure 13 shows a micro-CT image of a cross-section of the sample processed using binarization, where the white areas represent the pores and fractures (the length greater than 7.98 μm) and the black areas represent the oil shale matrix.

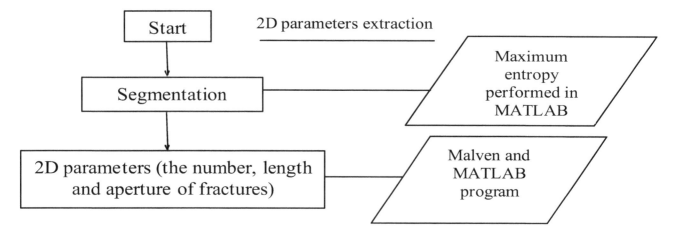

Figure 12. The workflow for processing images from CT scan.

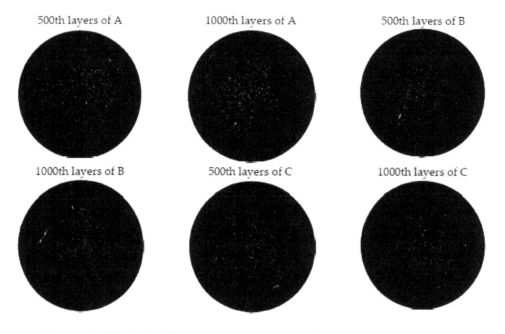

Figure 13. Oil shale CT-scan imaging after the binarization processing.

In order to quantitatively evaluate the development characteristics of pores and fractures of the oil shale after pyrolysis, the number, average length and average aperture of fractures with the length greater than 50 µm are determined. The statistical results are presented in Table 3.

Table 3. Parameters for the fracture of oil shale after pyrolysis.

Parameters of Fractures / Sample Number		Number	Average Length (µm)	Average Aperture (µm)
A	500th layers	48	54.4334	24.6902
	1000th layers	64	53.2752	24.4329
B	500th layers	30	59.4005	21.7064
	1000th layers	28	65.3627	27.0685
C	500th layers	25	57.5110	23.8858
	1000th layers	25	50.4162	23.0557

As can be seen from Table 3, the number of fractures in the 500th layer of oil shale at Location C is the lowest (only 25), while the number of fractures of oil shale at Location A is the highest (up to 56). The average length of fractures in oil shale at different locations lies within the range of 53.9636–62.3816 µm. Among them, the average length of fractures in the 1000th layer of oil shale at Location C is the smallest, whereas that at Location B is the largest. The average aperture of fractures in the 500th layer of oil shale at Location C is only 23.4708 µm and is the minimum, whereas that in the 1000th layer of oil shale at Location A is 24.5616 µm and is the largest. The fracture parameters of the 500th and 1000th layers of oil shale at different locations are averaged to obtain the variation characteristics of oil shale fractures, as shown in Figure 14.

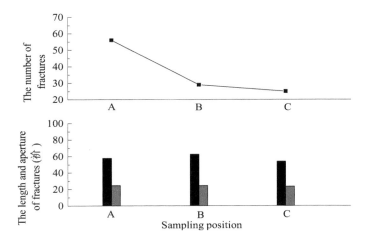

Figure 14. Variation characteristics of fracture parameters in oil shale.

In general, as the distance from the outlet of heat injection tube increases, the number of fractures continuously decreases, however the average length and the aperture of the fractures do not change significantly. This is due to the reason that, greater the distance from the outlet of the heat injection tube, lower is the temperature of superheated steam, and worse is the effect of thermal cracking of oil shale, which result in fewer number of fractures. The oil and gas generated by the pyrolysis of oil shale will both extend and expand the formed fractures in the process of migration, thus forming a more developed fracture channel. Furthermore, there will be little change in the length and aperture of fractures inside the oil shale at different locations. Generally speaking, oil shale is an anisotropic and heterogeneous rock, and the effect of thermal expansion of internal particles is different under the action of temperature. During the in-situ pyrolysis, many micron-scale cracks can be formed in oil shale. In this type of cracks, fluids, such as shale oil and shale gas, can migrate freely and obey the law of hydrostatic mechanics.

After CT scanning of the samples, a series of two-dimensional grayscale images are obtained, which can reflect the density distribution of different layers inside the samples. The 700th to 900th layers are selected and imported into the AVIZO 9.0 software. The threshold segmentation of these grayscale images is done through appropriate thresholds and a binary image characterizing fractures inside the oil shale is obtained [47]. Then, all the binary images obtained in the foregoing operation are successively stacked in the vertical direction, so that the reconstruction of the three-dimensional fracture structure can be achieved. In the process of three-dimensional reconstruction, a three-dimensional digital model of 200 × 200 × 200 voxel is obtained. In order to fully reflect the connectivity and distribution of fractures inside the oil shale, the computing load of the computer in the three-dimensional reconstruction is considered. Figure 15 shows a three-dimensional image of the distribution of fractures inside the oil shale at Locations A, B and C (the color of the fracture space is blue, while that of the matrix is gray), whereas the image is an 8-bit undefined grayscale image. The color range is 0–255.

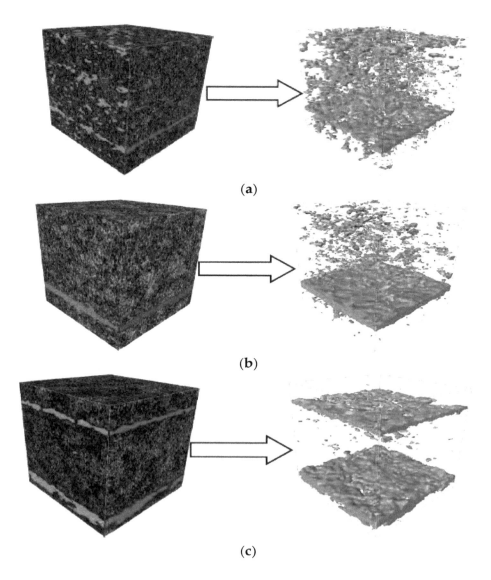

(a)

(b)

(c)

Figure 15. Distribution characteristics of internal holes and fractures in oil shale: (**a**) pore and fracture groups of oil shale at location A; (**b**). pore and fracture groups of oil shale at location B; (**c**) pore and fracture groups of oil shale at location C.

From Figure 15, it can be seen that the distribution of pore and fracture groups inside the oil shale is denser at Location A and the number of pores and fractures is large, which form a large permeate channel, which connects the two relative surfaces. The distribution of oil shale pore and fracture groups at Locations B and C is more scattered in the three-dimensional space and its size and number of oil shale fractures are smaller than that of oil shale at Location A. However, the seepage channel is still formed between the two relative surfaces, which is beneficial to the extraction of oil and gas. In general, the pyrolysis reaction of kerogen occurs obviously and a large number of pores and cracks are formed in the solid skeleton of oil shale, which constitute the entrance and exit channels of pyrolysis fluids and products in the pyrolysis process of oil shale. After pyrolysis using superheated steam, the oil shale can be regarded as a porous medium with high permeability. The results have shown that the in-situ oil shale pyrolysis technology using superheated steam is an efficient and feasible method for oil and gas production from shale oil.

4. Conclusions

The process of oil shale pyrolysis using superheated steam is a multi-field coupling process. The pyrolysis process involves the crack initiation, the decomposition of organic matter and the migration of the products. During the pyrolysis, the temperature distribution inside the oil shale will have serious inhomogeneity, which leads to the different pyrolysis effects of oil shale at different positions. On the basis of simulated experiments of in-situ oil shale pyrolysis using superheated steam, the variations in steam pressure and constrained stress during the pyrolysis are obtained. At the same time, the pyrolysis effect of the oil shale and the evolution of pores and fractures are studied after pyrolysis, which provide a certain level of guidance for the technological design of in-situ oil shale pyrolysis using superheated steam.

With the continuous development of pyrolysis, two forms of rupture occur inside the oil shale. These are the brittle fractures of the bedding plane and the shear failure of the rock mass between the bedding planes. After the pyrolysis, the rate of weight loss of oil shale residue was much lower than that of the original sample, indicating that the pyrolysis of the oil shale was more complete. After the pyrolysis, the pores and fractures inside the oil shale are widely distributed, whereas the oil shale can be regarded as a porous medium with high permeability. The feasibility of in-situ mining of oil shale using superheated steam is verified based upon an efficient pyrolysis process.

Author Contributions: All authors contributed to the research in the paper. Y.Z., D.Y. and L.W. conceived and designed the experiments; X.L. and G.W. performed the experiments; J.Z. analyzed the data; L.W. wrote the paper.

Acknowledgments: This work was supported by the National Natural Science Foundation of China (grant NOs. 11772213, U1261102), the National Youth Science Foundation of China (grant NOs. 51704206).

References

1. Dyni, J.R. Geology and resources of some world oil-shale deposits. *Oil Shale* **2003**, *20*, 193–252.
2. Yu, X.; Luo, Z.; Li, H.; Gan, D. Effect of vibration on the separation efficiency of oil shale in a compound dry separator. *Fuel* **2018**, *214*, 242–253. [CrossRef]
3. Hao, Y.; Gao, X.Q.; Xiong, F.S.; Zhang, J.L.; Li, Y.J. Temperature distribution simulation and optimization design of electric heater for in-situ oil shale heating. *Oil Shale* **2014**, *31*, 105–120. [CrossRef]
4. Wu, Y.; Lin, C.; Ren, L.; Yan, W.; An, S.; Chen, B. Reconstruction of 3D porous media using multiple-point statistics based on a 3D training image. *J. Nat. Gas Sci. Eng.* **2018**, *51*, 129–140. [CrossRef]
5. Rabbani, A.; Baychev, T.G.; Ayatollahi, S.; Jivkov, A.P. Evolution of Pore-Scale Morphology of Oil Shale during Pyrolysis: A Quantitative Analysis. *Transp. Porous Media* **2017**, *4*, 1–20. [CrossRef]
6. Eseme, E.; Urai, J.L.; Krooss, B.M.; Littke, R. Review of mechanical properties of oil shales: implications for exploitation and basin modelling. *Oil Shale* **2007**, *24*, 159–174.
7. Crawford, P.M.; Killen, J.C. New Challenges and Directions in Oil Shale Development Technologies. *Oil Shale A Solut. Liq. Fuel Dilemma* **2010**, *1032*, 21–60.

8.   Lin, L.; Lai, D.; Guo, E.; Zhang, C.; Xu, G. Oil shale pyrolysis in indirectly heated fixed bed with metallic plates of heating enhancement. *Fuel* **2016**, *163*, 48–55. [CrossRef]

9.   Kang, Z.; Yang, D.; Zhao, Y.; Hu, Y. Thermal cracking and corresponding permeability of Fushun oil shale. *Oil Shale* **2011**, *28*, 273–283. [CrossRef]

10.   Rangel-German, E.R.; Schembre, J.; Sandberg, C.; Kovscek, A.R. Electrical-heating-assisted recovery for heavy oil. *J. Petroleum Sci. Eng.* **2004**, *45*, 213–231. [CrossRef]

11.   Han, H.; Zhong, N.N.; Huang, C.X.; Liu, Y.; Luo, Q.Y.; Dai, N.; Huang, X.Y. Numerical simulation of in-situ conversion of continental oil shale in Northeast China. *Oil Shale* **2016**, *33*, 45–57.

12.   Lee, K.J.; Moridis, G.J.; Ehlig-Economides, C.A. Compositional simulation of hydrocarbon recovery from oil shale reservoirs with diverse initial saturations of fluid phases by various thermal processes. *Energy Explor. Exploit.* **2017**, *35*, 172–193. [CrossRef]

13.   Anonymous. Raytheon Technology Shows Promise in Extracting Oil from Shale. *Microw. J.* **2006**, *49*, 40.

14.   Yang, Z.; Zhu, J.; Li, X.; Luo, D.; Qi, S.; Jia, M. Experimental Investigation of the Transformation of Oil Shale with Fracturing Fluids under Microwave Heating in the Presence of Nanoparticles. *Energy Fuel* **2017**, *31*, 10348–10357. [CrossRef]

15.   Gerasimov, G.; Khaskhachikh, V.; Potapov, O. Experimental study of kukersite oil shale pyrolysis by solid heat carrier. *Fuel Process. Technol.* **2017**, *158*, 123–129. [CrossRef]

16.   Chen, C.; Gao, S.; Sun, Y.; Guo, W.; Li, Q. Research on Underground Dynamic Fluid Pressure Balance in the Process of Oil Shale In-Situ Fracturing-Nitrogen Injection Exploitation. *J. Energy Resour. Technol.* **2017**, *139*. [CrossRef]

17.   Yang, Y.; Liu, S.C.; Li, Q.; Sun, Y.H. Research on experiment of in-situ pyrolysisof oil shaleand production analysis. *J. Chem. Pharm. Res.* **2013**, *5*, 763–767.

18.   Yang, Y.; Sun, Y.H.; Li, Q. Experiment and simulation of oil shale pyrolysis. *Int. J. Earth Sci. Eng.* **2013**, *6*, 1311–1317.

19.   Allawzi, M.; Al-Otoom, A.; Allaboun, H.; Ajlouni, A.; Al Nseirat, F. CO_2 supercritical fluid extraction of jordanian oil shale utilizing different co-solvents. *Fuel Process. Technol.* **2011**, *92*, 2016–2023. [CrossRef]

20.   Lu, J.; Hawthorne, S.; Sorensen, J.; Pekot, L.; Kurz, B.; Smith, S. Advancing CO 2 enhanced oil recovery and storage in unconventional oil play-Experimental studies on Bakken shales. *Appl. Energy* **2017**, *208*, 171–183.

21.   Zhang, X.Q.; Li, Y.S. Changes in shale oil composition and yield after bioleaching by bacillus mucilaginosus and thiobacillus ferrooxidans. *Oil Shale* **2017**, *34*, 146. [CrossRef]

22.   Tong, J.; Han, X.; Wang, S.; Jiang, X. Evaluation of Structural Characteristics of Huadian Oil Shale Kerogen Using Direct Techniques (Solid-State13C NMR, XPS, FT-IR, and XRD). *Energy Fuel* **2011**, *25*, 4006–4013. [CrossRef]

23.   Mao, J.; Fang, X.; Lan, Y.; Schimmelmann, A.; Mastalerz, M.; Xu, L.; Schmidt-Rohr, K. Chemical and nanometer-scale structure of kerogen and its change during thermal maturation investigated by advanced solid-state 13C NMR spectroscopy. *Geochim. Cosmochim. Acta* **2010**, *74*, 2110–2127. [CrossRef]

24.   Sun, Y.; He, L.; Kang, S.; Guo, W.; Li, Q.; Deng, S. Pore Evolution of Oil Shale during Sub-Critical Water Extraction. *Energies* **2018**, *11*, 842. [CrossRef]

25.   Modica, C.J.; Lapierre, S.G. Estimation of kerogen porosity in source rocks as a function of thermal transformation: Example from the Mowry Shale in the Powder River Basin of Wyoming. *AAPG Bull.* **2012**, *96*, 87–108. [CrossRef]

26.   Chen, J.; Xiao, X. Evolution of nanoporosity in organic-rich shales during thermal maturation. *Fuel* **2014**, *129*, 173–181. [CrossRef]

27.   Geng, Y.; Liang, W.; Liu, J.; Cao, M.; Kang, Z. Evolution of pore and fracture structure of oil shale under high temperature and high pressure. *Energy Fuel* **2017**, *31*, 10404–10413. [CrossRef]

28.   Bai, F.; Sun, Y.; Liu, Y.; Li, Q.; Guo, M. Thermal and kinetic characteristics of pyrolysis and combustion of three oil shales. *Energy Convers. Manag.* **2015**, *97*, 374–381. [CrossRef]

29.   Sun, Y.; Bai, F.; Liu, B.; Liu, Y.; Guo, M.; Guo, W. Characterization of the oil shale products derived via topochemical reaction method. *Fuel* **2014**, *115*, 338–346. [CrossRef]

30.   Bai, F.; Sun, Y.; Liu, Y.; Guo, M. Evaluation of the porous structure of huadian oil shale during pyrolysis using multiple approaches. *Fuel* **2017**, *187*, 1–8. [CrossRef]

31.   Kang, Z.; Zhao, J.; Yang, D.; Zhao, Y.; Hu, Y. Study of the evolution of micron-scale pore structure in oil shale at different temperatures. *Oil Shale* **2017**, *34*, 42–45. [CrossRef]

32. Saif, T.; Lin, Q.; Bijeljic, B.; Blunt, M.J. Microstructural imaging and characterization of oil shale before and after pyrolysis. *Fuel* **2017**, *197*, 562–574. [CrossRef]

33. Saif, T.; Lin, Q.; Singh, K.; Bijeljic, B.; Blunt, M.J. Dynamic imaging of oil shale pyrolysis using synchrotron x-ray microtomography. *Geophys. Res. Lett.* **2016**, *43*, 6799–6807. [CrossRef]

34. Saif, T.; Lin, Q.; Butcher, A.R.; Bijeljic, B.; Blunt, M.J. Multi-scale multi-dimensional microstructure imaging of oil shale pyrolysis using X-ray micro-tomography, automated ultra-high resolution SEM, MAPS Mineralogy and FIB-SEM. *Appl. Energy* **2017**, *202*, 628–647. [CrossRef]

35. Liu, Z.; Yang, D.; Hu, Y.; Zhang, J.; Shao, J.; Song, S. Influence of In Situ Pyrolysis on the Evolution of Pore Structure of Oil Shale. *Energies* **2018**, *11*, 755. [CrossRef]

36. Pan, L.; Dai, F.; Huang, J.; Liu, S.; Li, G. Study of the effect of mineral matters on the thermal decomposition of Jimsar oil shale using TG-MS. *Thermochim. Acta* **2016**, *627*, 31–38. [CrossRef]

37. Pan, L.; Dai, F.; Huang, J.; Liu, S.; Zhang, F. Investigation of the gas flowdistribution and pressure drop in Xinjiang oil shale retort. *Oil Shale* **2015**, *32*, 172–185. [CrossRef]

38. Aboulkas, A.; Harfi, K.E.; Bouadili, A.E. Thermal degradation behaviors of polyethylene and polypropylene. Part I: Pyrolysis kinetics and mechanisms. *Energy Convers. Manag.* **2010**, *51*, 1363–1369. [CrossRef]

39. Zhao, Y.S.; Feng, Z.C.; Yang, D. the Method for Mining Oil & Gas from Oil Shale by Convection Heating. China Invent Patent CN200,510,012,473, 20 April 2005.

40. Liu, Y.; Xue, X.; He, Y. Solvent Swelling Behavior of Fushun Pyrolysis and Demineralized Oil Shale. *J. Comput. Theor. Nanosci.* **2011**, *4*, 1838–1841. [CrossRef]

41. Li, Y.; Feng, Z.; Xue, X.; He, Y.; Qiao, G. Ecological utilization of oil shale by preparing silica and alumina. *J. Chem. Ind. Eng.* **2008**, *59*, 1051–1057.

42. Coats, A.W.; Redfern, J.P. Kinetic parameters from thermogravimetric data. *Nature* **1964**, *201*, 68–69.

43. Moine, E.C.; Groune, K.; Hamidi, A.E.; Khachani, M.; Halim, M.; Arsalane, S. Multistep process kinetics of the non-isothermal pyrolysis of Moroccan Rif oil shale. *Energy* **2016**, *115*, 931–941. [CrossRef]

44. Zhang, Y.; Lebedev, M.; Al-Yaseri, A.; Yu, H.; Xu, X.; Iglauer, S. Characterization of nanoscale rockmechanical properties and microstructures of a Chinese sub-bituminous coal. *J. Nat. Gas Sci. Eng.* **2018**, *52*, 106–116. [CrossRef]

45. Luo, L.F.; Lin, H.; Li, S.C.; Lin, H.; Flu-hler, H.; Otten, W. Quantification of 3-d soil macropore networks in different soil types and land uses using computed tomo- graphy. *J. Hydrol.* **2010**, *393*, 53–64. [CrossRef]

46. Kapur, J.N.; Sahoo, P.K.; Wong, A.K.C. A new method for gray-level picture thresholding using the entropy of the histogram. *Comput. Vis. Graph. Image Process* **1985**, *29*, 273–285. [CrossRef]

47. Ni, X.; Miao, J.; Lv, R.; Lin, X. Quantitative 3D spatial characterization and flow simulation of coal macropores based on μCT technology. *Fuel* **2017**, *200*, 199–207. [CrossRef]

10

Resonant Pulsing Frequency Effect for Much Smaller Bubble Formation with Fluidic Oscillation

Pratik Devang Desai [1,2] , **Michael John Hines** [2], **Yassir Riaz** [1] and **William B. Zimmerman** [1,*]

[1]   Department of Chemical and Biological Engineering, University of Sheffield, Mappin Street, Sheffield S1 3JD, UK; pratik@perlemax.com (P.D.D.); y.riaz@shef.ac.uk (Y.R.)

[2]   Perlemax Ltd., Kroto Innovation Centre, 318 Broad Ln, Sheffield S3 7HQ, UK; michael@perlemax.com

*   Correspondence: w.zimmerman@sheffield.ac.uk

Abstract: Microbubbles have several applications in gas-liquid contacting operations. Conventional production of microbubbles is energetically unfavourable since surface energy required to generate the bubbles is inversely proportional to the size of the bubble generated. Fluidic oscillators have demonstrated a size decrease for a system with high throughput and low energetics but the achievable bubble size is limited due to coalescence. The hypothesis of this paper is that this limitation can be overcome by modifying bubble formation dynamics mediated by oscillatory flow. Frequency and amplitude are two easily controlled factors in oscillatory flow. The bubble can be formed at the displacement phase of the frequency cycle if the amplitude is sufficient to detach the bubble. If the frequency is too low, the conventional steady flow detachment mechanism occurs instead; if the frequency is too high, the bubbles coalesce. Our hypothesis proposes the existence of a resonant mode or 'sweet-spot' condition, via frequency modulation and increase in amplitude, to reduce coalescence and produce smallest bubble size with no additional energy input. This condition is identified for an exemplar system showing relative size changes, and a bubble size reduction from 650 μm for steady flow, to 120 μm for oscillatory-flow, and 60 μm for resonant condition (volume average) and 250 μm for steady-flow, 15 μm for oscillatory-flow, 7 μm for the resonant condition. A 10-fold reduction in bubble size with minimal increase in associated energetics results in a substantial reduction in energy requirements for all processes involving gas-liquid operations. The reduction in the energetic footprint of this method has widespread ramifications in all gas-liquid contacting operations including but not limited to wastewater aeration, desalination, flotation separation operations, and other operations.

Keywords: microbubbles; fluidics; flow oscillation; oscillators; energetics

1. Introduction

Gas-liquid contacting operations are arguably among the most important processing operations. The oil we produce, the air, the food, the drinks (fizzy drinks, beer, and fermented beverages), chemical dyeing processes, mixing operations, wastewater aeration (WWA). and remediation, and several operations require good gas-liquid contacting [1–4]. One way to achieve such contacting is by increasing the surface area of the system corresponding to its volume. This results in an increase in surface area with respect to volume, and for a bubble, results in a slower rise velocity and a substantial increase in contact time. The major problem with smaller bubble generation is the energy required due to the large surface energy involved in generating these smaller bubbles.

Two examples where the bubble size in terms of number and volume contribution matters are dissolved/dispersed air flotation (DAF) and wastewater aeration (WWA). Both these operations are important, mandatory for any wastewater remediation, and highly energy intensive. DAF uses 70–100 μm size bubbles, generated by involving high pressure nozzles (14–15 bar (g)) [5], in order to separate systems out for remediation or further processing. Variants of this method include froth flotation such as for mineral processing—copper ore processing, effluent removal from gas processing/petrochemical plants, paper mills, and drinking water plants. The second is WWA which accounts for nearly 0.25–0.4% of the UK's total energy consumption [6]. This is due to the barriers to implementing high energy microbubbles and the inability to produce these in an energy efficient manner.

If smaller bubbles could be generated with less energy, it could lead to a fundamental change in gas-liquid contacting equipment design. Of particular importance to the interplay between fluid flow and heat transfer are applications of microbubbles in phase change separations, particularly vaporisation and distillation [7–10]. Experiments and models [7,10] demonstrate the layer height of liquid is extremely important, with fixed microbubble size, for determining the degree of non-equilibrium separation and the overall rate of vaporization. The highest vaporization and the greatest enrichment for 100 μm diameter microbubbles occur with a contact time of ~1 ms. Since bubble rise rate at terminal velocity is well established to depend on bubble size, hitting the optimum vaporisation rates and enrichment (joint heat and mass transfer at the microbubble interface) strongly depend on tuning the microbubble size. The purpose of this paper is to explore how microbubble size depends on the fluid dynamics of fluidic oscillator induced microbubble generation.

There are several methods to generate microbubbles such as ablative technology, ultrasound, microfluidic devices, nozzles, and other techniques, but each faces problems either due to scalability or energy input [11–17]. Microbubbles have been divided into several size classes and depend on the application [18]. In this paper, microbubbles are gas-liquid interfaces (bubbles) ranging from 1 μm to 1000 μm in size. Several applications have been identified for them including microalgal separation [19], wastewater clean-up [20,21], theranostics [22–24], algal growth [25–28], oil emulsion separation [29] and for heat and mass transfer applications (due to the vastly increased surface area to volume ratios) [30–32]. A major advantage is gained if a different bubble generation regime can be formulated such that it does not specifically depend on the conventional form of detachment.

The fluidic oscillator is a fluidic device that creates hybrid synthetic jets which help engender microbubbles in an economical fashion via pulsatile flow through the aerator. The adherence of the jet to the wall, due to the Coandă effect, and its subsequent detachment to the other leg due to a switch over created by a geometric cusp [33–35] generates the pulsatile flow. The actual mechanism of bubble formation via fluidic oscillation resulting in the back flow into the membrane due to the net positive displacement is responsible for the smaller bubble size and has been demonstrated by Tesař [35]. Zimmerman et al. [9,30,35] have shown the efficacy of the fluidic oscillator with respect to bubble size reduction via aeration.

Although introduction of the negative feedback can be achieved using both the Warren and the Spyropoulos configuration for the fluidic oscillator [36], for the experiments herein, to facilitate the use of discrete frequencies, a Spyropoulos type feedback loop has been used. This type of negative feedback has several advantages. Firstly, frictional losses are kept to a minimum and secondly, the frequency of oscillation can be easily controlled by changing the negative feedback configuration i.e., with the use of different feedback loop lengths and volumes.

The Spyropoulos loop has been adequately described in Tesař et al. [37], which introduces a negative feedback to the system and in physical form is a single loop connecting the control terminals of the fluidic oscillator shown in Figure 1 as X_1 and X_2. S (supply nozzle), X_1 and X_2 (control terminals) and Y_1 and Y_2 (outlets). The incoming jet enters via the supply nozzle, is amplified at the throat via a constriction of appropriate size. Control terminals aid in switching the flow due to a pressure differential formed in order to switch on outputs Y_1 and Y_2 at relevant frequencies. Y_1 and Y_2 can be connected to microporous diffusers placed in a liquid stream which result in bubble formation when oscillatory gas exits Y_1 and Y_2 and into the microporous spargers. Typical widths for X_1 and X_2 are 2–4 mm. Fluidic oscillation is a nozzle free bubble generation method which also allows generation in the laminar flow mode. This is contrasted with conventional nozzles used for microbubbles in dissolved air flotation microbubble generation or droplet generation [38].

An added advantage of using a single feedback loop over the two required for the Warren type oscillator is that there is a reduction in the degrees of freedom, which results in a simpler system to control and quantitatively understand it.

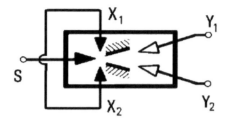

Figure 1. The Spyropoulos loop configuration for the fluidic oscillator, reproduced with permission [37]. S is the supply nozzle where gas enters, X_1 and X_2 are the control terminals where gas switches, Y_1 and Y_2 are the outlets where microporous spargers can be connected with oscillatory gas output. These spargers are placed in liquid and generate bubbles.

The usual method for bubble generation using fluidic oscillation is using two outlets connected to a set of bubble generating microporous membranes (spargers/aerators) placed under liquid of interest and gas entering via supply nozzle S. The entire flow can be utilised for the generation for bubbles and this results in an unvented condition for the fluidic oscillator.

Venting this jet, increases the momentum of the pulse post the fluidic oscillator as the oscillator is a flow amplifier. The flow switches from each leg, and if the momentum of the jet and therefore the pulse strength associated with the jet can be increased whilst maintaining the appropriate flow into the aerator.

Bubble formation conventionally requires the bubble to detach when the forces acting on it are balanced. According to Zimmerman et al., the proposed mechanism for bubble formation mediated by the fluidic oscillator typically takes place in the pulse cycle of the frequency switch. Therefore, technically this should lead to a bubble formed at every pulse and the throughput determined by the frequency of the oscillator. However, this does not take place as seen in the previous papers [9,30,35]. Each pulse of air is based on the frequency of oscillation. This means that higher the frequency, shorter the oscillatory pulse, and therefore in theory should lead to a smaller bubble. This has not been observed and led to one of the hypotheses proposed in this paper. The first part of the hypothesis is that the amplitude of the pulse must be high enough for detachment to occur. Post the bubble detachment, the bubbles may coalesce if the frequency is too high as they are close to each other. The frequency and amplitude of the fluidic oscillator are the two control parameters capable of producing the bubbles at the required frequencies. To increase the amplitude of the flow, higher flow rates will be used and then vented such that the actual flow into the aerator is as desired but the amplitude of the flow has increased. Different feedback conditions will be used in order to see the variations of the amplitude. Higher feedback conditions will have a higher amplitude whilst lower feedback conditions will introduce a higher friction loss and therefore lower resultant amplitude of the flow.

The second part of the hypothesis is that there is a resonant condition for the system which depends on a specific frequency which determines the bubble size. Even when the amplitude condition is met and bubbles are generated at each pulse, just increasing the frequency will not result in a smaller bubble size. At a specific frequency, there will be the presence of the resonant mode condition, where the bubbles are detached, but not too quickly so as to coalesce, and not too slowly so as to resemble a conventional steady flow bubble formation. The paper aims to explore this new regime of bubble formation and check if the hypothesis is supported with experimental evidence.

The flow has to be vented in order to generate higher amplitude of the jet and only partially diverted into the aerators used for bubble generation. This helps in two ways—it increases the effect of the oscillatory flow (by virtue of an increase in amplitude by increasing momentum of the wave) in order to observe the difference between discrete oscillatory flow conditions (frequencies) for the lab scale and minimises the bubble coalescence due to the increased momentum imparted to the newly engendered microbubble. This fluidic oscillator mediated oscillation has an amplitude and frequency dependence on the inlet flow rate to the oscillator provided that all other conditions are maintained. This proposed addition to the experiment enhances the fluidic oscillator mediated microbubble size reduction.

Several industrial applications can afford to have gas wastage for a significantly larger bubble throughput concomitant with reduced bubble size. Since air, in particular, is not that expensive with respect to the decreased energetics, this is justified by the increase in reaction surface area and the lack of maintenance requirements with the no-moving part fluidic oscillation.

Figure 2 shows the vented schematic of the fluidic oscillator with V_1 and V_2 acting as the vents to the system. Additional venting ensures that the appropriately controlled flow can pass through the aerator whilst maintaining the appropriate flow into oscillator in order for it to actuate the oscillation. Additionally, this increases the momentum of the jet and the amplitude of the oscillatory flow.

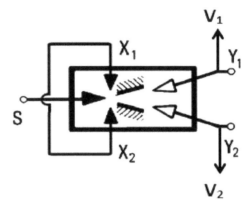

Figure 2. Vented Fluidic Oscillator (adapted from [37]) S is the supply nozzle where gas enters, X_1 and X_2 are the control terminals where gas switches , Y_1 and Y_2 are the outlets where microporous spargers can be connected with oscillatory gas output. These spargers are placed in liquid and generate bubbles, V_1 and V_2 (vents).

Conventional bubble formation depends on a host of other factors such as surface energy of the bubble engendering surface and the liquid, liquid and gas viscosity, momentum of the gas, height of liquid, pressure exerted by the system and acting upon it, and size of the bubble-engendering orifice. Fluidic devices such as the Tesař-Zimmerman fluidic oscillator generate a net positive hybrid synthetic oscillatory jet that results in a specific reduction in bubble size compared to conventional or steady flow as discussed previously.

2. Methods and Materials

2.1. Fluidic Oscillator

The fluidic oscillator is oscillated at 86 L per min (lpm) and 92 lpm (corrected for pressure and temperature) with most of the flow being vented and a frequency sweep is performed in order to support the hypothesis posited earlier. As discussed previously, vents have been introduced in order to increase the momentum of jet pulse and amplitude of oscillation whilst controlling appropriate flow into the aerator (MBD 75, Point Four Systems, Coquitlam, BC, Canada). Rotameters have been utilised to act as metered valves. The frequency of oscillation and the amplitude of the oscillation are measured using an Impress G-1000 pressure transducer (Impress Sensors and Systems, Ltd., Berkshire, UK) controlled and recorded using characterisation software developed in LabView (National Instruments, Austin, TX, USA). The pressure drop across the fluidic oscillator is 100 mbar. The total pressure drop depends on the combined pressure drop across the fluidic oscillator and aerator.

2.2. The Aerator

The proprietary Point Four Systems MBD 75 aerator produces a cloud of fine bubbles approximately 500 μm in size under steady flow. MBD 75 has ultrafine ceramic pores and a flat surface, thereby retarding bubble coalescence as compared to other types of material such as sintered glass or steel membranes. Ceramic, being inert, hydrophilic and robust, is a preferred surface for bubble generation in water.

2.3. System Set up

The system has been set up according to the schematic shown in Figure 3.

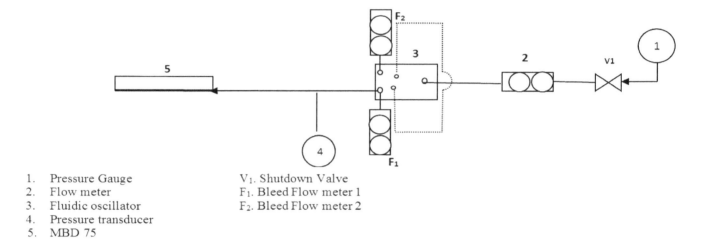

1. Pressure Gauge
2. Flow meter
3. Fluidic oscillator
4. Pressure transducer
5. MBD 75

V_1. Shutdown Valve
F_1. Bleed Flow meter 1
F_2. Bleed Flow meter 2

Figure 3. Schematic of the setup.

2.4. Pneumatic Set Up

Figure 3 shows the system schematic described here. The pressure regulator (Norgren, Littleton, CO, USA) controls the systemic pressure which is set at 2 bar(g)—the required pressure for the particular aerator to bubble. A different aerator would require lesser pressure. The flow controller (FTI Instruments, Sussex, UK) has been corrected for the pressure being used in the system. The air enters the system from the compressor via the pressure regulator and the flow controller regulates the global flow entering the fluidic oscillator. The fluidic oscillator is connected to two vent rotameters (F_1 and F_2) to act as metered valves and these are set up in order to vent appropriately and send the appropriate amount of flow into the aerator. The aerator is placed in a tank wherein the bubble size is measured using acoustic bubble spectrometry.

The frequency and amplitude of pulse from fluidic oscillator is measured simultaneously using a combination of pressure transducers and Fast Fourier Transform (FFT) code developed in LabView (*cf.* frequency measurement and FFT).

The aerator is kept in the centre with the hydrophones set around it as shown in Figure 4 with the set operational conditions. Bubble sizing is performed continuously and the distilled deionized (DDI) water in the tank is replaced after each reading.

The frequency is changed whilst all other conditions are kept constant and different feedback configurations are used coupled with an additional flow rate. This is an exemplar system and we are just demonstrating the ability for the system to change with different systems and the aim of this paper is to show that relative changes are possible for the system without any additional energy input. Several configurations of the Spyropoulos type fluidic oscillator being used are also able to engender the same frequency which helps observe the effect of the frequency.

2.5. Bubble Sizing Using Acoustic Bubble Spectrometry

The hydrophones are placed over at a height 5 cm above the diffuser and equidistant at 15 cm. These data were repeated 7 times and performed at 2 different flow rates—86 lpm and 92 lpm (standard conditions—293.15 K and 101,325 Pa). Twenty two frequencies were used in the study performing a frequency sweep and 3 different configurations of feedback for fluidic oscillator. Bubble sizing was performed using Acoustic Bubble Spectrometry commercially available from Dynaflow Inc.™ (Jessup, MD, USA). This has been found to be an effective method for visualising cloud bubble dynamics. The Acoustic Bubble Spectrometer (ABS) with 4 pairs of hydrophones—50, 150, 250, and 500 kHz were used in this study, with the capability to collate a size distribution from 3 μm to 600 μm in size (radius). The ABS is then set up along with an octaphonic set up and Figure 4 schematically represents the flow diagramme of the bubble visualisation set up.

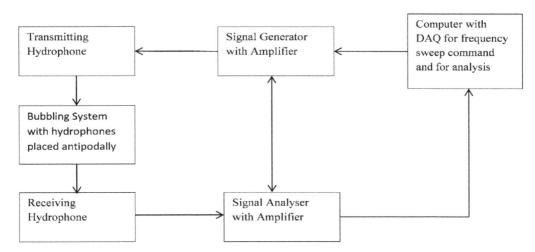

Figure 4. Information flow diagram of the Acoustic Bubble Spectrometer (ABS) and the hydrophone set up.

Two equiresponsive hydrophones are placed antipodal to each other with the aerator bubbling in the centre i.e., bubble cloud in the centre. The data acquisition module and computer (controlled using software) send out a frequency sweep via a signal generator with amplifier into the transmitting hydrophone. The signal passes through the bubble cloud (calibrated under no bubble condition) and into the receiving hydrophone which is demodulated using an amplifier too. This is used in conjunction with the no bubble condition in order to generate a bubble size distribution.

Acoustic bubble sizing relies on the principle of bubble resonance upon frequency insonation and the resonant bubble approximation. Upon insonation by a specific frequency, a bubble starts to oscillate and this frequency is specific to the bubble by a sixth power to the radius, i.e., [39]:

$$f \propto r^6 \qquad (1)$$

Hydrophone pairs are used, with one being a transmitter and the other acting as the receiver, and each hydrophone pair has a specific resonant frequency at which it works best and a range of frequencies that it can operate reasonably. When a hydrophone insonates a bubble cloud with a specific frequency, the bubble corresponding to that size resonates and therefore oscillates, thereby attenuating the signal due to the pressure change caused by the oscillating bubble as compared to a clear/bubble-free solution. This attenuation is then measured by the receiving hydrophone and the signal is inverted in order to garner a bubble size distribution. A frequency sweep is performed at equally spaced frequencies between 5 kHz to 950 kHz, from which the bubble size distribution is compiled. Chahine et al. [40–46] describe the methodology underpinning ABS and the algorithm for transformation of the raw data into a bubble size distribution.

2.6. Frequency Measurement and Fast Fourier Transform

Impress G-1000 pressure transducers were used in this experiment. Fast Fourier Transforms are simple algorithms designed to convert a signal from one domain (time or space) and convert it to the frequency domain and vice versa. Figure 5 shows an exemplar waveform with the FFT. This provides a quick and easy way to determine the frequency of the oscillatory flow from a fluidic oscillator. LabView is used to acquire the signal and process it. The oscillatory pulse from a fluidic oscillator is composed of the amplitude and frequency of oscillation.

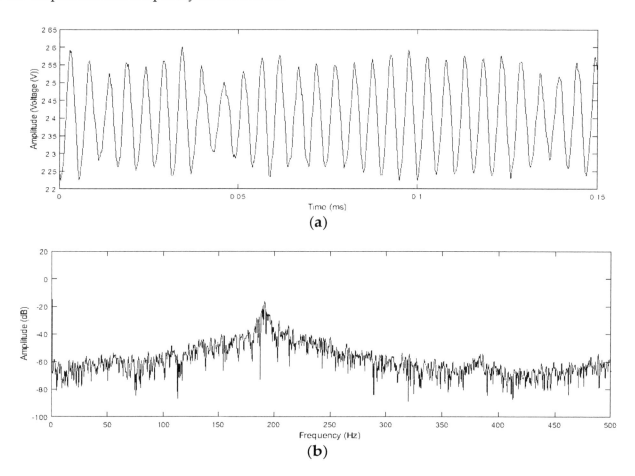

Figure 5. An exemple of a raw waveform from a fluidic oscillator (**a**) and its translation in frequency domain post FFT (**b**). Aliasing effects of FFT are mitigated by having a high acquisition rate (512 kHz with 512 k samples acquisition window).

2.7. Bubble Sizing Analyses

Two factors are readily computable for the bubble size analysis obtained from the ABS- average bubble size in terms of number of bubbles and average bubble size in terms of void fraction contribution (volume contribution) of the bubble. Depending on the application, bubble sizing is usually reported using either of these two factors. Since volume contributions (due to their association with increased mass transfer/heat transfer for microbubbles) are more relevant for a majority of industrial processes, this paper discusses bubble sizes in terms of volume contributions.

Table 1 provides an exemplar for a case wherein there are three classes of bubbles—Class A—with a size of 1 μm and 600 in number, Class B with size of 100 μm and 200 in number and Class C of size 500 μm and 200 in number. This would lead to a total of 1000 bubbles. The surface area of the bubbles is calculated and so is the volume. The bubble size can be computed by either using average bubble size in terms of numbers i.e. weighted bubble size divided by total numbers, or in terms of average bubble size in terms of volume contribution i.e., weighted bubble volume divided by total bubble volume:

$$N_{av} = \sum_{i=1}^{n} \frac{n_i x_i}{n} \tag{2}$$

$$N_{vc} = \sum_{i=1}^{n} \frac{n_i V_i}{nV} \tag{3}$$

with:

$$V_i = \frac{4}{3}\pi x_i^3 \tag{4}$$

where average bubble size in terms of number (2) volume (3) is shown, n is the total number of bubbles and n_i is the bubble contribution for number (2) volume (3) of each bubble of size x_i represented by V_i (4).

Table 1. Exemplar.

S.No.	Bubble	Size	Number	Volume of Individual Bubbles	Total Volume Contribution	Surface Area	Total Surface Area	Surface Area/Volume
1	A	1	600	5.24×10^{-1}	3.14×10^2	3.14	1.88×10^3	6.00
2	B	100	200	5.24×10^5	1.05×10^8	3.14×10^4	6.28×10^6	6.00×10^{-2}
3	C	500	200	6.54×10^7	1.31×10^{10}	7.85×10^5	1.57×10^8	1.20×10^{-2}
			1000		1.32×10^{10}			
N_{Av}	120.6 μm	N_{VC}		200 μm				

This also means that 1 million 1 μm bubbles would be required to occupy the same volume as a single 100 μm bubble. This brings about a massive disparity in bubble size in terms of volume contribution. However, the volume contribution would be a useful tool for estimating any transport phenomena exercise over number contribution. Generally speaking, size distributions collated from membranes are narrow and the difference in the two averages is lower. A large difference in bubble sizes is observed for a highly dispersed distribution and it is beneficial to the system to have a narrower size distribution. This exemplar demonstrates how these two values need not be the same and their dispersity results in the width of the bubble size distribution. Work by Allen [47] and Merkus [48] explain the nuances associated with particle sizing and statistical calculations performed for them in detail.

2.8. Results and Discussion

In order to prove the hypotheses, two experimental modalities were tested. Bubble sizes were measured at various frequencies and under different conditions. Bubbles were sized at 22 frequencies, for three feedback conditions (to test amplitude variations) and two higher flowrates (vented, so flow through the fluidic oscillator could be 86 lpm and 92 lpm). Bubble size for steady flow at the same conditions resulted in a bubble size of approximately 350 μm and 450 μm, respectively. This confirmed previous work performed in literature [19,30,49] that fluidic oscillation resulted in a significant decrease in bubble size when compared to conventional methods of microbubble generation.

An approximate 60% reduction in bubble size than the average bubble size estimated from oscillatory flow at other frequencies was observed for all configurations, supporting the proposed hypothesis. The variations between the amplitudes provided additional information on this new bubble formation dynamic under the resonant mode regime.

With the configuration that induced the highest negative feedback, there is a suggestion that two dips observed and this is probably due to the higher feedback introduced for oscillatory control thereby changing the effect of the system.

Figure 6 shows the length of feedback loop compared to the average bubble size at two amplitudes. This shows that although the conditions have been slightly changed, the resonant condition observed is at the same frequency as can be seen in Figure 7. The frequency of the fluidic oscillator can be changed by changing the feedback loop length. This changes the amount of feedback introduced into the system. Figure 6 shows that although the dip is at the same frequency, the actual feedback loop is different for both cases for different conditions of feedback loop lengths. This causes the slight shift in the frequency observed for the dip.

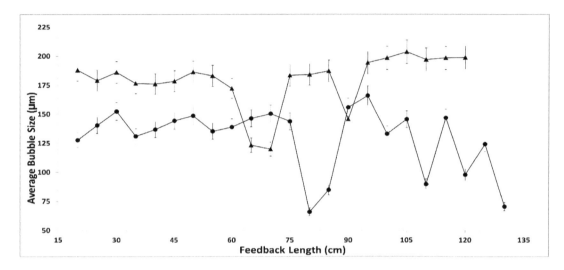

Figure 6. Feedback length vs. average bubble size (●—92 lpm OD4 mm, ▲—0 86 lpm OD 4 mm).

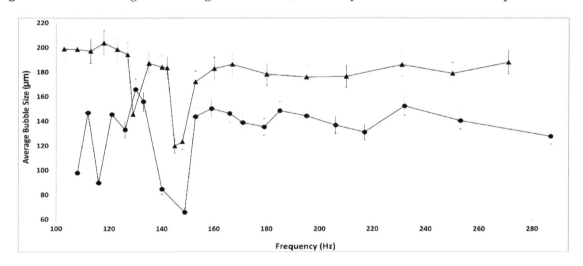

Figure 7. Frequency vs. average bubble size at different flow rates (●—92 lpm OD4 mm, ▲—86 lpm OD 4 mm).

Figure 7 shows the average bubble size at different flowrates when the frequency is modulated for the smallest feedback configuration. At these two conditions, it is observed that the primary dip occurs circa 150 Hz. The dip is large and observable and there is a suggestion of another dip prior to that. This is shifted slightly for different amplitudes. This is what is seen herein. The frequency remains the same for different configurations of feedback, resulting in a change in bubble size even for different conditions.

Varying the lengths of the tube were used to introduce a change in the feedback. Figures 6 and 7 show that even though the feedback loop lengths were different, the frequency change observed due to the change in flow rate with respect to feedback loop length resulted in a minimum bubbles size at the same frequency. Figures 6 and 7 are plots from a reading taken for single tubing, lowest feedback configuration, at two different inlet global flow rates at 22 different frequencies. The bubble size distribution showed the dip in bubble size and the resonant mode—'sweet spot' was observed here.

Figure 6 shows the presence of the 'sweet spot' at different lengths of the feedback loop. Figure 7 shows that although the feedback loop lengths were different for the different dips in bubble size, the actual frequency remained the same and was approximately 150 Hz.

Figure 8 shows the resonant mode for the medium feedback condition. The dip is quite close to each other for this scenario. The frequency for the sweet spot is similar to the low feedback condition and is at 150 Hz. There is a significant dip observed due to potential matching of the flow rate with the frequency of the system and bubble formation.

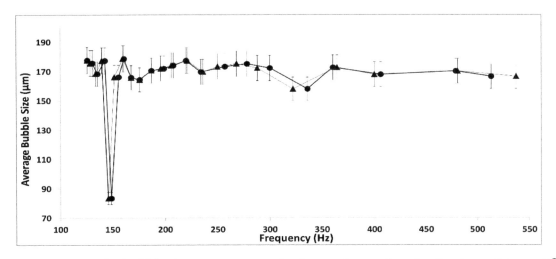

Figure 8. Frequency for bubble size—resonant mode observed—medium feedback condition—(● — 92 lpm OD 6 mm, ▲ —86 lpm OD 6 mm).

Figure 9 shows the bubble size versus frequency at the higher feedback configuration. Resonant condition dips are observed for all three conditions as seen in Figures 7–10.

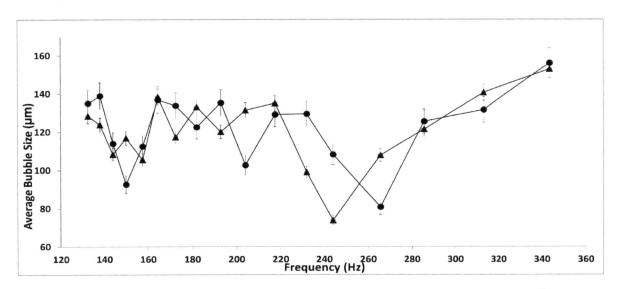

Figure 9. Bubble Size vs. frequency for two flow rates (higher feedback configuration) (● —92 lpm OD 10mm, ▲ —86 lpm OD 10 mm).

Under higher feedback condition, i.e., 10 mm OD, the frequency is higher for the dip aside from the initial dip. This is probably due to the higher amplitude observed for the higher feedback condition which results in lesser coalescence for the system. The dip observed is still consistent with the average bubble size formed at other conditions of the resonant modes.

The presence of the resonant condition or 'sweet spot' was observed for the higher feedback condition. However, the magnitude of the decrease in bubble was not as significant as for the lower feedback condition (i.e., feedback loop length is constant but volume is lower) when considered in a relative manner. However, due to the greater amplitude, there is a higher frequency and several smaller bubbles being formed which results in the dampening of the 'sweet spot'. There is another 'sweet spot' formed at 250 Hz.

These figures also show a skew observed in terms of frequency and flow rate, with the skew being less for the larger feedback condition than for the lower feedback configuration indicating that even the shift observed in the frequency is due to the combination of the length of the feedback tube and the flow rate and this skews the bubble size and the resonant condition.

These three conditions (amplitude variations) consistently show the resonant mode condition. The different vent flows also show the resonant mode condition and the resonant condition changes based on the variations introduced. The flow rate is an additional variable influencing the frequency [49] can be adjusted to achieve the same average bubble size. This is because the change in flow rate has a much larger impact on the frequency of the smaller feedback configuration than that of the larger one due to the difference in the frictional losses for both conditions. The dip in bubble size is observed at different feedback conditions when the global flow rate is different. However, it is the same frequency at which bubble size reduction is observed, demonstrating that the change in global flow results in no significant shift in the frequency 'sweet spot', ~150 Hz. Of note is that not only is a similar trend being observed at these different flow rates but there is also a clear indication of the dip in bubble size at the specific flow rate and frequency that seems to be characteristic of this newly discovered 'sweet spot'. It is interesting to note that the extent of the dip in bubble size varies with the flow rate and therefore on the amplitude of fluidic oscillator. The 'sweet spot' depends on the fluidic circuit i.e., aerator, liquid, fluidic oscillator, and gas aside from incoming flow rate. Amplitude is one of the major causes for a bubble size reduction which results in higher frequencies and bubble throughput.

Figure 10 shows the amplitudes and the differences seen for the different resonant conditions and shows the resonant condition for these configurations. It is seen that there is a greater decrease in bubble size for the higher feedback condition (i.e., higher amplitude), as also observed for higher fluidic oscillator incoming flow rate. This happens at a higher frequency because of the change in the 'sweet spot' condition. This means that the amplitude is higher for the same condition. This results in the lowest feedback condition having the 'largest' bubble size for its sweet spot condition. There is a smaller difference between OD 4 mm and OD 6 mm condition due to smaller differences in the feedback introduced as compared to the OD 10 mm. This increases the condition substantially resulting in an increase in the effect observed.

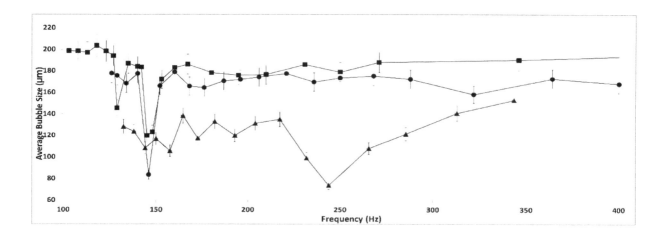

Figure 10. Comparison between amplitudes for the resonant conditions or 'sweet spot' (■—OD 4 mm, ●—OD 6 mm , ▲—OD 10 mm).

2.9. Mechanism for the Resonant Mode Condition or 'Sweet Spot'

Bubble size is directly proportional to the rise velocity and inversely proportional to the bubble wake and liquid present in the system. There are two ways that can be used to increase the amplitude to test the hypothesis we have proposed—increase it by imposing a higher feedback (i.e., 10 mm OD volume) or by increasing the flow through the fluidic oscillator (86/92 lpm). Both conditions have been used in this study and this provides several conditions to observe for fluidic oscillation mediated bubble formation. The change between the OD 4 mm and OD 6 mm is not as significant as between the two and OD 10 mm. This is due to the large feedback introduced in the system in order to cause an improved performance regime in terms of momentum of the jet as well as amplitude of the oscillatory pulse.

This results in different systems being imposed whilst keeping the rest of the system as constant as possible. The reason why different aerators were not used was because it would be difficult to compare two aerators (they can have several different properties—wettability, porosity, thickness, mesoporosity, polydispersity of orifice sizes, material of fabrication, and pressure drop across the membrane.

3. Dimensionless Analysis

The dynamics of bubbles generated in an oscillatory system can be defined by dimensionless quantities such as the Weber Number—*We*, Stokes Number—*Sk*, Strouhal Number—*Sh* and Reynolds Number—*Re*.

We is defined as the ratio of the inertia of fluid to its surface tension and determines the curvature of the bubble which means smaller the bubble greater is the curvature and higher is the surface tension and higher the *We*. *Sk* relates the bubble size to the rise velocity of the system whereas the *Sh* is used for oscillatory systems and for the fluidic oscillator, helps determine the frequency of bubble generation and the characteristics of the fluidic oscillator, especially the oscillatory flow and bistability.

Conjunctions occur when either of these values is unbalanced, leading to a bigger bubble size being observed. Greater unbalance leads to coalescence in the system. Tesař et al. [50] describes the case of the smallness of microbubble being limited due to coalescence:

$$Sh_t = \frac{fL}{V} \tag{5}$$

$$Re = \frac{\rho V D_h}{\mu} = \frac{Vb}{v} = v\frac{\dot{M}}{hv} \tag{6}$$

$$Sk = \frac{fb^2 L}{V} \tag{7}$$

$$We = \frac{w^2 D_b}{2v\sigma} \tag{8}$$

$$We = \frac{f^2 D^3_b}{2v\sigma} n \tag{9}$$

f = oscillation frequency (Hz), L = length of feedback loop (m), V = supply nozzle bulk exit velocity (m/s), b = Constriction width (m), ρ = density (kg/m³), η = viscosity (Pa.s), D_h = Hydraulic diameter (m), v = specific fluid volume (m³), w = bubble rise velocity (m/s), D_b = bubble diameter (m), and σ = surface tension (N/m).

Sanada et al., [51] discussed the coalescence observed in rising bubbles and the interactions between two rising bubbles. This interaction depends on the rise velocity, (therefore size) and the rate of formation of the bubble (oscillation velocity in case of fluidic oscillator). Therefore this means that coalescence is directly related to the *We*, (bubble formation and size), *Sk* (bubble rise and size), *Sh* (bubble formation via oscillation and oscillatory flow) and *Re* (determining the momentum carried by the bubble due to the pulse). This, coupled with the resultant bubble wake and zeta potential (if any due to presence of ions/ surfactant layers on the system), defines the ultimate bubble size. This does not account for the fact that the surface of the membrane generating the bubble (involves the We and $We_{Oscillatory}$ since surface tension is involved) and the orifice size (dependent on Sh since it is the constriction that determines the bulk exit velocity of the orifice).

A conclusion to be drawn from this is that if there is the appropriate balance in the system, (with respect to rise velocity, frequency of bubble generation, bubble size and orifice diameter), analogous to a resonant mode of the system where these conditions are balanced just about correctly, then bubble conjunctions would be avoided leading to a significant size reduction in the bubbles. This is a substantial reduction in size by optimising the parameters of an existing system without any modifications or retrofitting. The amplitude of the pulse is also important as it increases the Sk and Sh which results in lesser conjunctions and coalescence.

- *Force Balance:*

Balancing the forces on the bubble being formed as seen in Figure 11.

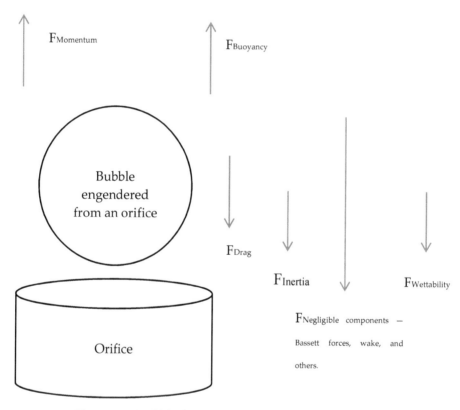

Figure 11. Bubble forces resolved (adapted from [52]).

Figure 11 shows the forces on a bubble detaching and being formed. The buoyancy force and momentum force act upwards, the cosine component of the wettability (anchoring force), drag force, surface tension act downwards. For an orifice of diameter D, density of liquid—ρ_L, density of gas—ρ_g and volume of bubble former—V_b, the following equations are obtained for the forces acting upwards:

- *Bouyancy Force*

$$F_{Bouyancy} = V_b \left(\rho_L - \rho_g \right) g \tag{10}$$

$$V_b = \frac{1}{6} \pi d_b^3 \tag{11}$$

resulting in:

$$F_{Bouyancy} = \frac{1}{6} \pi d_b^3 \left(\rho_L - \rho_g \right) g \tag{12}$$

- *Momentum Force*

$$F_{Momentum} = \frac{1}{4} \pi D_o^2 u_o^2 \rho_g \tag{13}$$

taking downward forces into account.

- *Wettability Force*

$$F_{Wettability} = \pi D_o \sigma \cos \theta \tag{14}$$

where θ is the wetting angle / contact angle made by the engendering bubble and the surface and σ is the surface tension force / anchoring force/ wetting force and D_o is the diameter of the orifice.

- *Drag Forces*

$$F_{Drag\ Force} = \frac{1}{2} C_d \rho_L \frac{\pi}{4} d_b^2 u_b^2 \tag{15}$$

with the rise velocity denoted by u_b.

- *Bubble Inertial Force*

$$F_{IF} = \frac{d\left(u_b V_b \rho_g\right)}{dt} = \frac{\rho_g Q^2 V_b^{\frac{-2}{3}}}{12 \pi \left(\frac{3}{4\pi}\right)^{\frac{2}{3}}} \tag{16}$$

where V_b is the bubble volume, ρ_g is the gas density, ρ_l is the liquid density, g is the acceleration due to gravity, D_b is the bubble diameter, D is the orifice diameter, u_b is the rise velocity of the bubble centre , C_d is the drag coefficient and Q is the volumetric gas flow rate.

Our hypothesis comes from the fact that the bubble formation is most dependent on frequency of the system for oscillatory flow and the amplitude associated with it. If these two are appropriate, such that the bubbles are formed at regular intervals, fast enough to have substantial size reduction (reduction of pulse length and increase in throughput) and not too fast as to coalesce, with the increase in amplitude such that the bubble will cut off instantaneously and not coalesce. Balancing these forces together would result in pinch off. Compensation must be provided for the bubble rise and pinch off due to the oscillatory flow. The force balance turns out to be complicated due to the oscillatory waves and the hybrid synthetic jet engendered by the oscillator, resulting in a highly non-linear system.

- *Prediction of Bubble Size at Resonant Frequency (Volume-based Bubble Size)*

It is seen in Tesař et al. [53] that the bubble rise is dominated by the coalescence and each individual coalescing bubble leads to an increased probability for another staged coalescence which leads to largeness in bubble size as compared to the orifice. Once two bubbles merge together, due to the increase in size and the change in the surface energy as well as the energy associated with the ascent, it is easier for the other bubble to catch up with it. This is better explained by the concept of bubble wake. Each bubble creates a wake (region of lower pressure) upon being created and this allows other bubbles to catch up, coalesce and result in larger bubbles.

Therefore, the smaller the bubble that is created will result in a smaller wake. However, it is easier for the small bubble to be affected by a wake of another bubble (especially a larger one—which is why a smaller bubble generated after a larger one usually results in coalescence.

The active diffusing area of the aerator is 0.15×0.03 m^2. The bubble flux recorded by the acoustic bubble spectrometry is based on the capture rate or acquisition rate that is set for the system at 200 ms.

The 200 ms ensures that the flow in the system is captured at a fixed rate and this results in a delimiting information due to the resultant lower acquisition observed. The capture results in determining the flow to be 0.5 lpm. Using these results, a size of 74 μm approximately is observed. This is thus the minimum size achieved for this. Taking this value and placing it in the equations (10–16) results in the bubble formation force, and taking a frequency of 150 Hz as the bubbling frequency, and the bubble flux from the ABS (120,000) it results in a pulse requirement of 0.007 bar to detach the bubble. This has been achieved by the oscillator as observed in Figure 5 and therefore explains the sweet spot possibility for these conditions of resonance.

Figure 5 shows that the amplitude of the pulse is approximately 0.2 V, which is equivalent to 0.02 bar. This is roughly twice the required pulse strength for the bubble formation and this is why the bubble is detached at the frequency. Judging by that, the frequency band is slightly wide due to the lack of coalescence at this point.

The amplitude of the pulse reduces as the frequency increases for the same flow resulting in intermittent pinch off (since the amplitude is important for imparting sufficient force for bubble detachment). Lower frequency has a larger amplitude but the rate of generation is not fast enough. Additionally as can be seen in Tesař ([34,50]), the coalescence perpetuates the higher rise velocity of the larger bubble. This results in larger bubbles for the non-resonant conditions at lower frequencies. Therefore, the resonant condition or 'sweet spot' is possibly the only condition where the amplitude, frequency, and size are balanced so as not to have bubble coalescence.

There is greater amplitude for the higher feedback condition which results in the 'sweet spot' occurs at a higher frequency (250 Hz). This is what results in the larger increase in the number of bubbles throughput.

- *Prediction of Bubble Size at Resonant Frequency (Number of Bubbles-Based Bubble Size)*

This paper has placed a lot of emphasis on bubble size suitable for transport phenomena but for finding out what the actual bubble size is, when compared to those that have been reported previously, such as Hanotu et al. [19], the size distribution based on number of bubbles calculation—N_{Av}, provides a better idea for those applications concerned with size of the bubbles (flotation/DAF for example) in Figure 12.

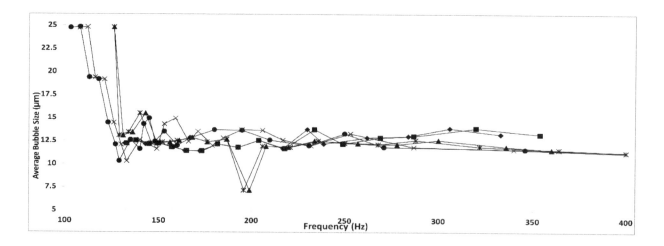

Figure 12. This is an example of the sweet spot for the average bubble size when one considers number of bubbles as opposed to the volume fraction of bubbles. The sweet spot changes slightly but there is still a dip observed at the higher frequencies (◆—OD 10 mm (92 lpm), ▲—OD 6 mm (92 lpm), ✳—OD 4 mm (92 lpm), ■—OD 10 mm (86 lpm), ✳—OD 6 mm (86 lpm), ●—OD 4 mm (86 lpm)).

This brings down the average bubble size reduction producing about 7 µm size bubbles in terms of average size distribution which makes it suitable for generating the bubbles for flotation studies. The 'sweet spot' changes to 200 Hz. This is because the size distribution changes due to the change in flow dynamics. The flow results in this size change as the effect of the number of bubbles cannot be discounted just as the volume average bubble size depends on the dispersity of the size distribution.

An example is provided herewith: The outlet flow of the diffuser measured at 0.1 lpm (1.67×10^{-6} m³/s) for which, when corrected for the 200 ms acquisition rate of the ABS is $Q = 3.33 \times 10^{-7}$ m³/200 ms. N = No of Bubbles— Approximating total number of bubbles as 120,000 as an average measured from the readings, assuming that the distribution is narrow (for the purposes of a general understanding).

Assuming that since bubble pinch-off force (by using Equations (10)–(16)) results in 0.007 bar, and the pulse of the oscillator is an average of 0.02 bar, it exceeds the force required for bubble pinch-off (*cf.* Figure 5). As seen from the FFT of the pulse for the oscillator, the pulse strength is at 0.02 bar per pulse which ensures that there is sufficient momentum and amplitude to generate the bubbles and ensure that the bubble detaches at each oscillatory pulse, so it is an accurate assumption providing no coalescence takes place. This is why the sweet spot is likely at 200 Hz as for the same momentum the force of the bubble pinch-off seems to be at the appropriate level. Figure 5 is shown herein for reference. This has been done in order to reduce the bubble size by matching the bubble formation characteristics to achieve the lowest possible bubble size and by increasing the amplitude of the oscillation in order to impart momentum to the jet and therefore the bubble so that it has a higher rise velocity. This reduces the bubble coalescence when the appropriate conditions are met. Too slow, and the bubble does not detach quickly enough and therefore coalesces, too fast, and the bubble cannot detach due to reduced amplitude and pinch-off. At the right frequency and amplitude, what we term—resonant condition or 'sweet spot', one can detach the bubble at the smallest possible size for that system and it is significantly smaller than what was originally possible via conventional steady flow. This is achieved for the frequency at 200 Hz. Since f = 200 Hz means that it is per second, for 200 ms, the equivalent frequency to be considered would be 40 Hz.

Using the A = diffusing area (0.15×0.03 m²), and assuming that all the flow forms a bubble (there are no leaks and the bubble sizes are small enough to form bubbles rather than slugs:

$$V_{av} = Average\ Volume\ of\ bubble\ formed = \frac{Q}{ANf_{eq}}$$

which results in D_B i.e., $(2r)$ = 6.8 µm for 200 Hz.

This is in close agreement to the size obtained in the system. Depending on changes introduced to the system, these values will change. Whilst these values will change depending on the system, the work supports the presence of the resonant bubbling condition that can occur in any oscillatory flow mediated bubble generating system.

Figure 13 shows the calculations for a sweep for bubble formation via frequencies juxtaposed with the size for the bubbles that would be formed. There is an extension but this is likely due to the fact that at lower frequencies, steady bubble formation dominates as it is results in faster bubble pinch-off.

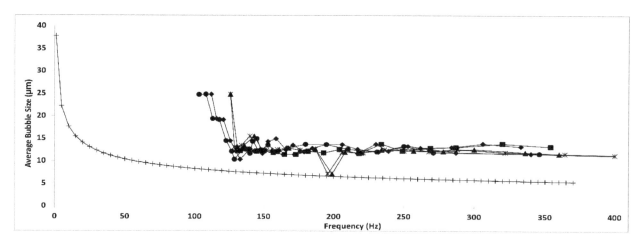

Figure 13. Calculated value as compared to the average bubble size garnered experimentally. Deviation from the sweet spot is less than 0.0001%. ╋—Calculated ◆—OD 10 mm (92 lpm), ▲—OD 6 mm (92 lpm), ✳—OD 4 mm (92 lpm), ■—OD 10 mm (86 lpm), ✳—OD 6 mm (86 lpm), ●—OD 4 mm (86 lpm).

As discussed earlier, 0.25–0.4% of the U.K.'s energy use is utilised for WWA operations. This is not discussed earlier These energy requirements could be substantially reduced by bubble size reduction. The 10 fold size reduction would provide a significant increase in the transport phenomena associated with the bubbles. The cost of adding the fluidic oscillator is based on the pressure drop of the oscillator, which for an industrial plant would typically be about 400 mbar for a large scale system. The cost of frequency modulation in terms of energy is negligible. This is also applicable for several other applications and remediation steps including aeration of waste streams, oxidation of volatiles, advanced oxidation processes, and other treatment techniques. Several other processes exist where gas-liquid operations are involved and bubble sizes are required to be small.

4. Conclusions

This paper reports several scientific results. One of the major ones is the exploration of a new regime for bubble generation mediated by oscillatory flow which reduces coalescence and results in the smallest bubble formation for a given system (number averaged or volume averaged) at specific conditions of detachment with no additional energy.

The ability to generate substantially smaller bubbles, useful for several applications, by simply engineering the conditions for bubble formation under oscillatory flow such that the frequency and amplitude are maintained is a shift in understanding for bubble formation dynamics. Whilst oscillatory flow results in bubble size reduction over the gamut of frequencies, at a particular frequency, specific to a particular set up (but valid for other set ups as seen with different configurations), a substantial reduction in bubble size (≈60%) is observed. It has been hypothesised that tuning of all the different parameters controlling bubble formation will result in a smaller overall bubbles size, presumably by the combined effect of more efficient pinch off and reduced coalescence. This has resulted in a large size reduction from a typical bubble size distribution produced in a system. Conventional bubble generation would have produced bubbles of sizes: 650 μm (volume averaged) and 250 μm (number averaged), whereas the fluidic oscillator would result in an average size of 120 μm (volume averaged) and 14 μm (number averaged), and the sweet spot would result in 60 μm (volume averaged) and 7 μm (number averaged).

The addition of the fluidic oscillator reduces the bubble size to a certain extent but coalescence prevents the size being smaller than a certain range. This adds an additional pressure drop across the system by 20–150 mbar in the system (system dependent) which is insignificant compared to the actual savings in terms of surface energy. A further reduction of bubble size is achieved, without any additional cost to the system, which then brings about a step change in any gas liquid operation and overcomes any of the limitations that had been posed earlier for using microbubbles for mass transfer in gas-liquid operations. A 60% reduction in bubble size from what was already a reduced bubble size, resulting in a 10-fold reduction in bubble size with suitable tuning and adding a pressure drop of 150 mbar maximum, is paradigm shifting and has the potential to change the processing industry.

An example can be seen in DAF, which has been discussed at the start. In order to achieve a 70–100 μm bubble size (number averaged), 12–14 bar of pressure is used to generate the flux and throughput required. We can achieve a 10-fold reduction for this size (7 μm) with a tuned frequency, fluidic oscillator, and aerator totalling 2.15 bar(g) total pressure which means that the energetics involved in order to get 10 times larger size bubbles will result in substantial reduction. A simplistic approach will be to relate energy with pressure as a proxy and the energetics for a specific process can be taken into account. The fact the frequency modulation for a bistable diverter valve (the fluidic oscillator) does not require any additional energy input, but results in a 60% reduction in bubble size (increase in associated transport phenomena) [9,11,29], and is a 10-fold reduction in bubble size (100–fold increase in interfacial area) by using this tuned oscillator as opposed to a conventional steady flow system. The substantial energy saving and increase in transfer efficiencies for any gas-liquid operation involved results in widespread ramifications. This is not limited to only bistable diverters but all oscillatory flow systems as it is actually dependent on oscillatory flow and the control of frequency and amplitude as opposed to using a specific system.

These wider implications are not only limited to DAF, but for all operations that can be thought of as gas-liquid contacting or processing including aeration, gas-liquid contacting operations which constitute a large proportion of industrial processes, biochemical reactors, fermenters, remediation processes, digesters, incubators, bioreactors, and others.

Author Contributions: W.B.Z. and P.D.D. jointly proposed the central hypothesis. P.D.D. augmented the hypothesis with the inclusion of higher amplitude by artificial venting, created the experimental design and setup, and conducted the first three experiments and the design for the analysis, hence is the principal author. Y.R. provided the individual bias removal, substantial experimental work via repeats (5 in total), equation input, and collated the data for analysis. M.J.H. and W.B.Z. provided organisational and logistical support and compositional critique.

Acknowledgments: This work was carried out as part of the "4CU" programme grant, aimed at sustainable conversion of carbon dioxide into fuels, led by The University of Sheffield and carried out in collaboration with The University of Manchester, Queens University Belfast and University College London. The authors acknowledge gratefully the Engineering and Physical Sciences Research Council (EPSRC) for supporting this work financially (Grant no. EP/K001329/1). M.J.H. and P.D.D. would like to thank R3Water FP7 grant. W.B.Z. and P.D.D. would like to thank IUK IBCatalyst and Energy Catalyst award. P.D.D. would like to thank Professor Ray Allen for discussions, Audrey YY Tan for bias studies and GK for inspiration.

References

1. Bird, R.B.; Stewart, W.E.; Lightfoot, E.N. *Transport Phenomena*; Wiley: Hoboken, NJ, USA, 2007.
2. Incropera, F.P.; Lavine, A.S.; Bergman, T.L.; DeWitt, D.P. *Fundamentals of Heat and Mass Transfer*; Wiley: Hoboken, NJ, USA, 2007.
3. Green, D.; Perry, R. *Perry's Chemical Engineers' Handbook*, 8th ed.; McGraw-Hill Education: New York, NY, USA, 2007.
4. Treybal, R.E. *Mass-Transfer Operations*; McGraw-Hill Education: New York, NY, USA, 1980.
5. Tesař, V. What can be done with microbubbles generated by a fluidic oscillator? (survey). *EPJ Web Conf.* **2017**, *143*. [CrossRef]
6. Metcalf, E.I.; Tchobanoglous, G.; Burton, F.; Stensel, H.D. *Wastewater Engineering: Treatment and Reuse*; McGraw-Hill Education: New York, NY, USA, 2002.
7. Abdulrazzaq, N.; Al-Sabbagh, B.; Rees, J.M.; Zimmerman, W.B. Purification of Bioethanol Using Microbubbles Generated by Fluidic Oscillation: A Dynamical Evaporation Model. *Ind. Eng. Chem. Res.* **2016**, *55*, 12909–12918. [CrossRef]
8. Abdulrazzaq, N.; Al-Sabbagh, B.; Rees, J.M.; Zimmerman, W.B. Separation of zeotropic mixtures using air microbubbles generated by fluidic oscillation. *AIChE J.* **2015**, *62*, 1192–1199. [CrossRef]

...

9.   Al-Yaqoobi, A.; Zimmerman, W.B. Microbubble distillation studies of a binary mixture. In *USES—University of Sheffield Engineering Symposium 2014*; The University of Sheffield: Sheffield, UK, 2014.

10. Zimmerman, W.B.; Al-Mashhadani, M.K.H.; Bandulasena, H.C.H. Evaporation dynamics of microbubbles. *Chem. Eng. Sci.* **2013**, *101*, 865–877. [CrossRef]

11. Fiabane, J.; Prentice, P.; Pancholi, K. High Yielding Microbubble Production Method. *BioMed Res. Int.* **2016**, 3572827. [CrossRef] [PubMed]

12. Kantarci, N.; Borak, F.; Ulgen, K.O. Bubble column reactors. *Process Biochem.* **2005**, *40*, 2263–2283. [CrossRef]

13. Tesař, V. What can be done with microbubbles generated by a fluidic oscillator? (survey). *EPJ Web Conf.* **2017**, *143*. [CrossRef]

14. Makuta, T.; Takemura, F.; Hihara, E.; Matsumoto, Y.; Shoji, M. Generation of micro gas bubbles of uniform diameter in an ultrasonic field. *J. Fluid Mech.* **2006**, *548*, 113–131. [CrossRef]

15. Shirota, M.; Sanada, T.; Sato, A.; Watanabe, M. Formation of a submillimeter bubble from an orifice using pulsed acoustic pressure waves in gas phase. *Phys. Fluids* **2008**, *20*. [CrossRef]

16. Stride, E.; Edirisinghe, M. Novel microbubble preparation technologies. *Soft Matter* **2008**, *4*, 2350–2359. [CrossRef]

17. Zimmerman, W.B.; Tesař, V.; Bandulasena, H.C.H. Towards energy efficient nanobubble generation with fluidic oscillation. *Curr. Opin. Colloid Interface Sci.* **2011**, *16*, 350–356. [CrossRef]

18. Parmar, R.; Majumder, S.K. Microbubble generation and microbubble-aided transport process intensification—A state-of-the-art report. *Chem. Eng. Process. Process Intensif.* **2013**, *64*, 79–97. [CrossRef]

19. Hanotu, J.; Bandulasena, H.C.; Zimmerman, W.B. Microflotation performance for algal separation. *Biotechnol. Bioeng.* **2012**, *109*, 1663–1673. [CrossRef] [PubMed]

20. Agarwal, A.; Ng, W.J.; Liu, Y. Principle and applications of microbubble and nanobubble technology for water treatment. *Chemosphere* **2011**, *84*, 1175–1180. [CrossRef] [PubMed]

21. Rehman, F.; Medley, G.; Bandalusena, H.C.H.; Zimmerman, W.B. Fluidic oscillator-mediated microbubble generation to provide cost effective mass transfer and mixing efficiency to the wastewater treatment plants. *Environ. Res.* **2015**, *137*, 32–39. [CrossRef] [PubMed]

22. Mulvana, H.; Eckersley, R.J.; Tang, M.X.; Pankhurst, Q.; Stride, E. Theoretical and experimental characterisation of magnetic microbubbles. *Ultrasound Med. Biol.* **2012**, *38*, 864–875. [CrossRef] [PubMed]

23. Lukianova-Hleb, E.Y.; Hanna, E.Y.; Hafner, J.H.; Lapotko, D.O. Tunable plasmonic nanobubbles for cell theranostics. *Nanotechnolohy* **2010**, *21*. [CrossRef] [PubMed]

24. Cai, X. Applications of Magnetic Microbubbles for Theranostics. *Theranostics* **2012**, *2*, 103–112. [CrossRef] [PubMed]

25. Zimmerman, W.B.; Hewakandamby, B.N.; Tesař, V.; Bandulasena, H.C.H.; Omotowa, O.A. Design of an airlift loop bioreactor and pilot scales studies with fluidic oscillator induced microbubbles for growth of a microalgae Dunaliella salina. *Appl. Energy* **2011**, *88*, 3357–3369. [CrossRef]

26. Zimmerman, W.B.; Hewakandamby, B.N.; Tesař, V.; Bandulasena, H.C.H.; Omotowa, O.A. On the design and simulation of an airlift loop bioreactor with microbubble generation by fluidic oscillation. *Int. Sugar J.* **2010**, *112*, 90–103. [CrossRef]

27. Ying, K.; Gilmour, D.J.; Shi, Y.; Zimmerman, W.B. Growth Enhancement of Dunaliella salina by Microbubble Induced Airlift Loop Bioreactor (ALB)—The Relation Between Mass Transfer and Growth Rate. *J. Biomater. Nanobiotechnol.* **2013**, *4*, 1–9. [CrossRef]

28. Ying, K.; AlMashhadani, K.H.; Hanotu, J.O.; Gilmour, D.J.; Zimmerman, W.B. Enhanced Mass Transfer in Microbubble Driven Airlift Bioreactor for Microalgal Culture. *Engineering* **2013**, *5*, 735–743. [CrossRef]

29. Hanotu, J.; Bandulasena, H.C.H.; Chiu, T.Y.; Zimmerman, W.B. Oil emulsion separation with fluidic oscillator generated microbubbles. *Int. J. Multiph. Flow* **2013**, *56*, 119–125. [CrossRef]

30. Zimmerman, W.B.; Tesař, V.; Butler, S.; Bandulesena, H.C.H. Microbubble Generation. *Recent Pat. Eng.* **2008**, *2*. [CrossRef]

31. Zimmerman, W.B.; Hewakandamby, B.N.; Tesař, V.; Bandulasena, H.C.H.; Omotowa, O.A. On the design and simulation of an airlift loop bioreactor with microbubble generation by fluidic oscillation. *Food Bioprod. Process.* **2009**, *87*, 215–227. [CrossRef]

32. Tesař, V. *Pressure-Driven Microfluidics*; Artech House: Norwood, UK, 2007.

33. Tesař, V.; Bandalusena, H. Bistable diverter valve in microfluidics. *Exp. Fluids* **2011**, *50*, 1225–1233. [CrossRef]

34. Tesař, V. Mechanisms of fluidic microbubble generation part Ii: Suppressing the conjunctions. *Chem. Eng. Sci.* **2014**. [CrossRef]

35. Zimmerman, W.B.; Tesař, V.; Bandulasena, H.C.H. *Efficiency of an Aerator Driven by Fluidic Oscillation. Part I: Laboratory Bench Scale Studies*; The University of Sheffield: Sheffield, UK, 2010.

36. Warren, R.W. Negative Feedback Oscillator. U.S. Patent 3,158,166, August 1962.

37. Tesař, V. Microbubble generation by fluidics. part I: Development of the oscillator. In Proceedings of the Fluid Dynamics 2012, Prague, Czech Republic, 24–26 October 2012.

38. Kooij, S.; Sijs, R.; Denn, M.M.; Villermaux, E.; Bonn, D. What determines the drop size in sprays? *Phys. Review X* **2018**, *8*. [CrossRef]

39. Leighton, T.G. *The Acoustic Bubble*; Academic Press: Cambridge, MA, USA, 1994.

40. Chahine, G.L. Numerical Simulation of Bubble Flow Interactions. In *Cavitation: Turbo-machinery and Medical Applications-WIMRC FORUM 2008*; Warwick University: Coventry, UK, 2008.

41. Chahine, G.L.; Duraiswami, R.; Frederick, G. *Detection Of Air Bubbles In Hp Ink Using Dynaflow's Acoustic Bubble Spectrometer (Abs) Technology*; Hewlett Packard: Palo Alto, CA, USA, 1998.

42. Chahine, G.L.; Gumerov, N.A. An inverse method for the acoustic detection, localization and determination of the shape evolution of a bubble. *Inverse Probl.* **2000**, *16*, 1–20.

43. Duraiswami, R.; Prabhukumar, S.; Chahine, G.L. Bubble counting using an inverse acoustic scattering method. *J. Acoust. Soc. Am.* **1998**, *104*, 2699–2717. [CrossRef]

44. Chahine, G.L.; Kalamuck, M.K.; Cheng, J.Y.; Frederick, G.S. Validation of Bubble Distribution Measurements of the ABS with High Speed Video Photography. In Proceedings of the CAV 2001: Fourth International Symposium on Cavitation, Pasadena, CA, USA, 20–23 June 2001.

45. Tanguay, M.; Chahine, G.L. Acoustic Measurements of Bubbles in Biological Tissue. In *Cavitation: Turbo-machinery and Medical Applications*; Warwick University: London, UK, 2008.

46. Wu, X.-J.; Chahine, G.L. Development of an acoustic instrument for bubble size distribution measurement. *J. Hydrodyn. Ser. B* **2010**, *22*, 330–336. [CrossRef]

47. Allen, T. *Particle Size Measurement: Volume 1: Powder Sampling and Particle Size Measurement*; Springer: New York, NY, USA, 1996.

48. Merkus, H.G. *Particle Size Measurements: Fundamentals, Practice, Quality*; Springer: New York, NY, USA, 2009.

49. Tesař, V. Microbubble smallness limited by conjunctions. *Chem. Eng. J.* **2013**, *231*, 526–536. [CrossRef]

50. Sanada, T.; Sato, A.; Shirota, M.; Watanabe, M. Motion and coalescence of a pair of bubbles rising side by side. *Chem. Eng. Sci.* **2009**, *64*, 2659–2671. [CrossRef]

51. Pinczewski, W.V. The formation and growth of bubbles at a submerged orifice. *Chem. Eng. Sci.* **1981**, *36*, 405–411. [CrossRef]

52. Tesař, V. Shape Oscillation of Microbubbles. *Chem. Eng. J.* **2013**, *235*, 368–378. [CrossRef]

53. Tesař, V.; Hung, C.-H.; Zimmerman, W.B. No-moving-part hybrid-synthetic jet actuator. *Sens. Actuators A Phys.* **2006**, *125*, 159–169. [CrossRef]

11

POD Analysis of Entropy Generation in a Laminar Separation Boundary Layer

Chao Jin and Hongwei Ma *

School of Energy and Power Engineering, Beihang University, Beijing 100191, China; jin_chao@buaa.edu.cn
* Correspondence: mahw@buaa.edu.cn

Abstract: Separation of laminar boundary layer is a great source of loss in energy and power machinery. This paper investigates the entropy generation of the boundary layer on the flat plate with pressure gradient. The velocity of the flow field is measured by a high resolution and time related particle image velocimetry (PIV) system. A method to estimate the entropy generation of each mode extracted by proper orthogonal decomposition (POD) is introduced. The entropy generation of each POD mode caused by mean viscous, Reynolds normal stress, Reynolds sheer stress, and energy flux is analyzed. The first order mode of the mean viscous term contributes almost 100% of the total entropy generation. The first three order modes of the Reynolds sheer stress term contribute less than 10% of the total entropy generation in the fore part of the separation bubble, while it reaches to more than 95% in the rear part of the separation bubble. It indicates that the more unsteady that the flow is, the higher contribution rate of the Reynolds sheer stress term makes. The energy flux term plays an important role in the turbulent kinetic energy balance in the transition region.

Keywords: POD; entropy generation; boundary layer; laminar separation bubble

1. Introduction

For the energy and power machinery, such as gas turbine and other turbomachinery, the efficiency must be the most important performance parameters [1]. An effective way to improve efficiency is to reduce the loss, namely to avoid the generation of entropy [2–4]. In thermodynamics, any irreversible physical process will inevitably lead to the increase of entropy [5]. In this paper, it mainly focuses on the entropy generation in the boundary layer.

The entropy generation has been the subject of many past studies. Bejan [6], Rotta [7], and McEligot [8] studied the entropy generation in the viscous layer and analyzed the generation rates in diffident Y-plus layer. Moore [9] attempted to develop a numerical model for turbulent flow entropy generation, but Kramer-Bevan [10] verified it is not consistent near the wall due to a small temperature gradient. Adeyinka et al. [11] investigated the error of entropy generation model affected by the mesh grids in the fully-developed laminar flow.

The entropy generation is difficult to be accurately calculated [12]. Since it is difficult to predict the small fluctuating velocity and temperature in the turbulent flow through numerical and experimental [13] few experiments have had sufficient measurements to calculate the entropy generation. Thus, most studies try to estimate the entropy generation with a simplified formulation, which will be introduced in Section 3.2.

To describe and extract coherent structures in boundary layer, researchers attempt to develop new data analysis methods, such as proper orthogonal decomposition (POD), dynamic mode decomposition (DMD), and spectral POD (SPOD). POD is a method that identifies coherent structures by decomposing the flow field into orthogonal modes in space. Moreover, the dominant features in the flow is identified based on the energy rank [14]. Dynamic mode decomposition is a method that is orthogonal in time in the sense that the dynamic mode information from a given flow field is based on the Koopman analysis [15]. In contrast to POD, it is a method that gives the energy of the fluctuations at distinct frequencies meaning the modes are arranged in descending order of energy content. Spectral POD is a method that combines proper orthogonal decomposition with a spectral method to analyze and extract reduced order models of flows from time data series of velocity fields [16].

Since this paper mainly focuses on the entropy generation in the boundary layer. It is more convenient to identify the flow field based on the energy rank and care little about the time series and frequencies. Thus, the POD method is applied to the measurement to estimate the entropy generation. The POD has been widely applied for the experimental and numerical data to identify the coherent structure [17,18], such as the plate boundary layer [19], cylinder engine flow [20], and turbine rotor-stator interaction [21]. Particularly, in the case of laminar separation bubbles, POD can extract the different scale coherent structures [22,23].

While few works have been done to quantify the entropy generation of different coherent structures in the boundary layer. Calculating the entropy generation of those coherent structure helps to explore the mechanisms of entropy generation in the boundary layer. It can also identify the loss resource of the flow flied. In this paper, the entropy generation rate is analyzed by proper orthogonal decomposition (POD) applied to the PIV measurements.

2. Experimental Facility

The measurement is taken in a transparent circulating water tunnel with a 700 mm (width) × 500 mm (depth) experimental section, and the whole length of the water tunnel is 6.8 m, just as Figure 1 shows. The inlet velocity in this experiment is 0.065 m/s. The mean velocity profile in the empty experimental section was uniform except for the thin boundary layers on the walls. The turbulence of the water tunnel is less than 1%, and the Reynolds number is based on the total length of the flat plate and the inlet velocity is about 3×10^4. The flat plate is mounted horizontally in the middle of the water tunnel to ensure zero flow incidence. Total length of the flat plate is 390 mm and the geometrical structure of the leading edge is an ellipse with a ratio of 3:1 to the semi-minor axis length, just as Figure 2 shows.

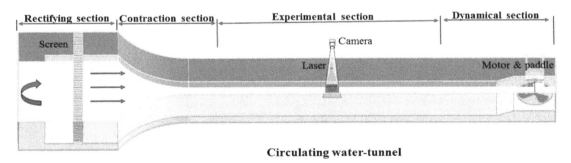

Circulating water-tunnel

Figure 1. Water tunnel and experimental layout.

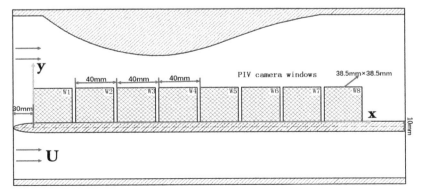

Figure 2. Placement of the PIV camera windows.

The instantaneous velocity field of eight streamwise planes (W1–8) is measured by a PIV system. The distance from the first window to the leading edge of the flat plate is 30 mm, and the distance between each windows is 40 mm. The streamwise plane is illuminated by a light sheet provided by a double cavity Nd: YAG laser, which has the maximum illumination energy of 200 mJ/pulse and a maximum repetition rate of 15 Hz. To ensure the PIV system operates in stably, the sampling rate is set as 12 Hz. One thousand pairs of signal-exposure images are continuously captured by a CCD camera with a resolution of 2072 pixels × 2072 pixels, and the reality view window is about 38.5 mm × 38.5 mm. The PIV measurement uncertainty can refer to the paper [24], in which the same PIV system is used. A more detailed investigation of the uncertainty in the turbulent boundary layer can be found in [25].

3. Dara Processing Method

3.1. POD Method

Since any turbulent flow can be viewed as a superposition of a small number of coherent structures, the equations describing these structures can be considered in a low-dimensional description of turbulence. The POD is an effective method to extract these coherent structures. The time related flow field snapshots can be decomposed into a linear basis set consisting of N basis function $\phi^{(k)}$ (POD mode) and the corresponding coefficients $\chi^{(k)}$ (POD coefficients or time coefficients). The POD mode provides the spatial information on coherent structures, and the POD coefficients retain the temporal information. In addition to those two parameters, the eigenvalues $\lambda^{(k)}$ represents the total kinetic energy that each POD mode captured. The original flow field can be reconstructed by POD mode and POD coefficients.

$$\nu = \sum_{n=1}^{N} \chi_n \phi_n \tag{1}$$

The instantaneous velocity of kth POD mode is:

$$\nu_n^{(k)} = \chi_n^{(k)} \phi^{(k)} \quad n = 1, 2 \ldots N, \tag{2}$$

The mean velocity of kth POD mode is:

$$V^{(k)} = \frac{1}{N} \sum_{n=1}^{N} \chi_n^{(k)} \phi^{(k)}, \tag{3}$$

The velocity fluctuation of kth POD mode is:

$$v'^{(k)}_n = v_n^{(k)} - V^{(k)} \quad n = 1, 2 \ldots N, \tag{4}$$

Thus, the Reynolds stresses of kth POD mode can be computed as:

$$u'^{(k)}_n v'^{(k)}_n = \left(u_n^{(k)} - U^{(k)} \right) \left(v_n^{(k)} - V^{(k)} \right) \quad n = 1, 2 \ldots N, \tag{5}$$

3.2. Estimation of Entropy Generation

Basic thermodynamics tells us that entropy generates when the process is irreversible such as viscous friction [1]. The viscous dissipation is a main frictional irreversibility in boundary layers, which includes mean viscous dissipation and turbulent dissipation. So the entropy generation can be expressed as:

$$S = \frac{\mu \Phi + \rho \varepsilon}{T}, \tag{6}$$

where the mean viscous dissipation $\mu\Phi$ is [2]:

$$\mu\Phi = 2\mu\left[\left(\frac{\partial U}{\partial x}\right)^2 + \left(\frac{\partial V}{\partial y}\right)^2\right] + \mu\left(\frac{\partial U}{\partial y} + \frac{\partial V}{\partial x}\right)^2, \tag{7}$$

The turbulent dissipation $\rho\varepsilon$ is:

$$\rho\varepsilon = 2\mu\left[\left(\frac{\partial u'}{\partial x}\right)^2 + \left(\frac{\partial v'}{\partial y}\right)^2\right] + \mu\left(\frac{\partial u'}{\partial y} + \frac{\partial v'}{\partial x}\right)^2, \tag{8}$$

Rotta [7] gives a simple estimation function of entropy generation:

$$S = \frac{\left[\mu\left(\frac{\partial U}{\partial y}\right)^2 - \rho\left(\overline{u'v'}\right)\left(\frac{\partial U}{\partial y}\right)\right]}{T}, \tag{9}$$

Since, the energy flux term plays an important role in the turbulent energy balance in the transition, so it should also take the energy flux term into account in entropy generation. Thus, the entropy generation can be calculated as following form [7,26,27]:

$$S\{\delta\} \approx \frac{\rho}{T}\left[v\int_0^\delta\left(\frac{\partial U}{\partial y}\right)^2 dy - \int_0^\delta\left(\overline{u'v'}\right)\left(\frac{\partial U}{\partial y}\right)dy - \int_0^\delta\left[\left(\overline{u'^2}\right) - \left(\overline{v'^2}\right)\right]\left(\frac{\partial U}{\partial x}\right)dy - \frac{d}{dx}\int_0^\delta\frac{1}{2}U\left(\overline{q^2}\right)dy - \frac{1}{2}v'_\delta\left[\left(u'^2_\delta\right) + \left(v'^2_\delta\right) + \left(w'^2_\delta\right)\right] - \overline{v'_\delta p_\delta}\right], \tag{10}$$

where the first term on the right side of Equation (10) is the viscous term, the second term is the Reynolds shear stress term, the third term is the Reynolds normal stress term; the fourth term is the energy flux term, the fifth term is the turbulent diffusion term, and the last term is the pressure diffusion term. As the last two terms are small, comparing to the magnitudes of other terms, so it can be neglected [26]. All above equations are widely used to estimate the entropy generation especial in experimental work. Since it does not involve the pressure and temperature, only the instant velocity should be measured. It is easy to measure the instant velocity through PIV, hotline probe, and LDV.

3.3. Effect of Decomposition Region Size on POD

In order to explore the effect of decomposition region size on POD results, different decomposition region size has been investigated. The results of four different height cases are shown in Figure 3. It shows that, when the height of the decomposition region is less than 30 mm, the spatial structure of different windows is consecutive, but the symbols of the velocity in W5 and W6 are reverse at different cases. When the height reaches to 35 mm, the POD can even not extract the structure of boundary layer in window 5. This is mainly because the POD extracts a coherent structure according to the total kinetic energy of the flow field. The first-order mode captures more than 99% of the energy at W5 and W6, as shown in Figure 4; the higher-order modes naturally capture a very low energy. Therefore, the higher order modes are easily to be effect by the disturbance in main flow. In W7 and W8, the energy distribution is more uniform because of a more unsteady boundary layer, so it shows more stabilization with different decomposition region sizes. For the same reasons, the size of streamwise decomposition region has the same effect on the POD results, which is not shown in this paper. Thus, in order to extract more flow field structure and avoid the effect of main flow disturbance on the POD mode, the decomposition region size in following work is all 30 mm × 38.5 mm (only shows 20 mm × 38.5 mm).

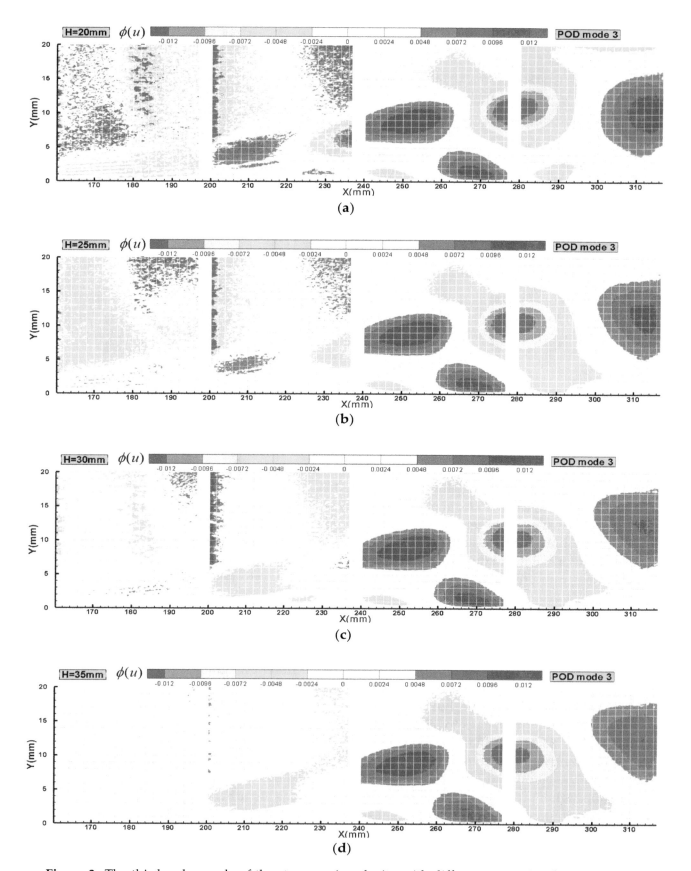

Figure 3. The third-order mode of the streamwsie velocity with different spanwise decomposition region size: (**a**) $H = 20$ mm; (**b**) $H = 25$ mm; (**c**) $H = 30$ mm; and (**d**) $H = 35$ mm.

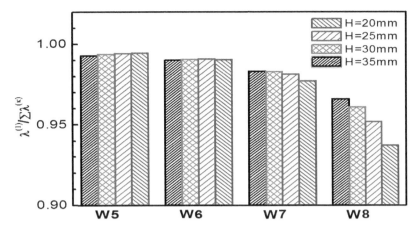

Figure 4. Contribution to total energy of the first order POD mode.

4. Results and Discussion

4.1. Time-Mean Flow Field

Figure 5 shows the time-mean flow field on the flat plate. The thickness of boundary layer decreases gradually with the velocity increase under the favorable pressure gradient, and then increases rapidly under the adverse gradient. The boundary layer separates at about $X = 200$ mm, and it reattaches again at the end of the outlet, where the adverse gradient is disappearance. It forms a time-mean separation bubble in the separated boundary layer, while for the temporary flow field, it consists of a series of vortex. In [28], some criteria can be used to assess if flow history (upstream evolution of the streamwise pressure gradient) have an impact on the development of the boundary layer. The following analyses concentrates on the aft portion of the flat plate boundary layer downstream of the transition.

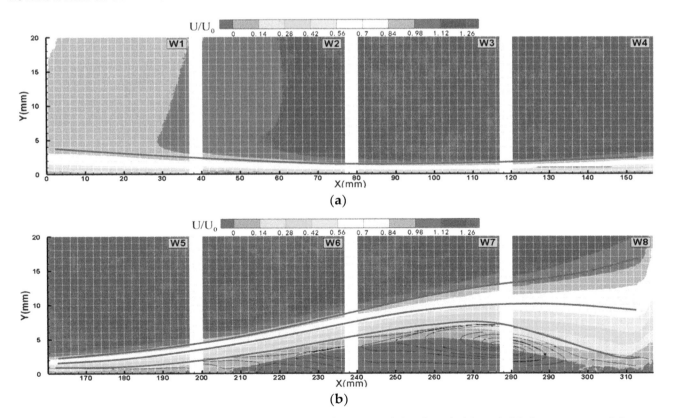

Figure 5. Time mean streamwise velocity: (**a**) the fore part of the flow field; and (**b**) the rear part of the flow field.

Figure 6 shows the normalized velocity pattern of the boundary layer along the streamwise. It is easy to judge the state of the boundary layer through the velocity pattern. The boundary layer at $X/\delta_{x=150} = 41$–46 is a typical attached boundary layer. There is no inflexion point in the velocity pattern, which indicates that the flow at those positions has not yet separated. From the position of $X/\delta_{x=150} = 47.5$, the velocity pattern has obvious inflection point. It indicates that the boundary layer has separated. The height of the inflection point also represents the boundary of the separation bubble shown as the dotted line in Figure 6.

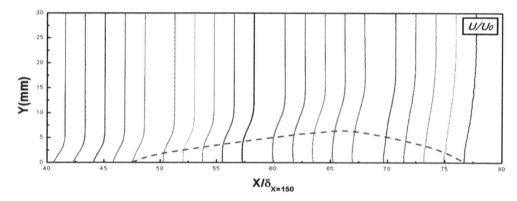

Figure 6. Normalized velocity of the boundary layer.

Figure 7 shows the variation of boundary layer thickness, displacement boundary layer thickness, momentum boundary layer thickness and the shape factor along the streamwise. What should be noted that strong pressure gradient may lead to an inconsistent boundary layer edge by the common techniques to define the boundary layer edge [29]. The shape factor maintains about 2 to 3 in W1–3, where the boundary layer keeps laminar. The boundary layer separates at the position of $X = 200$ mm, the value of the shape factor just consistent with the typical value of the separation laminar boundary. Between the laminar region and the separation region, there is a transition region. Downstream of the transition region, the shape factor increase rapidly due to the increase of the displacement boundary layer thickness. The shape factor start to decrease from the location of $X = 260$ mm, the position corresponds the maximum thickness of the separation bubble.

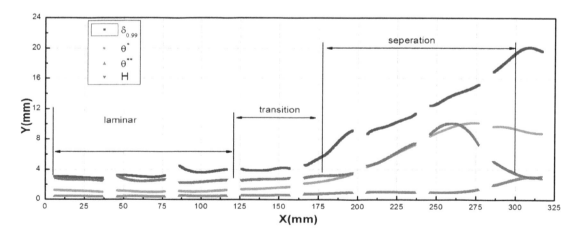

Figure 7. Streamwise variation of the boundary layer parameters.

4.2. POD Analysis of Flow Field

Figure 8 shows the different POD modes of the streamwise velocity. The flow structure of the first order mode is consistent with the time-mean flow field. Since the flow field is almost steady in W5 and W6, the flow structure of different modes is a little similar to each other, while in W7 and W8, the flow structure of different modes is much different from each other because of strong unsteady.

The second order mode is still the large-scale coherent structures that affected by the mean flow. The third–fourth-order modes represent the small coherent structures that induced by the large-scale vortex, and the coherent structure paired increases with the increase of the order number [30].

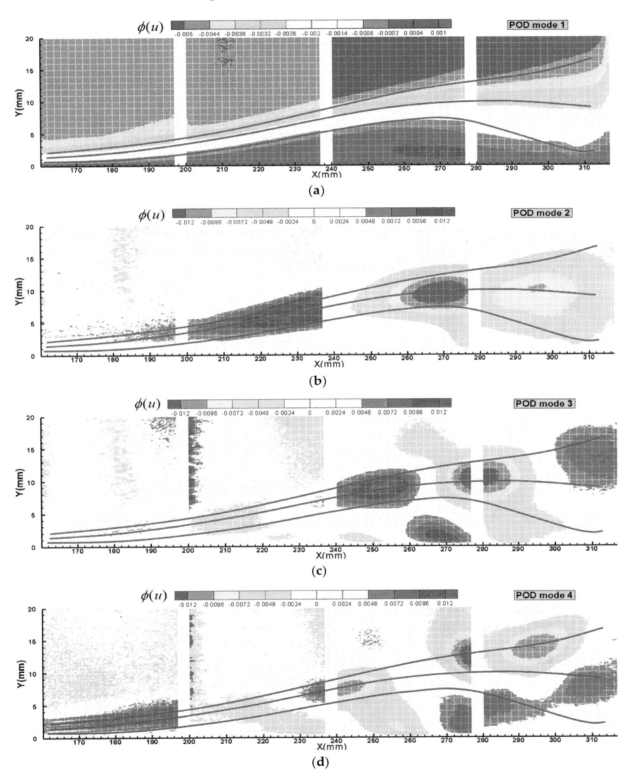

Figure 8. POD modes of streamwise velocity: (**a**) the first order mode; (**b**) the second order mode; (**c**) the third order mode; and (**d**) the forth order mode.

Figure 9 shows the power spectrum density of the POD coefficient, which represents the turbulent kinetic energy of the POD mode. There is a basic frequency (0.024 Hz) in the power spectrum of the POD coefficient, which corresponds with the frequency of the main flow turbulence. The energy mainly concentrates in a low frequency band for those four windows. The amplitude of the energy spectrum increases gradually from W5 to W8 with the increasing unsteady of the flow. The energy at high frequency decreases faster and faster from W5 to W8. The slope of the forth order mode at W8 tends to be $-3/5$, which indicates that the flow field of the forth order mode tends to be fully developed turbulent flow.

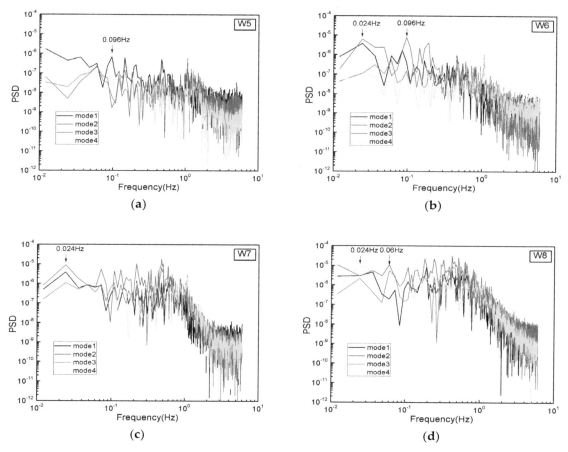

Figure 9. Power spectrum density of the POD coefficient: (**a**) W5; (**b**) W6; (**c**) W7; and (**d**) W8.

According to the Equations (2)–(5), the Reynolds stress of different POD modes are calculated and shown in Figure 10. The distribution of the first order mode Reynolds stress is basically consistent with the original flow field (not given in this paper), which also verifies the correctness of the above equations. The extracted Reynolds normal stress of mode 2 and mode 3 mainly distributes above the separation bubble, while the Reynolds shear stress distributes in the whole boundary layer.

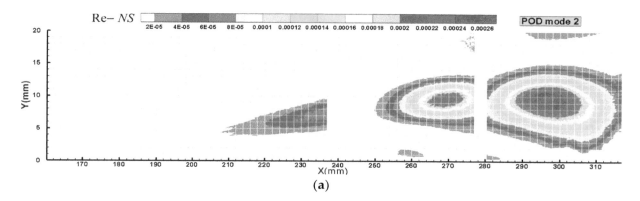

Figure 10. *Cont.*

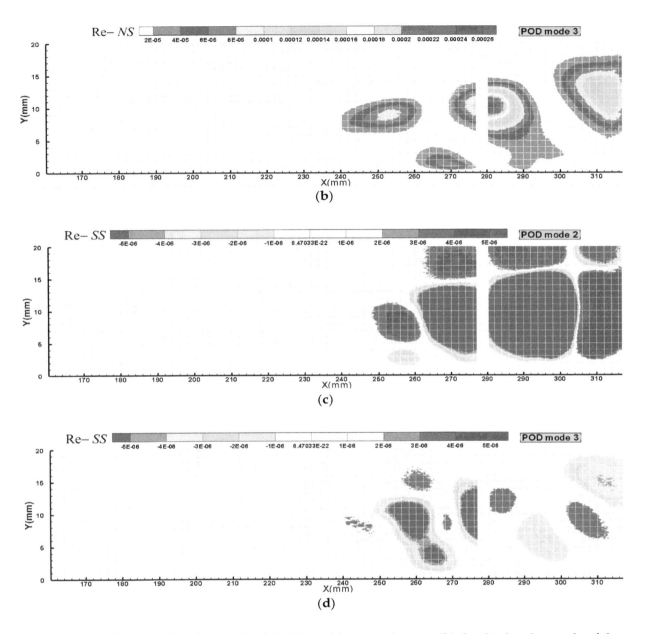

Figure 10. (**a**) The second-order mode of the Reynolds normal stress; (**b**) the third-order mode of the Reynolds normal stress; (**c**) the second-order mode of the Reynolds sheer stress; and (**d**) the third-order mode of the Reynolds sheer stress.

4.3. Entropy Generation Analysis

4.3.1. Entropy Generation of Original Flow Field

According to the Equation (10), the different entropy generation terms are calculated and shown in Figure 11. It shows that the magnitude of the mean viscous dissipation term is much smaller than other terms. The entropy of the mean viscous dissipation only generates in laminar boundary layer and the edge of the separation bubble. Since the velocity in the separation bubble is very small, the entropy generation tends to be very small in the separation bubble. Nevertheless, it does not mean that the separation bubble will not cause any loss. The separation bubble will induce a rapidly increase of the boundary layer thickness and the Reynolds stress dissipation loss in the separated boundary layer is much higher than that of the attached boundary layer, just as Figure 11 shows. The energy flux term distributes in the whole main flow and there is not a clear boundary between the positive area with the negative area.

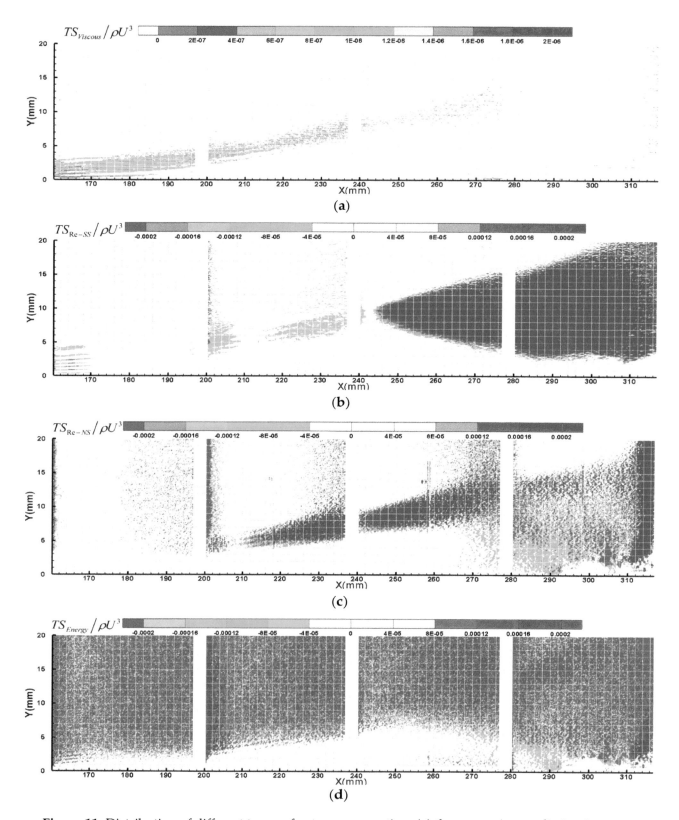

Figure 11. Distribution of different terms of entropy generation: (**a**) the mean viscous dissipation term; (**b**) the Reynolds sheer stress term; (**c**) the Reynolds normal stress term; and (**d**) the energy flux term.

Figure 12 shows the variation of the integrated entropy generation in the boundary along the streamwise. In the laminar region (W1–3), the entropy generation of the laminar boundary layer keeps at a very low level. In the transition region (W4), the amplitude of the energy flux term increases significantly. It indicates that the energy flux term plays an important role in the turbulent kinetic energy balance. The integrated energy flux shows discontinuity between windows and data sparsity, which results from the discrete distribution of energy flux term in the counter map just as the Figure 11 shows. Figure 13 shows the total integrated entropy generation in the boundary layer. It shows that the total entropy generation increases significantly when the boundary layer separated. The Reynolds shear stress term tends to be positive growth and the energy flux term is negative growth, which is consistent with the trend described in [27].

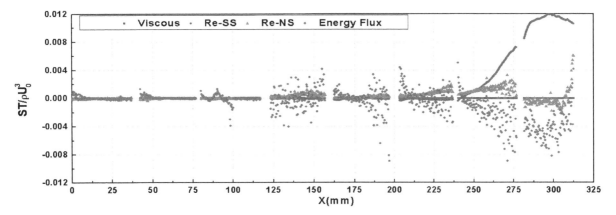

Figure 12. Streamwise variation of integral entropy generation.

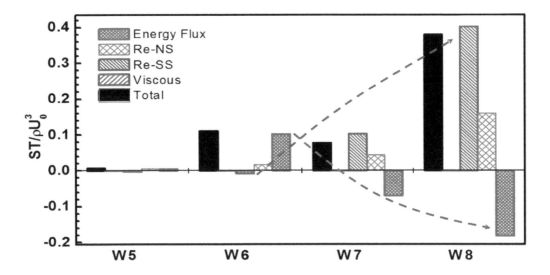

Figure 13. Total integral entropy generation of different terms.

4.3.2. Entropy Generation of POD Mode

As described above, the POD is an effective method to extract the coherent structures. The entropy generation of the coherent structures extracted by POD can be calculated through Equations (2)–(5) and (10). It significant to quantize the entropy generation of the coherent structures. Once you can identify the source of the coherent structures, then the source of the loss production can be confirmed [31].

Figures 14–16 show the entropy generation of different POD modes. The first-order mode of the mean viscous dissipation term is almost equal to that of the original flow field and the value of higher order mode tends to be zero. It also demonstrates that the first order mode represents the mean flow. The entropy generation distribution of the second and third-order mode of the Reynolds stress term and the energy flux term is similar to the distribution of the related POD mode.

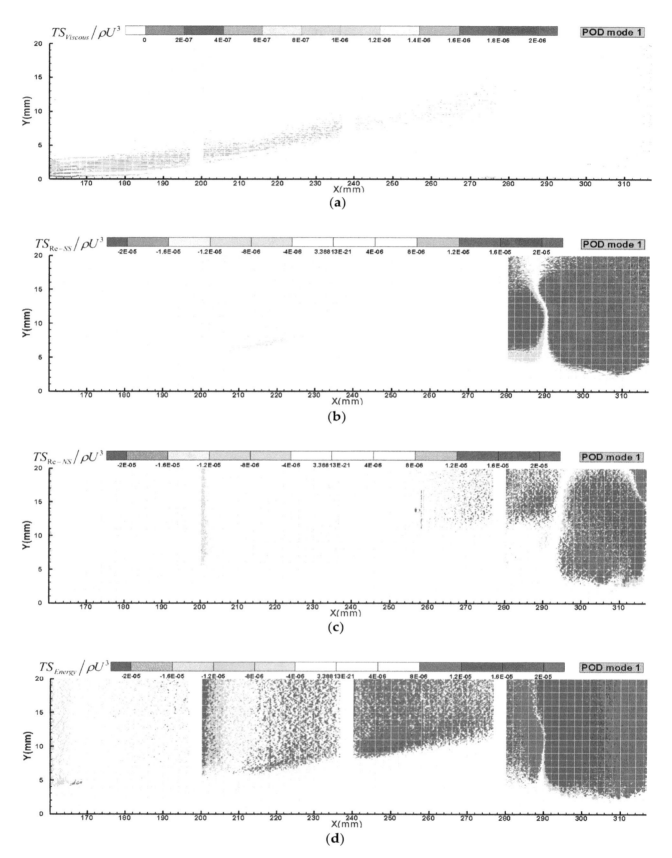

Figure 14. Entropy generation of first order mode: (**a**) the mean viscous dissipation term; (**b**) the Reynolds sheer stress term; (**c**) the Reynolds normal stress term; and (**d**) the energy flux term.

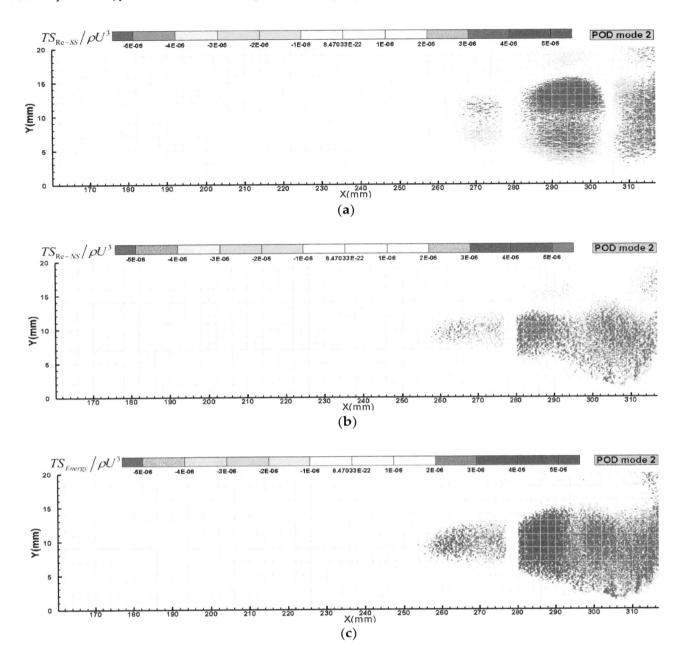

Figure 15. Entropy generation of second order mode: (**a**) the Reynolds sheer stress term; (**b**) the Reynolds normal stress term; and (**c**) the energy flux term.

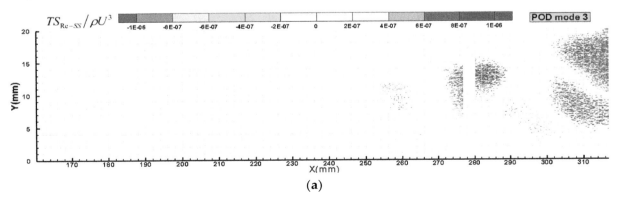

Figure 16. *Cont.*

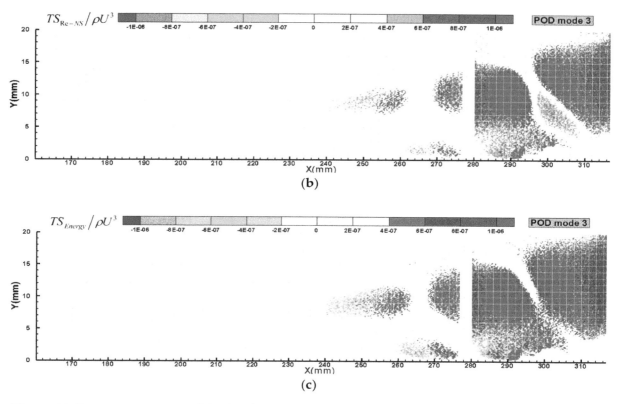

Figure 16. Entropy generation of third order mode: (**a**) the Reynolds sheer stress term; (**b**) the Reynolds normal stress term; and (**c**) the energy flux term.

Figure 17 shows the cumulative contribution to the total integral entropy generation of each POD mode in the boundary layer. The contribution rate of the first order mode of the mean viscous term reaches 100% in all of those windows. In W5, the contribution rate of the first two modes of the Reynolds stress (normal and sheer) is about 10%, the magnitude of the higher modes is almost at the same level. Thus, the contribution of each mode seems to be the same and the cumulative contribution line increases with a constant slope. The Reynolds sheer stress term in W6 has the same situation with that in W5, while the first third-order mode of the Reynolds normal stress in W6 contributes about 80% of the total entropy generation. The contribution rate of the energy flux term reaches to 90% until the cumulative order number of the POD mode is about 400 in both W5 and W6.

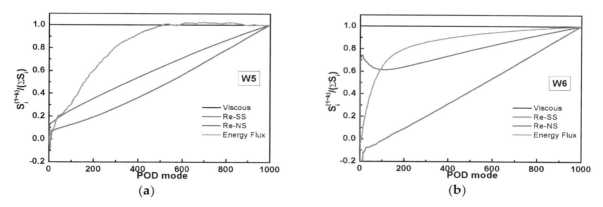

Figure 17. *Cont.*

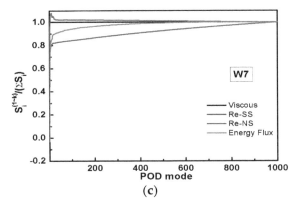
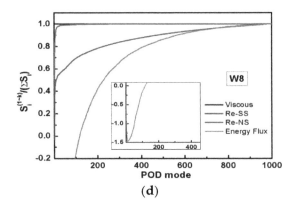

Figure 17. Cumulative contribution to the total integral entropy generation of each POD mode: (**a**) W5; (**b**) W6; (**c**) W7; and (**d**) W8.

In W7, the first third-order modes of Reynolds stress (normal and sheer) term contributes more than 80% of the total entropy generation, the higher order contributes less than 20%. In W8, the contribution of the first 30th-order modes of the Reynolds stress (normal and sheer) is about 60%. It indicates that the contribution of the Reynolds stress term of the low-order mode in the rear part of the separation bubble (W7 and W8) is much higher than that in the fore part of the separation bubble (W5). The low order mode of the energy flux term in W8 contributes the negative value, which results from the negative energy captured by the first-order mode just as shown in Figure 16.

5. Conclusions

In this paper, the boundary layer of the flat plate with a pressure gradient in the water tunnel has been studied using high-resolution particle image velocimetry (PIV). The entropy generation rate is analyzed by proper orthogonal decomposition (POD) applied to the measurements. Several conclusions can be made.

The separation bubble will dramatically increase the thickness of the boundary layer and result in a sharp increase of the loss. The loss due to the Reynolds normal stress mainly distributes above the separation bubble, and the loss due to the Reynolds shear stress distributes in the whole boundary layer. The decomposition region size has a significantly effects on the POD result of the laminar boundary layer. It should reduce the proportion of the main flow area as much as possible when carrying out the POD analysis. In the transition region, the energy flux terms plays an important role in transiting the energy from the mean flow to the turbulent flow and contributes a large ratio to the entropy generation. Estimation of the entropy generation of the POD mode helps to explore the mechanism of entropy generation and identify the source of the loss production. Additionally, it is easy to promote this method to some other complex flows, such as wake-induced separation on the suction side of blade, or the leakage flow on the blade tip from either numerical or experimental data, which is only based on the instantaneous velocity field.

Author Contributions: Conceptualization: C.J.; data curation: C.J.; investigation: C.J.; methodology: C.J.; software: C.J.; supervision: H.M.; validation: C.J.; writing—original draft: C.J.; writing—review and editing: H.M.

Nomenclature

H	Shape factor
Re	Reynolds number
S	Entropy generation
T	Temperature
u	Instantaneous streamwise velocity
u'	Streamwise velocity fluctuation $(u - U)$
U	Averaged streamwise velocity

U_0	Streamwise velocity of main flow
v	Instantaneous spanwise velocity
v'	Spanmwise velocity fluctuation $(v - V)$
V	Averaged spanwise velocity
χ	POD coefficients
λ	Eigenvalue of POD
ϕ	Basis function of POD
ρ	Density
μ	Kinetic viscosity
υ	Kinematic viscosity
δ	Boundary layer thickness
θ^*	Displacement boundary layer thickness
θ^{**}	Momentum boundary layer thickness

References

1. Denton, J.D. The 1993 IGTI Scholar Lecture: Loss Mechanisms in Turbomachines. *J. Turbomach.* **1993**, *115*, 621–656. [CrossRef]
2. Bejan, A.; Kestin, J. Entropy Generation through Heat and Fluid Flow. *J. Appl. Mech.* **1983**, *50*, 475. [CrossRef]
3. Naterer, G.F.; Camberos, J.A. *Entropy Based Design and Analysis of Fluids Engineering Systems*; CRC Press: Boca Raton, FL, USA, 2008.
4. Bejan, A. *Entropy Generation Minimization*; Advanced Engineering Thermodynamics; John Wiley & Sons, Inc.: Hoboken, NJ, USA, 1995.
5. Enrico, S. Calculating entropy with CFD. *Mech. Eng.* **1997**, *119*, 86–88.
6. Suder, K.L.; Obrien, J.E.; Reshotko, E. *Experimental Study of Bypass Transition in a Boundary Layer*; National Aeronautics & Space Administration Report; NASA: Washington, DC, USA, 1988.
7. Rotta, J.C. Turbulent boundary layers in incompressible flow. *Prog. Aerosp. Sci.* **1962**, *2*, 1–95. [CrossRef]
8. Mceligot, D.M.; Walsh, E.J.; Laurien, E.; Spalart, P.R. Entropy Generation in the Viscous Parts of Turbulent Boundary Layers. *J. Fluids Eng.* **2008**, *130*, 61205. [CrossRef]
9. Moore, J.; Moore, J.G. Entropy Production Rates from Viscous Flow Calculations: Part I—A Turbulent Boundary Layer Flow. In Proceedings of the ASME 1983 International Gas Turbine Conference and Exhibit, Phoenix, AZ, USA, 27–31 March 1983.
10. Kramer-Bevan, J.S. A Tool for Analyzing Fluid Flow Losses. Master's Thesis, University of Waterloo, Waterloo, ON, Canada, 1992.
11. Adeyinka, O.B.; Naterer, G.F. Apparent Eentropy Production Difference with Heat and Fluid Flow Irrevesibilities. *Numer. Heat Transf. Part B Fundam.* **2002**, *42*, 411–436. [CrossRef]
12. Adeyinka, O.B.; Naterer, G.F. Modeling of Entropy Production in Turbulent Flows. *J. Fluids Eng.* **2005**, *126*, 893–899. [CrossRef]
13. Hanjalić, K.; Launder, B. *Modelling Turbulence in Engineering and the Environment*; Cambridge University Press: Cambridge, UK, 2000.
14. Lumley, J.L. *Stochastic Tools in Turbulence*; Dover Publications: Mineola, NY, USA, 2007.
15. Schmid, P.; Sesterhenn, J. Dynamic mode decomposition of experimental data. In Proceedings of the 8th International Symposium on Particle Image Velocimetry, Melbourne, Australia, 25–28 August 2009.
16. Cammilleri, A.; Gueniat, F.; Carlier, J.; Pastur, L.; Mémin, E.; Lusseyran, F.; Artana, G. POD-spectral decomposition for fluid flow analysis and model reduction. *Theor. Comput. Fluid Dyn.* **2013**, *27*, 787–815. [CrossRef]
17. Zhu, J.; Huang, G.; Fu, X.; Fu, Y.; Yu, H. Use of POD Method to Elucidate the Physics of Unsteady Micro-Pulsed-Jet Flow for Boundary Layer Flow Separation Control. In Proceedings of the ASME Turbo Expo 2013: Turbine Technical Conference and Exposition, San Antonio, TX, USA, 3–7 June 2013.
18. Hammad, K.J. Coherent Structures in Turbulent Boundary Layer Flows over a Shallow Cavity. In Proceedings of the ASME 2017 International Mechanical Engineering Congress and Exposition, Tampa, FL, USA, 3–9 November 2017.

19. Anbry, N. The dynamics of coherent structures in the wall region of a turbulent boundary layer. *J. Fluid Mech.* **1988**, *192*, 115–173.

20. Chen, H.; Reuss, D.L.; Sick, V. On the use and interpretation of proper orthogonal decomposition of in-cylinder engine flows. *Meas. Sci. Technol.* **2012**, *23*, 085302. [CrossRef]

21. Cizmas, A.; Paul, G.; Palacios, A. Proper Orthogonal Decomposition of Turbine Rotor-Stator Interaction. *J. Propuls. Power* **2003**, *19*, 268–281. [CrossRef]

22. Lengani, D.; Simoni, D. Recognition of coherent structures in the boundary layer of a low-pressure-turbine blade for different free-stream turbulence intensity levels. *Int. J. Heat Fluid Flow* **2015**, *54*, 1–13. [CrossRef]

23. Lengani, D.; Simoni, D.; Ubaldi, M.; Zunino, P.; Bertini, F. Experimental Investigation on the Time–Space Evolution of a Laminar Separation Bubble by Proper Orthogonal Decomposition and Dynamic Mode Decomposition. *J. Turbomach.* **2016**, *139*, 31006. [CrossRef]

24. Tian, Y.; Ma, H.; Wang, L. An Experimental Investigation of the Effects of Grooved Tip Geometry on the Flow Field in a Turbine Cascade Passage Using Stereoscopic PIV. In Proceedings of the ASME Turbo Expo 2017: Turbomachinery Technical Conference and Exposition, Charlotte, NC, USA, 26–30 June 2017.

25. Vinuesa, R.; Schlatter, P.; Nagib, H.M. Role of data uncertainties in identifying the logarithmic region of turbulent boundary layers. *Exp. Fluids* **2014**, *55*, 1751. [CrossRef]

26. Walsh, E.J.; Mc Eligot, D.M.; Brandt, L.; Schlatter, P. Entropy Generation in a Boundary Layer Transitioning under the Influence of Freestream Turbulence. *J. Fluids Eng.* **2011**, *133*, 61203. [CrossRef]

27. Skifton, R.S.; Budwig, R.S.; Crepeau, J.C.; Xing, T. Entropy Generation for Bypass Transitional Boundary Layers. *J. Fluids Eng.* **2017**, *139*, 041203. [CrossRef]

28. Vinuesa, R.; Orlu, R.; Vila, C.S.; Ianiro, A.; Discetti, S.; Schlatter, P. Revisiting History Effects in Adverse-Pressure-Gradient Turbulent Boundary Layers. *Flow Turbul. Combust.* **2017**, *99*, 565–587. [CrossRef] [PubMed]

29. Vinuesa, R.; Bobke, A.; Örlü, R.; Schlatter, P. On determining characteristic length scales in pressure-gradient turbulent boundary layers. *Phys. Fluids* **2016**, *28*, 55101. [CrossRef]

30. Vila, C.S.; Orlu, R.; Vinuesa, R.; Schlatter, P.; Ianiro, A.; Discetti, S. Adverse-Pressure-Gradient Effects on Turbulent Boundary Layers: Statistics and Flow-Field Organization. *Flow Turbul. Combust.* **2017**, *99*, 589–612. [CrossRef] [PubMed]

31. Lengani, D.; Simoni, D.; Ubaldi, M.; Zunino, P.; Bertini, F.; Michelassi, V. Accurate Estimation of Profile Losses and Analysis of Loss Generation Mechanisms in a Turbine Cascade. *J. Turbomach.* **2017**, *139*, 121001–121007. [CrossRef]

12

Investigation of the Concepts to Increase the Dew Point Temperature for Thermal Energy Recovery from Flue Gas, using Aspen ®

Nataliia Fedorova [1,2]🆔, Pegah Aziziyanesfahani [1], Vojislav Jovicic [1,2,*]🆔, Ana Zbogar-Rasic [1], Muhammad Jehanzaib Khan [1] and Antonio Delgado [1,2]

[1] Institute of Fluid Mechanics (LSTM), Friedrich-Alexander University (FAU), 91058 Erlangen, Germany; nataliia.fedorova@fau.de (N.F.); pegah.aziziyan@gmail.com (P.A.); ana.zbogar-rasic@fau.de (A.Z.-R.); muhammad.j.khan@fau.de (M.J.K.); antonio.delgado@fau.de (A.D.)

[2] Erlangen Graduate School in Advanced Optical Technologies (SAOT), 91054 Erlangen, Germany

* Correspondence: vojislav.jovicic@fau.de

Abstract: Thermal energy of flue gases (FG) dissipating from industrial facilities into the environment, constitute around 20% of the total dissipated thermal energy. Being part of the FG, water vapour carries thermal energy out of the system in the form of the latent heat, which can be recovered by condensation, thus increasing the overall efficiency of an industrial process. The limiting factor in this case is the low dew point temperature (usually 40–60 °C) of the water vapour in the FG. The increase of the dew point temperature can be achieved by increasing the water content or pressure. Taking these measures as a basis, the presented work investigated the following concepts for increasing the dew point temperature: humidification of the flue gas using water, humidification using steam, compression of the FG and usage of the steam ejector. Modelling of these concepts was performed using the commercial software Aspen®. The humidification of the FG using water resulted in the negligible increase in the dew point (3 °C). Using steam humidification the temperatures of up to 92 °C were reached, while the use of steam ejector led to few degrees higher dew point temperatures. However, both concepts proved to be energy demanding, due to the energy requirements for the steam generation. The FG compression enabled the achievement of a 97 °C dew point temperature, being both energy-efficient and exhibiting the lowest energy cost.

Keywords: thermal energy recovery; flue gas; dew point temperature; condensation; Aspen®

1. Introduction

The dissipation of thermal energy from industrial facilities during production processes is and has been a challenging issue worldwide. Flue gases from industrial processes constitute around 20% of the total dissipated energy [1–3]. Water vapour, as a part of the flue gas, carries latent heat, which can be recovered. Depending on the industry, flue gas temperatures vary from 120 to 200 °C [4–6], whereas water vapour content can be up to 90 %vol. For example, as presented in Table 1, in the flue gas of natural gas-fired boilers 20 %vol accounts for the water vapour [3,6], while in the potato crisps manufacturing its content is around 40 %vol [5]. Unlike the flue gas from the drying processes, which can contain up to 90% of the water vapour [7], most of the flue gases from the other processes have a much lower water vapour content. Baking, textile, pulp and paper industries have a potential for the recovery of thermal energy from the flue gas, due to the large quantity of water vapour in their flue gases and massive production rates [8–10].

Table 1. Composition of different industrial flue gases.

Industry	H_2O [%]	CO_2 [%]	N_2 [%]	O_2 [%]
Potato crisps manufacturing [5]	41.1	5	50.6	3.3
Natural gas-fired boiler [11]	18–20	8–10	67–72	2–3
Coal-fired boiler [11]	8–10	12–14	72–77	3–5

The difference in the concentration levels of water vapour in flue gases of different processes is in general less influenced by the type of the fired fuel and more by the amount of the water vapour originating from the process itself. In many of the processes, especially in the food industry, a single fan at the outlet of the system evacuates all the gases from the system. In this way, the flue gas originating from the burners is commonly mixed with the gases originating from the processed product and also often with the excess air, sucked into the system from the surrounding. In the processes like drying or baking, the raw material (paper, potato, dough, etc.) can release significant amounts of the water vapour leading to the increase of the overall water vapour concentration in the flue gas at the outlet of the system.

The higher the amount of water vapour in the flue gas, the more latent heat can be recovered from it. Therefore, this issue is essentially relevant for industries with water vapour-rich flue gases. Recovering sensible and latent heat by water vapour condensation from the flue gas has been reported by different sources as a promising way to improve the total energy efficiency by around 10% [3,7,12]. Herewith come economic and environmental benefits.

A traditional unit for thermal energy recovery from the flue gas is a gas-to-liquid condensing heat exchanger. During the heat exchange with the cooling liquid, the flue gas temperature is reduced below the dew point, so that the water vapour is condensed and the release of the latent and sensible heat occurs. Porous and non-porous gas separation membranes have been actively developed for the simultaneous heat and water recovery from the flue gas [7,13]. The growing interest for the thermal energy recovery application is connected to the Organic Rankine Cycle (ORC) technology for generation of electricity and absorption refrigerator (AR) technology for driving cooling processes [5,14–16].

One of the factors limiting the application of the aforementioned recovery systems is the low dew point temperature of the water vapour in the flue gas, usually in the range of 40–60 °C, i.e., the temperature increase of the working liquid is not sufficient to be used as process heat within the production cycle. Increasing the dew point temperature can help to upgrade the temperature level of working fluids and to reach higher energy efficiencies. For example, in AR the increase of the thermal source temperature from 80 °C to 86 °C, increases the coefficient of performance by 7% [17,18].

On the example of the condensing heat exchangers (HE) as a recovery technology, the increase of the dew point temperature of the water vapour contained in the flue gas leads to obtaining the condensate and the cooling water at the higher temperature level. In this way, these otherwise waste streams can be further used within an industry as process heat for a wide range of applications, e.g., space heating, heating, ventilation and air-conditioning (HVAC) of office areas, as sanitary water, for washing and cleaning in the production areas, for preparation of dough in the case of a baking industry, for preheating of fuel and combustion air in the case of a power generation industry [19–21], etc. Other industry-dependent in-plant demands for recovered thermal energy should be generally determined by conducting a full energy audit.

Process modelling is one of the ways to analyse the thermal energy recovery from industrial processes. The licensed software Aspen Plus® and Aspen® HYSYS® find applications in modelling of chemical, biological and physical processes. It enables to model complex processes using simple models with built in unit operation models (e.g., heat exchangers, columns, reactors, mixers, splitters, etc.) and property methods.

The Aspen® software is a widely used tool for analysing system performances. Jana et al. [22] used it to model the utilisation of the waste heat by means of a condensing HE for the post-combustion CO_2 capture. Luyben [23] simulated HEs with phase changes, namely a condenser and an evaporator, for low-level energy recovery with n-hexane as a working fluid. Duan et al. [24] analysed coal gasification, which included a gasifier and a boiler to recover steam and blast furnace slag. Ishaq et al. [25] modelled a trigeneration system for electricity, hydrogen and fresh water production from the flue gas of a glass melting furnace. Mazzoni et al. [26] proposed ORC plant arrangements, based on the turbo-expander pumping system and internal regeneration processes for low grade waste thermal energy recovery, aiming to improve the plant efficiency and to reduce the cooling load on the condenser.

The motivation behind the presented research was to investigate the concepts, leading to the increase of the dew point temperature of the water vapour contained in the flue gas (preferably above 80 °C), in order to recover its latent and sensible heat. It is generally known, that the increase of the dew point temperature can be achieved through the increase of the water vapour share in the flue gas or through the increase of the flue gas pressure. Taking these measures as a basis, the presented work investigated the following concepts for increasing the dew point temperature: humidification of the flue gas using water, humidification of the flue gas using steam, compression of the flue gas and the usage of the steam ejector.

The investigated models were developed in Aspen Plus® V8.8 and Aspen® HYSYS® V8.8. The condensing shell-and-tube HE was chosen as a thermal energy recovery technology. Under examination was the influence of the dew point temperature of the water vapour in the flue gas and the flow rate of the cooling water in the HE on the amount of the recovered thermal energy, the temperature of the cooling water at the HE outlet and the flow rate of the condensate. In order to analyse each concept from the economic point of view, the energy cost was estimated.

2. Model Description and Methodology

Within the presented work, four concepts leading to the increase of the dew point temperature of the water vapour in the flue gas were investigated, by means of the aspenONE® Engineering Suite:

1.   Humidification of the flue gas using water (Aspen Plus® V8.8)
2.   Humidification of the flue gas using steam (Aspen Plus® V8.8)
3.   Compression of the flue gas (Aspen Plus® V8.8)
4.   Usage of the steam ejector (Aspen® HYSYS® V8.8)

The steady state process modelling based on the energy balance was performed for each concept. The Peng-Robinson equation of state was chosen as a physical property method, commonly used in the gas processing industry. The investigated models were established under the following assumptions: the flue gas follows the ideal gas behaviour, the flue gas is composed of noncondensable (dry air) and condensable gas (water vapour), which is the approach followed in many studies [3,6,27].

In the tested concepts, the parameters of the flue gas at the outlet of the industrial baking oven were measured and used for calculations. The flue gas at the exit of an industrial baking oven had the temperature of 120 °C and the pressure of 1.01 bar, with the dew point temperature of 67 °C. The flue gas mass flow rate was 276.2 kg/h, with the air mass flow rate of 229.4 kg/h (83 %mas) and 46.8 kg/h for water vapour (17 %mas).

2.1. Concept 1: Humidification of the Flue Gas Using Water

Modelling of the concept with humidification of the flue gas using water, shown in Figure 1, is realised by adding WATER (95 °C and 1.01 bar) into the flue gas stream EXGAS. During perfect mixing (MIXER unit model) water evaporates using the thermal energy of the flue gas. Evaporation of the sprayed water increases the concentration of the water vapour in the flue gas stream and consequently its partial pressure and the GASIN dew point temperature.

The humidified flue gas stream GASIN leaves the mixer and is cooled in the counter-current heat exchanger HE (HEATX model) by the flow of WATERIN, supplied at a temperature of 60 °C and pressure of 1.01 bar. During the heat exchange between the two streams, the water vapour from the humidified flue gas stream condenses at constant pressure, and the condensate CONDENS occurs. The heated cooling water WATEROUT and the cooled dried flue gas GASOUT leave the HE.

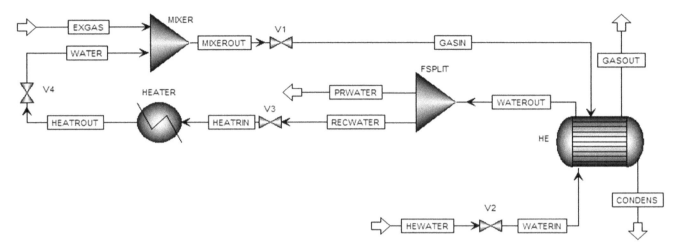

Figure 1. Flowsheet for the humidification of the flue gas using water.

The heated cooling water is then separated in the stream splitter (FSPLIT model), depending on the required amount of water for the GASIN dew point temperature increase. The first part, the water for further recycling RECWATER, is heated in the HE (HEATER model), in order to be sprayed into the MIXER. The second part, the process water PRWATER, can be used in other applications required within the industry, and is considered as the useful process stream, obtained from the thermal energy recovery cycle of the flue gas.

2.2. Concept 2: Humidification of the Flue Gas Using Steam

In the process model of the second investigated concept, the steam generator SG (HEATER model) is provided for the production of STEAM (120 °C and 1.01 bar). This steam is further used for humidification of the flue gas stream EXGAS, as presented in Figure 2. The recycling water in the concept RECWATER passes through the SG, in order to be evaporated. The other components are the same as in the previous model.

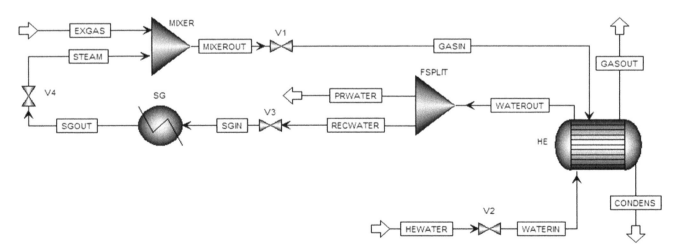

Figure 2. Flowsheet for the humidification of the flue gas using steam.

It is expected that in the case of the steam humidification, the higher GASIN dew point temperature can be reached, since there is no need to use the thermal energy of the flue gas for the evaporation of water, as in concept 1.

2.3. Concept 3: Compression of the Flue Gas

The third investigated concept is the compression of the flue gas, demonstrated in Figure 3. The flue gas stream EXGAS is directed to the compressor (COMPR model), where during the isentropic compression its temperature and pressure increase. The compressed flue gas stream GASIN is then led to the heat exchanger HE, as in the previously described cases. The heated cooling water WATEROUT at the outlet of the HE is not needed further in the recovery process, thus it can be fully utilized as process heat for different applications within the industry.

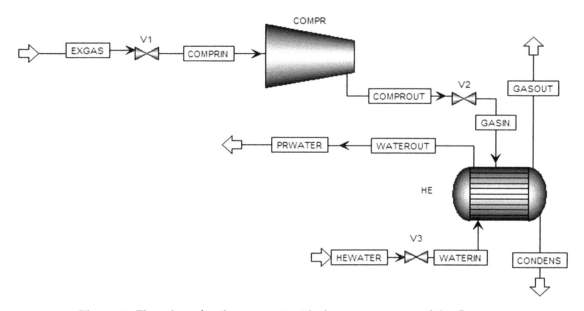

Figure 3. Flowsheet for the concept with the compression of the flue gas.

2.4. Concept 4: Usage of the Steam Ejector

Using the steam ejector leads to the GASIN dew point temperature increase, due to the simultaneous increase in the water content of the flue gas and its pressure, resulting from the geometry of the device. The flowsheet for concept 4 is given in Figure 4. The modelling was performed in the Aspen® HYSYS® software, since Aspen Plus® does not incorporate the steam ejector model. The other units are the same as in the previously described concepts.

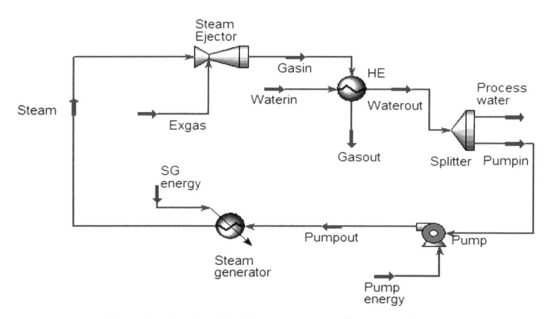

Figure 4. Flowsheet for the concept with the steam ejector.

The flue gas stream EXGAS is introduced from the low pressure side of the STEAM EJECTOR (Ejector model), whereas STEAM as a motive fluid is supplied from the high-pressure side. The discharged stream GASIN is directed to the HE (Heat Exchanger model), where the water vapour condenses from the flue gas during water cooling by WATERIN. The heated cooling water WATEROUT is separated in the SPLITTER (Tee model) into two parts: (1) PUMPIN, recirculated to generate the high pressure steam, and (2) PROCESS WATER, utilized for the needs of the industry. Pressure of the water PUMPIN is increased in the PUMP (Pump model), and then the high pressure water PUMPOUT is directed to the STEAM GENERATOR (Heater model) to be evaporated at the constant pressure. The generated high pressure STEAM is led to the STEAM EJECTOR to humidify the flue gas.

2.5. Heat Exchanger Design

The shell-and-tube heat exchanger for extracting the thermal energy from the flue gas was designed using the Aspen Shell & Tube Exchanger^{TM}. The following inlet parameters for the HE were used: physical properties of the inlet streams (T, p, mass flow and composition), temperature change, pressure drops in the shell and the tube side and the fouling factors for each inlet fluid. The possible HE design, according to the given inlet parameters, coupled with the failure analysis was provided by the software. The suggested design was improved by trial-and-error calculations following Kern's method [28]. The HE with the key design parameters, shown in Table 2, was used for modelling of all the previously described concepts.

Table 2. Design parameters of the HE.

Shell OD (mm)	Tube OD (mm)	Tube Length (mm)	Tube Pattern	Number of Tubes	Baffle Spacing (mm)	Number of Baffles
219	19	2700	Triangular	49	0.15	15

The effectiveness of the heat exchanger was calculated using the Number of Transfer Units (NTU) method. The typical values for the reference case with the dew point temperature of 67 °C lay in the range of 83–93% for the flow rate of cooling water of 400–2000 kg/h, respectively.

The allowable pressure drops were determined, based on the standard values given by Aspen®, as 0.1 bar and 0.2 bar for the shell and the tube side, respectively. The fouling factor of 0.0003 m²·°C/W for the cooling water and 0.0005 m²·°C/W for the flue gas were taken from the literature [29].

2.6. Theoretical Calculations

All the calculations and considerations in this work are related to one specific industrial facility, namely the natural gas fired, tunnel baking oven. The limitations related to the investigated facility (composition, temperature and flow rate of the flue gas, fluid parameters at the inlet/outlet of the process components, etc.) influence most of the obtained results. Therefore, it is imperative to perform the similar analysis for each specific system of interest with its unique properties.

Based on the parameters of the flue gas, previously listed in the preface to chapter two of this work, the inlet values for process modelling were determined by thermodynamic calculations. Few major ones are presented below with Equations (1)–(4). The used terminology can be found in Nomenclature.

Humidity ratio of an air-vapour mixture:

$$x = \frac{m_w}{m_{air}} \tag{1}$$

Partial pressure and density of water vapour in the flue gas:

$$P_w = \frac{x \cdot P}{0.622 + x} \tag{2}$$

$$\rho_w = \frac{P_w}{R_w \cdot T} \tag{3}$$

Dalton's law for air-vapour mixtures:

$$P = y_{air}P + y_{vap}P \tag{4}$$

The dew point of the water vapour was taken from the thermodynamic tables for the saturated vapour pressure values (P_i^s), calculated using Raoult's law:

$$y_i P = x_i P_i^s \tag{5}$$

The amount of the recovered thermal energy, representing the enthalpy difference between the flue gas at the HE inlet (GASIN stream) and at HE outlet (GASOUT stream), was calculated with Equations (6)–(8):

$$Q_{RE} = m_{air} \cdot (h_{in} - h_{out}) \tag{6}$$

$$h_{in} = \left(C_{p,air} \cdot t_{gas,in}\right) + x_{in}\left(C_{p,vap} \cdot t_{gas,in} + \Delta h_{vap}\right) \tag{7}$$

$$h_{out} = \left(C_{p,air} \cdot t_{gas,out}\right) + x_{out}\left(C_{p,vap} \cdot t_{gas,out} + \Delta h_{vap}\right) \tag{8}$$

For each investigated concept described above (Figures 1–4), the energy gain and the energy demand were estimated. The energy dissipation in the pipelines between the system components was neglected. As energy gain (Equations (9)–(12)) was considered the case, when the temperature of the cooling water, of the flue gas and of the condensate at the HE outlet exceeded $t_{washing}$. The temperature $t_{washing}$ was set to be 60 °C, based on the average temperature level useful for washing purposes within the production.

$$Q_{gain} = Q_{w,out} + Q_{gas,out} + Q_{condens} \tag{9}$$

$$Q_{w,out} = m_{prwater} \cdot C_{p,w} \cdot \left(t_{w,out} - t_{washing}\right) \tag{10}$$

$$Q_{gas,out} = m_{gas,out} \cdot C_{p,gas} \cdot \left(t_{gas,out} - t_{washing}\right) \tag{11}$$

$$Q_{condens} = m_{condens} \cdot C_{p,condens} \cdot \left(t_{condens} - t_{washing}\right) \tag{12}$$

The energy demand for increasing the dew point temperature in each concept was calculated as the difference between the thermal energy of a stream at the outlet and the inlet of a device, taking the corresponding device's efficiency into account (90% for the heater, 90% for the steam generator, 85% for the compressor and 90% for the pump [30,31]):

$$Q_{dem} = \frac{Q_{d,out} - Q_{d,in}}{\eta} \tag{13}$$

The energy cost was estimated for each tested concept, assuming that: (1) the flue gas compressor (concept 3) is driven by the electro motor, while (2) the heater and the steam generator (concepts 1, 2 and 4) use natural gas as fuel. The average electricity and the natural gas prices for German industries in 2017 were 0.127 €/kWh and 0.026 €/kWh, respectively [32].

3. Results and Discussion

The effects of the dew point temperature of the water vapour in the flue gas (Θ) and the flow rate of the cooling water in the HE ($m_{w,in}$) on the recovered thermal energy (Q_{RE}), the water temperature at the HE outlet ($t_{w,out}$) and the condensate flow rate ($m_{condens}$) were investigated for each concept described above. The fitting curves were created using the Origin® 2019 by polynomial fit of the second order with $R^2 \in [0.9882-1]$.

3.1. Concept 1: Humidification of the Flue Gas Using Water

In this concept, water (t = 95 °C and p = 1.01 bar) for humidification of the flue gas is sprayed directly into the flue gas stream. The computational results of the calculations, obtained for concept 1, are presented in Figure 5.

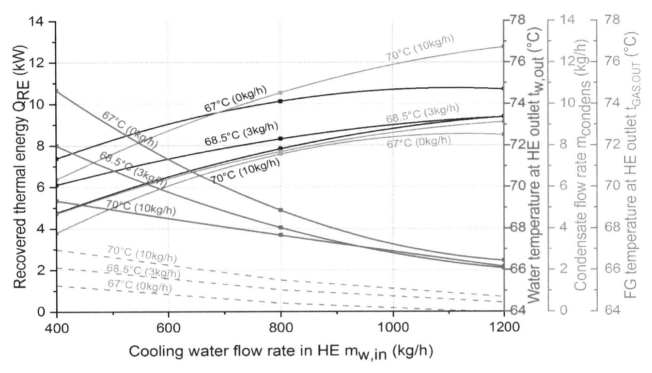

Figure 5. Influence of the cooling water flow rate in the heat exchanger (HE) and the dew point temperature on the recovered thermal energy (black lines), the water temperature at the HE outlet (red lines), the condensate flow rate (green lines) and the flue gas temperature at the HE outlet (gray lines) for concept 1.

Increasing the flow rate of sprayed water (0, 3, 10 kg/h) leads to an increase of the dew point temperature from 67 °C (no water added) to 70 °C (water addition of 10 kg/h). The relatively low temperature increase of $\Delta T = 3$ °C can be attributed to the negligible rise in the water content of the flue gas. On the other hand, a further increase of the sprayed water amount above 10 kg/h is limited by the ability of the flue gas to evaporate the sprayed water.

The increase of the dew point temperature from 67 °C to 70 °C results in the 8% reduction of the water temperature at the outlet of the HE ($t_{w,out}$) for 400 kg/h flow rate of cooling water. This is due to the fact that in case of the low flow rate of cooling water, the transferred heat is mostly sensible and, therefore, higher for the flue gas with the lower dew point temperature. By increasing the flow rate of cooling water more condensate is generated, meaning that the transferred heat is both sensible and latent. The same explanation is applicable to the thermal energy recovery trend line, which increases with the increase of the cooling water flow rate.

Nevertheless, due to the low amount of the recovered thermal energy (maximum of 10.5 kW) and unattainability of the high dew point temperature, this concept does not meet the goals of the research and is not recommended for the practical use in the investigated facility.

3.2. Concept 2: Humidification of the Flue Gas Using Steam

The computational results of the calculations, obtained for concept 2, are presented in Figure 6. Steam for the humidification purpose is supplied in this case at the temperature of 120 °C and pressure of 1.01 bar.

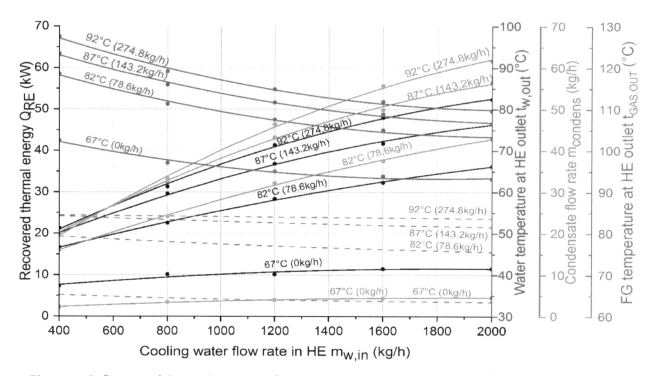

Figure 6. Influence of the cooling water flow rate in the HE and the dew point temperature on the recovered thermal energy (black lines), the water temperature at the HE outlet (red lines), the condensate flow rate (green lines) and the flue gas temperature at the HE outlet (gray lines) for concept 2.

The increase of the dew point temperature (67, 82, 87 and 92 °C) is achieved by increasing the flow rate of the introduced steam (0, 78.6, 143.2 and 274.8 kg/h, respectively). As expected, this results in the increase of all investigated parameters. For the flow rate of cooling water of 2000 kg/h, the recovered thermal energy is increased by 80% (from 10 to 52 kW), due to the enthalpy increase of the flue gas at the inlet of the HE caused by the steam humidification. The flow rate of condensate is increased by around 75% and reaches the maximum absolute value of 62 kg/h for the case with 92 °C dew point and 2000 kg/h flow rate of cooling water. Since by steam introduction, the enthalpy of the flue gas at the HE inlet increases, water which is passing through the HE is increasingly heated up. The temperature of water at the HE outlet increases on average by around 25%, independent of the cooling water flow rate at the HE, with the increase of the dew point temperature.

The flow rate of the cooling water in the HE has the similar influence on investigated parameters compared to the previous concept with water humidification. Supplying more cooling water to the HE results in the release of latent heat from the flue gas, so the condensate flow rate and the recovered thermal energy show the intensive growth. On the other hand, the higher flow rate of the cooling water results in the decrease of the water temperature at the HE outlet. This follows from the fact that the same amount of the flue gas at the same conditions has to warm up the significantly (five times) increased amount of the cooling water in the HE.

3.3. Concept 3: Compression of the Flue Gas

In the third investigated concept the increase of the flue gas pressure (1.01, 1.92, 2.33, 2.83 and 3.41 bar) results in the increase of the dew point temperature (67, 82, 87, 92 and 97 °C, respectively), whereas the water vapour content of the flue gas remains constant (17 %mas). The computational results, obtained for concept 3 are shown in Figure 7.

The dependences of Q_{RE}, $t_{w,out}$ and $m_{condens}$ are qualitatively similar to the ones obtained in the concept with the steam humidification. However, the achieved values are higher in the case of steam humidification. For instance, the recovered thermal energy displays the rise while increasing the dew point temperature, but the maximum absolute value is 30 kW in the case of the flue gas compression, which is ca. 20 kW less than in the case of the steam humidification. This is because the flue gas contains four times less water vapour: 46.8 kg/h in the case of the flue gas compression concept and (46.8 + 143.2) kg/h in the case of the steam based humidification.

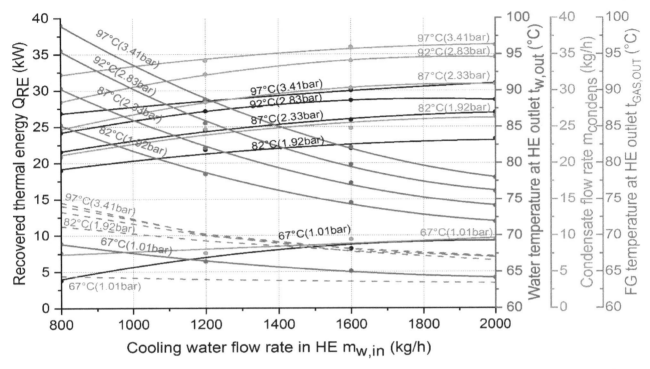

Figure 7. Influence of the cooling water flow rate in the HE and the dew point temperature on the recovered thermal energy (black lines), the water temperature at the HE outlet (red lines), the condensate flow rate (green lines) and the flue gas temperature at the HE outlet (gray lines) for concept 3.

Nevertheless, looking at the condensation rate, the pressure rise of the flue gas of around 2 bar (from 1.01 to 3.41 bar) leads to the 70% increase in the condensate flow rate.

The temperature of the flue gas is significantly increased during the isentropic compression and the high amount of the sensible heat is transferred to the water supplied to the HE. For this reason the water temperature at the outlet of the HE rises with the increase of the dew point temperature by around 30% for the low flow rate of the cooling water (800 kg/h) and by around 17% for the high flow rate (2000 kg/h).

3.4. Concept 4: Usage of the Steam Ejector

The modelling results for concept 4 (steam supply via the steam ejector) in comparison to concept 2 (steam supply via the steam generator) are presented in Figure 8. The comparison is eligible, due to the same flow rate of steam introduced in both cases, which amounts to 143.2 kg/h. While the humidification via the steam generator was performed by adding steam at 120 °C and 1.01 bar, the steam ejector was supplied with steam at 160 °C and 4 bar. In concept 2, the dew point temperature was 87 °C, whereas in concept 4 the higher dew point temperature was reached, namely 91 °C.

The process tendencies, corresponding to concepts 2 and 4, are qualitatively similar, but the target temperatures (dew point and cooling water outlet temperature) are higher in the case of the steam ejector. This is attributed to the fact that the usage of the steam ejector leads to the increase in both the water content and the pressure of the flue gas.

The condensate flow rate and the recovered thermal energy are almost the same in the case of the low flow rate of the cooling water (400–1200 kg/h) for both concepts. In this case the condensate flow rate dependencies are similar and therefore, the recovered thermal energy is mainly influenced by the sensible heat of the flue gas.

Increasing the cooling water flow rate further demonstrates the more noticeable difference between the two concepts. For 2000 kg/h the recovered thermal energy and the condensate flow rate are around 8% and 12% higher in the case of the steam ejector usage, because steam is introduced at higher thermodynamic parameters.

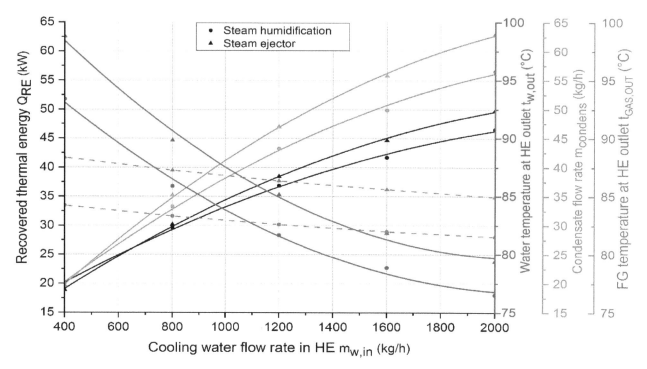

Figure 8. Comparison between the results calculated for concepts 2 and 4. Influence of the cooling water flow rate in the HE on the recovered thermal energy (black lines), the water temperature at the HE outlet (red lines), the condensate flow rate (green lines) and the flue gas temperature at the HE outlet (gray lines).

The water temperature at the outlet of the HE by using the steam ejector is 10 °C higher compared to the steam humidification case in the whole range of the cooling water flow rate. This follows from the fact that the flue gas at the inlet of the HE has higher temperature in the case of the steam ejector concept, therefore the flue gas transfers more sensible heat to the cooling water passing through the HE.

3.5. Comparison of the Results of the Investigated Concepts

The energy balance with the gain and the demand trend lines for concepts 2, 3 and 4, at the cooling water flow rate of 1200 kg/h, are shown in Figure 9. Concept 1 (humidification of the flue gas using water) is not considered, due to the relatively low increase in the dew point temperature.

During the humidification of the flue gas using steam (concept 2), additional energy input is required for the steam generator. In the concept with the compression of the flue gas (concept 3), additional energy input is required to run the flue gas compressor. In the case of the steam ejector usage (concept 4), additional energy is required by the pump, for the pressure rise of water, and by the steam generator, for the production of steam.

The analysis indicates that the humidification of the flue gas using steam (concept 2) is the most energy demanding process for increasing the dew point temperature. The energy demand of the steam generator rises with the dew point temperature increase, since the greater mass flow rate of water has to be evaporated in order to get the required amount of steam for injection into the flue gas. The concept with the steam ejector (concept 4) has the lower energy demand than concept 2, because the desirable dew point temperature is reached using the lower amount of steam but at the higher pressure. Moreover, the pump has a relatively high efficiency and is considered as the low-energy consuming equipment for increasing the pressure. In case of the third investigated concept, where the flue gas is compressed, the energy demand is the lowest of all three concepts. The reason is that the mass flow rate of the flue gas to the compressor remains constant for all of the investigated dew point temperatures. The maximum considered pressure increase by the compressor is about 2 bar (for increasing the dew point temperature from 67 °C to 97 °C), which requires a relatively low isentropic work.

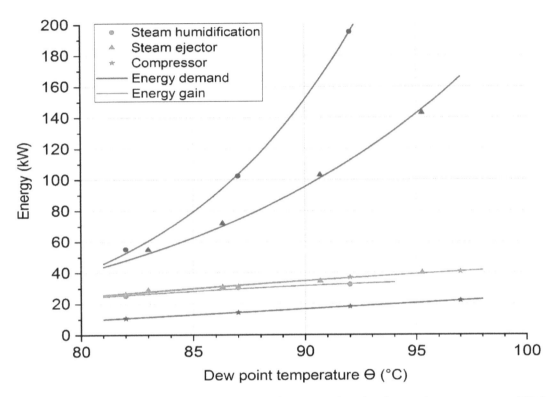

Figure 9. The energy demand and the energy gain for increasing the dew point temperature (Θ) for the investigated concepts.

Further analysis of the obtained results is related to the ratio of the recovered and additionally introduced energy. Since the energy gain stays almost the same for all investigated concepts, the conclusion is that only the third investigated concept, based on the flue gas compression, is of practical interest. In this case the energy gain exceeds the energy demand, therefore, the concept with compression is considered as the most promising for the energy recovery in the systems with the dew point temperature increase.

Cost analysis of the investigated concepts was conducted based on the current energy costs and additional energy demands of different concepts (Figure 10). The prices of different energy sources were introduced in chapter 2.6. The required investment costs including the prices of equipment were not taken into consideration.

As it was expected, the energy cost follows the trend of the energy demand, and grows as the dew point temperature increases. The growth of the energy cost per temperature degree for the steam humidification concept is especially steep, since the generation of the high amount of steam is coupled with the high energy demand. The energy cost can be reduced up to 35% (for the dew point temperature of 92 °C) by applying the steam ejector, which uses the lower amount of steam, but at higher pressure.

The energy cost in the case of the flue gas compression concept also increases, since the greater pressure is required for increasing the dew point temperature. Compared to two other concepts, concept 3 exhibits the lowest energy cost (20,000 €/year), which is half of the costs of concept 2 in the case of the dew point temperature of 92 °C. Nevertheless, the installation, the maintenance and the repair costs of the compressor are normally higher than the equipment used in the other proposed concepts and should be taken into account in the future work.

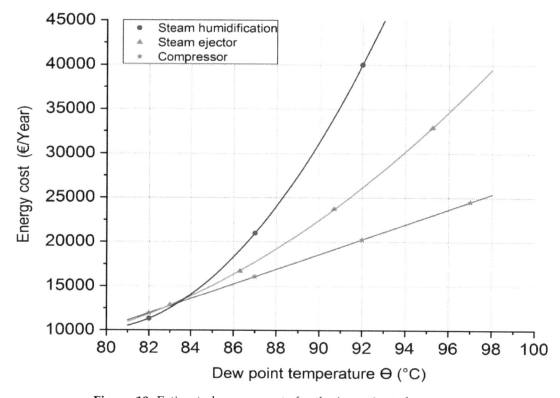

Figure 10. Estimated energy costs for the investigated concepts.

The potential practical application of the theoretical concepts investigated in this work is also related to several major challenges. The first one is an appropriate selection of the optimal method based on the actual size, requirements and properties of each specific facility. For example, all the calculations and considerations in this work are related to one specific industrial facility, namely the natural gas fired, tunnel baking oven. The limitations related to the investigated facility (composition, temperature and flow rate of the flue gas, fluid parameters at the inlet/outlet of the process components, etc.) influence most of the obtained results. Therefore, it is imperative to perform the similar analysis for each specific system of interest with its unique properties. The next technical challenge is to compensate for the eventual dynamic changes within the process. The presented work is based on the continuous industrial process with time independent process parameters which is often not the case. Although compensation of the dynamic process behavior is out of the scope of this work, the use of cognitive algorithms for the process management in connection to the dew point temperature increase, could result in the optimal use of the available potentials for the thermal energy recovery. Nevertheless, one of the major challenges for the practical implementation of the investigated techniques would probably be of the financial nature and related to the investment costs for the additional equipment. Taking into account the investment costs, the additional maintenance costs and the price of the low-temperature thermal energy in Germany, some of the simplified financial calculations, not presented in this work, indicate that most of the investigated processes are financially justified only for the larger facilities and pay-off periods of 20–30 years. Therefore, a full economic analysis for the considered concepts will be part of a future detailed investigation.

4. Conclusions

Recovering both the sensible and latent heat is part of the energy efficiency optimization of every process that has thermal energy dissipating with the flue gas. Essential relevance is related to the industrial processes where flue gases are rich in water vapour (e.g., baking, textile, pulp and paper industries). Cooling the flue gas below the water vapour dew point temperature in a shell-and-tube heat exchanger, leads to the condensation of the water vapour and to the release of the sensible and latent heat. In order to use this energy as process heat, the increase of the usually low dew point temperatures (40–60 °C) to the higher levels (80–95 °C) is of special interest.

The increase of the dew point temperature of the water vapour contained in the flue gas, leads to the increase of the temperature level of the cooling fluid and the condensate. In this way, these otherwise waste streams can be further used within an industrial process, thus improving its energy efficiency.

The present study was focused on the investigation of four concepts for the increase of the dew point temperature of the industrial flue gas: humidification of the flue gas using water, humidification of the flue gas using steam, compression of the flue gas and the usage of the steam ejector. All the calculations and considerations in this work are related to one specific industrial facility, namely the natural gas fired, tunnel baking oven. The process modelling was performed using the commercial software Aspen®. For each considered concept the effects of the dew point temperature and the cooling water flow rate in the HE on the recovered thermal energy, the water temperature at the HE outlet and the condensate flow rate were investigated.

The major conclusions for the considered concepts are summarized as follows:

- The increase of the dew point temperature above 90 °C is possible by the steam humidification, compression of the flue gas and using the steam ejector.

- The humidification of flue gas using water is not recommended, due to the relatively low thermal energy recovery level (maximum of 10 kW is achieved) and the negligible increase in the dew point temperature (up to 70 °C) in comparison to other tested concepts.

- Although the steam humidification shows the highest potential for the thermal energy recovery, it is also the most energy demanding and, consequently, the most expensive process, in the case when the steam has to be specially produced for the dew point temperature increase. When this low parameter steam is available as a waste product, the investigated concept gains on its importance.

- In the steam humidification concept the maximum of 52 kW of the recovered thermal energy is achieved and the flow rate of the condensate is increased by app. 75% at $m_{w,in}$ = 2000 kg/h. The maximal calculated dew point temperature is 92 °C. The energy demand of the steam generator increases with the dew point temperature increase, since the greater water mass flow rate has to be heated and evaporated, in order to get the required amount of steam.

- Using the steam ejector leads to the increase in both the water content and pressure of the flue gas, due to the geometry of the device. Therefore, the recovered thermal energy and the condensate flow rate are around 8% and 12% higher compared to the steam humidification concept. Yet, the energy demand exceeds the energy gain, making this concept not suitable for the use in a thermal energy recovery cycle, in the case when there is no waste steam available in the facility.

- Both from the energetic and economic point of view, the compression of the flue gas has the highest potential, as the energy gain exceeds the energy demand and the energy cost is the lowest out of all the investigated methods. Although the low required pressure increase of app. 2 bar corresponds to the relatively low isentropic work by the compressor, a detailed economic analysis should be performed, taking into account the equipment price, the installation and the maintenance costs for each concrete industrial facility.

- The concept with the flue gas compression will be the subject of further investigations, in combination with the Organic Rankine Cycle and the absorption refrigerator.

Author Contributions: Conceptualization, N.F., V.J., A.Z.-R. and A.D.; Methodology, N.F., V.J. and P.A.; Validation, P.A.; Investigation, P.A., N.F.; Resources, A.D.; Data Curation, P.A., N.F., V.J., A.Z.-R. and M.J.K.; Writing—Original Draft Preparation, N.F., P.A., A.Z.-R. and V.J.; Writing—Review and Editing, N.F., V.J., A.Z.-R., M.J.K. and P.A.; Software, P.A. and M.J.K.; Visualization, P.A. and N.F.; Supervision, N.F., V.J.; Funding acquisition, N.F., V.J., A.D.

Acknowledgments: The authors gratefully acknowledge the financial support of the German Academic Exchange Service (DAAD), and the funding of the Erlangen Graduate School in Advanced Optical Technologies (SAOT) by the German Research Foundation (DFG) in the framework of the German excellence initiative. We also acknowledge the support by Deutsche Forschungsgemeinschaft and Friedrich-Alexander-Universität Erlangen-Nürnberg (FAU) within the funding programme Open Access Publishing.

Nomenclature

C_p	Specific heat capacity (kJ kg^{-1} K^{-1})
h	Specific enthalpy (kJ kg^{-1})
Δh_{vap}	Vaporization enthalpy of water (kJ kg^{-1})
m	Mass flow rate (kg h^{-1})
P	Total pressure (bar)
P_i^s	Saturation vapour pressure (bar)
Q	Heat (kW)
R	Gas constant (kJ kg^{-1} K^{-1})
t	Temperature (°C)
x	Humidity ratio of an air-vapour mixture (kg$_w$ kg$_{air}$$^{-1}$)
x_i	Mole fraction of a component in liquid phase (-)
y	Mole fraction of a component (-)
y_i	Mole fraction of a component in gas phase (-)

Greek symbols

ρ	Density (kg m^{-3})
Θ	Dew point temperature (°C)
η	Efficiency (-)

Abbreviations

AIR	Dry air
COMPR	Compressor
CONDENS	Condensate
D	Device
DEM	Energy demand
EXGAS	Flue gas at the exit of an industrial flue before a concept modification
FG	Flue gas
FSPLIT	Splitter
GAIN	Energy gain
GAS	Flue gas after a concept modification
HE	Heat exchanger
HEWATER	Cooling water for a heat exchanger
IN	Inlet
OD	Outer diameter
OUT	Outlet
PRWATER	Useful water for processes in a company
RE	Recovered thermal energy
RECWATER	Recycling water to be used in a concept
SG	Steam generator
VAP	Vapour
W	Water

References

1. Hu, Y.; Gao, Y.; Lv, H.; Xu, G.; Dong, S. A New Integration System for Natural Gas Combined Cycle Power Plants with CO_2 Capture and Heat Supply. *Energies* **2018**, *11*, 3055.

2. Schwabe, K.; Walsdorf-Maul, M.; Schaudienst, F.; Vogdt, F.U. Using Waste Heat for Sustainable Manufacturing Based on the Example of a Conventional Industrial Bakery. *Int. J. Mater. Mech. Manuf. (IJMMM)* **2013**, 274–277. [CrossRef]

3. Terhan, M.; Comakli, K. Design and economic analysis of a flue gas condenser to recover latent heat from exhaust flue gas. *Appl. Therm. Eng.* **2016**, *100*, 1007–1015. [CrossRef]

4. Xu, G.; Huang, S.; Yang, Y.; Wu, Y.; Zhang, K.; Xu, C. Techno-economic analysis and optimization of the heat recovery of utility boiler flue gas. *Appl. Energy* **2013**, *112*, 907–917. [CrossRef]

5. Aneke, M.; Agnew, B.; Underwood, C.; Wu, H.; Masheiti, S. Power generation from waste heat in a food processing application. *Appl. Therm. Eng.* **2012**, *36*, 171–180. [CrossRef]

6. Che, D.; Da, Y.; Zhuang, Z. Heat and mass transfer characteristics of simulated high moisture flue gases. *Heat Mass Transf.* **2005**, *41*, 250–256. [CrossRef]

7. Wang, D.; Bao, A.; Kunc, W.; Liss, W. Coal power plant flue gas waste heat and water recovery. *Appl. Energy* **2012**, *91*, 341–348. [CrossRef]

8. Bajpai, P. Pulp and Paper Production Processes and Energy Overview. In *Pulp and Paper Industry*; Elsevier: Amsterdam, The Netherlands, 2016.

9. Hasanbeigi, A.; Price, L. A review of energy use and energy efficiency technologies for the textile industry. *Renew. Sustain. Energy Rev.* **2012**, *16*, 3648–3665. [CrossRef]

10. Jank, R.; Schulte, S. Verfahren zur Energiegewinnung aus Wasserdampf enthaltenden Schwaden und Vorrichtung zur Durchführung dieses Verfahrens. European Patent No. WO2017064036, 20 April 2017.

11. Song, C.; Pan, W.; Srimat, S.T.; Zheng, J.; Li, Y.; Wang, Y.-H.; Xu, B.-Q.; Zhu, Q.-M. Tri-reforming of Methane over Ni Catalysts for CO_2 Conversion to Syngas With Desired H_2/CO Ratios Using Flue Gas of Power Plants Without CO_2 Separation. In *Carbon Dioxide Utilization for Global Sustainability, Proceedings of the 7th the International Conference on Carbon Dioxide Utilization, Seoul, Korea, 12–16 October 2003*; Elsevier: Amsterdam, The Netherlands, 2004; pp. 315–322.

12. Osakabe, M. Heat exchanger for latent heat recovery. *Mech. Eng. Rev.-Bull. JSME* **2015**, *2*, 1–24. [CrossRef]

13. Zhao, S.; Yan, S.; Wang, D.K.; Wei, Y.; Qi, H.; Wu, T.; Feron, P.H.M. Simultaneous heat and water recovery from flue gas by membrane condensation: Experimental investigation. *Appl. Therm. Eng.* **2017**, *113*, 843–850. [CrossRef]

14. Law, R.; Harvey, A.; Reay, D. Opportunities for low-grade heat recovery in the UK food processing industry. *Appl. Therm. Eng.* **2013**, *53*, 188–196. [CrossRef]

15. Hung, T.C.; Shai, T.Y.; Wang, S.K. A review of organic rankine cycles (ORCs) for the recovery of low-grade waste heat. *Energy* **1997**, *22*, 661–667. [CrossRef]

16. Tchanche, B.F.; Lambrinos, G.; Frangoudakis, A.; Papadakis, G. Low-grade heat conversion into power using organic Rankine cycles—A review of various applications. *Renew. Sustain. Energy Rev.* **2011**, *15*, 3963–3979. [CrossRef]

17. Ziegler, F. EnEff Wärme: Absorptionskältetechnik für Niedertemperaturantrieb—Grundlagen und Entwicklung von Absorptionskältemaschinen für die fernwärme-und solarbasierte Kälteversorgung. In *Abschlussbericht*; Technische Universität Berlin Fakultät III-Prozesswissenschaften: Berlin, Germany, 2013.

18. Offizielles Stadtportal für die Hansestadt Hamburg. Solares Kühlen für Büro- und Dienstleistungsgebäude, Österreichisches Forschungs-und Prüfzentrum Arsenal Ges.m.b.H. Available online: https://www.hamburg.de/contentblob/1356374/8760da9f3d940e22e7f926723591eae7/data/solare-kuehlung.pdf (accessed on 25 March 2019).

19. U.S. Department of Energy. Waste Heat Reduction and Recovery for Improving Furnace Efficiency, Productivity and Emissions Performance. Available online: https://www.energy.gov/sites/prod/files/2014/05/f15/35876.pdf (accessed on 25 March 2019).

20. Mukherjee, S.; Asthana, A.; Howarth, M.; Mcniell, R. Waste heat recovery from industrial baking ovens. *Energy Procedia* **2017**, *123*, 321–328. [CrossRef]

21. EcoStep. Energieeffizienz in Bäckereien-Energieeinsparungen in Backstube und Filialen, ttz Bremerhaven. Available online: http://www.ecostep-online.de/cms_uploads/files/eneff_baeckerei_-_leitfaden_-_juli_2014. pdf (accessed on 25 March 2019).

22. Jana, K.; De, S. Utilizing waste heat of the flue gas for post-combustion CO_2 capture—A comparative study for different process layouts. *Energy Sources Part A Recovery Util. Environ. Eff.* **2016**, *38*, 960–966. [CrossRef]

23. Luyben, W.L. Heat exchanger simulations involving phase changes. *Comput. Chem. Eng.* **2014**, *67*, 133–136. [CrossRef]

24. Duan, W.; Yu, Q.; Wang, K.; Qin, Q.; Hou, L.; Yao, X.; Wu, T. ASPEN Plus simulation of coal integrated gasification combined blast furnace slag waste heat recovery system. *Energy Convers. Manag.* **2015**, *100*, 30–36. [CrossRef]

25. Ishaq, H.; Dincer, I.; Naterer, G.F. New trigeneration system integrated with desalination and industrial waste heat recovery for hydrogen production. *Appl. Therm. Eng.* **2018**, *142*, 767–778. [CrossRef]

26. Mazzoni, S.; Arreola, M.J.; Romangoli, A. Innovative Organic Rankine arrangements for Water Savings in Waste Heat Recovery Applications. *Energy Procedia* **2017**, *143*, 361–366. [CrossRef]

27. Hu, H.W.; Tang, G.H.; Niu, D. Experimental investigation of convective condensation heat transfer on tube bundles with different surface wettability at large amount of noncondensable gas. *Appl. Therm. Eng.* **2016**, *100*, 699–707. [CrossRef]

28. Towler, G.; Sinnott, R. *Chemical Engineering Design: Principles, Practice, and Economics of Plant and Process Design*, 2nd ed.; Butterworth-Heinemann: Waltham, MA, USA, 2013.

29. Incropera, F.P.; Dewitt, D.P.; Bergman, T.L.; Lavine, A.S. *Principles of Heat and Mass Transfer*, 7th ed.; John Wiley & Sons: Hoboken, NJ, USA, 2011.

30. Baehr, H.D.; Kabelac, S. *Thermodynamik*; Springer Vieweg: Berlin, Germany, 2012.

31. Spirax Sarco. Grundlagen der Dampf-und Kondensattechnologie. Available online: http://www.spiraxsarco. com/global/de/Resources/Documents/Grundlagen-der-Dampf-und-Kondensattechnologie.pdf (accessed on 25 March 2019).

32. Statistisches Bundesamt (Destatis). Data on Energy Price Trends, Long-Time Series from January 2000 to March 2018. Available online: https://www.destatis.de (accessed on 25 March 2019).

Permissions

List of Contributors

Aysan Shahsavar Goldanlou
Institute of Research and Development, Duy Tan University, Da Nang 550000, Vietnam
Faculty of Electrical−Electronic Engineering, Duy Tan University, Da Nang 550000, Vietnam

Mohammad Badri
Department of Mechanical Engineering, University of Kashan, Kashan 8731753153, Iran

Behzad Heidarshenas
College of Mechanical of Electrical Engineering, Nanjing University of Aeronautics and Astronautics, Nanjing 210016, China

Ahmed Kadhim Hussein
College of Engineering−Mechanical Engineering Department, University of Babylon, Babylon 51001, Iraq

Sara Rostami
Laboratory of Magnetism and Magnetic Materials, Advanced Institute of Materials Science, Ton Duc Thang University, Ho Chi Minh City 758307, Vietnam
Faculty of Applied Sciences, Ton Duc Thang University, Ho Chi Minh City 758307, Vietnam

Mostafa Safdari Shadloo
CORIA-UMR 6614, CNRS & INSA of Rouen, Normandie University, 76000 Rouen, France

Meng-Ge Li, Feng Feng and Wei-Tao Wu
School of Mechanical Engineering, Nanjing University of Science & Technology, Nanjing 210094, China

Mehrdad Massoudi
U.S. Department of Energy, National Energy Technology Laboratory (NETL), Pittsburgh, PA 15236, USA

Zhongchao Zhao, Yimeng Zhou, Xiaolong Ma, Xudong Chen, Shilin Li and Shan Yang
School of Energy and Power, Jiangsu University of Science and Technology, Zhenjiang 212000, China

Hyoung Tae Kim
Thermal Hydraulic and Severe Accident Research Division, Korea Atomic Energy Research Institute, 989-111 Daedeok-daero, Yuseong-gu, Daejeon 34057, Korea

Se-Myong Chang and Young Woo Son
School of Mechanical Convergence Systems Engineering, Kunsan National University, 558 Daehak-ro, Gunsan, Jeonbuk 54150, Korea

Naveed Ahmed, Fitnat Saba and Syed Tauseef Mohyud-Din
Department of Mathematics, Faculty of Sciences, HITEC University, Taxila Cantt 47080, Pakistan

Umar Khan
Department of Mathematics and Statistics, Hazara University, Mansehra 21300, Pakistan

Ilyas Khan
Faculty of Mathematics and Statistics, Ton Duc Thang University, Ho Chi Minh City 736464, Vietnam

Tawfeeq Abdullah Alkanhal
Department of Mechatronics and System Engineering, College of Engineering, Majmaah University, Majmaah 11952, Kingdom of Saudi Arabia

Imran Faisal
Department of Mathematics, Taibah University, Universities Road, Medina, Kingdom of Saudi Arabia

Xiangyu Su, Xiaodong Ren, Xuesong Li and Chunwei Gu
Institute of Gas Turbine, Department of Energy and Power Engineering, Tsinghua University, Beijing 100084, China

Tarek A. Ganat
Department of Petroleum Engineering, Universiti Teknologi PETRONAS, Seri Iskandar, Perak 32610, Malaysia

Meftah Hrairi
Department of Mechanical Engineering, International Islamic University Malaysia, Kuala Lumpur 50728, Malaysia

Fei Zhang, Zhenxia Liu, Zhengang Liu and Yanan Liu
School of Power and Energy, Northwestern Polytechnical University, Xi'an 710129, China

Lei Wang, Jing Zhao and Yangsheng Zhao
College of Mining Engineering, Taiyuan University of Technology, Taiyuan 030024, China
Key Laboratory of In-situ Property Improving Mining of Ministry of Education, Taiyuan University of Technology, Taiyuan 030024, China

Dong Yang, Xiang Li and Guoying Wang
Key Laboratory of In-situ Property Improving Mining of Ministry of Education, Taiyuan University of Technology, Taiyuan 030024, China

Yassir Riaz and William B. Zimmerman
Department of Chemical and Biological Engineering, University of Sheffield, Mappin Street, Sheffield S1 3JD, UK

Michael John Hines
Perlemax Ltd., Kroto Innovation Centre, 318 Broad Ln, Sheffield S3 7HQ, UK

Pratik Devang Desai
Department of Chemical and Biological Engineering, University of Sheffield, Mappin Street, Sheffield S1 3JD, UK
Perlemax Ltd., Kroto Innovation Centre, 318 Broad Ln, Sheffield S3 7HQ, UK

Chao Jin and Hongwei Ma
School of Energy and Power Engineering, Beihang University, Beijing 100191, China

Nataliia Fedorova, Vojislav Jovicic and Antonio Delgado
Institute of Fluid Mechanics (LSTM), Friedrich-Alexander University (FAU), 91058 Erlangen, Germany
Erlangen Graduate School in Advanced Optical Technologies (SAOT), 91054 Erlangen, Germany

Pegah Aziziyanesfahani, Ana Zbogar-Rasic and Muhammad Jehanzaib Khan
Institute of Fluid Mechanics (LSTM), Friedrich-Alexander University (FAU), 91058 Erlangen, Germany

Index